Immunocytochemical Methods and Protocols

METHODS IN MOLECULAR BIOLOGY™

John M. Walker, SERIES EDITOR

METHODS IN MOLECULAR BIOLOGY™

Immunocytochemical Methods and Protocols
Second Edition

Edited by

Lorette C. Javois

Catholic University of America, Washington, DC

Humana Press ✳ Totowa, New Jersey

Dedication

To Larry, who wonders whether this "counts" as another publication, and to Alexandra, who decided to skip the midnight sessions down the home stretch this time.

© 1999 Humana Press Inc.
999 Riverview Drive, Suite 208
Totowa, New Jersey 07512

This publication is printed on acid-free paper. ∞
ANSI Z39.48-1984 (American Standards Institute)
Permanence of Paper for Printed Library Materials.

Cover illustration: The head and upper body column of the freshwater coelentrate, *Hydra oligactis,* labeled with the head-specific monoclonal antibody CP8, using the immunofluorescent technique described in Chapter 18 (ref. 10). Labeling on the whole mount preparation is restricted to the hypostome (mouth) and surrounding tentacles but not the body column beneath. Photograph by Daniel R. Bessette.
Cover design by Patricia F. Cleary.
For additional copies, pricing for bulk purchases, and/or information about other Humana titles, contact Humana at the above address or at any of the following numbers: Tel.: 973-256-1699; Fax: 973-256-8341; E-mail: humana@humanapr.com; or visit our Website: http://humanapress.com

Library of Congress Cataloging in Publication Data

Main entry under title:
Methods in molecular biology™.

Immunocytochemical methods and protocols / edited by Lorette C. Javois. -- 2nd ed.
    p.   cm. -- (Methods in molecular biology™ ; v. 115)
    Includes bibliographic references and index.
    ISBN 978-0-89603-813-4 (alk. paper)
    1. Immunocytochemistry--Laboratory manuals.   2. In situ hybridization--Laboratory manuals.
    I. Javois, Lorette, C. II. Series: Methods in Molecular Biology (Totowa, NJ) ; 115.
    QR187.I45I45 1999
    571.9--dc21   98-46930
                                                                                CIP

Preface

The principle that antibodies can be used as cytochemical agents provided they are tagged with suitable markers has been evident for over 55 years. During this time the use of immunocytochemical methods has spread to a wide array of biological disciplines. Early applications focused on the detection of microbial antigens in tissues while more recent applications have used monoclonal antibodies to study cell differentiation during embryonic development. For a select few disciplines volumes have been published focusing on the specific application of immunocytochemical techniques to that discipline.

What distinguished *Immunocytochemical Methods and Protocols* from earlier books when it was first published four years ago was its broad appeal to researchers in all disciplines, including those in both research and clinical settings. The methods and protocols presented here are designed to be general in their application and the accompanying "Notes" provide invaluable assistance in adapting or troubleshooting the protocols. These strengths hold true for the second edition. In addition, the last four years have seen the ever-expanding application of immunocytochemical techiniques in the clinical laboratory and the widespread use of immunocytochemistry in combination with *in situ* hybridization. Both of these subjects are given more detailed treatment in this second edition. As with the first edition, interspersed throughout the book are chapters providing overviews of select topics related to immunocytochemistry.

Immunocytochemical Methods and Protocols, Second Edition begins with a series of protocols geared to purify and prepare antibodies for use in subsequent staining procedures. The next series of protocols focuses on the fixation and preparation of tissues for light microscopic analyses, including antigen retrieval methods. Then, the use of fluorescent- and enzyme-conjugated antibodies to localize antigens in a variety of preparations is considered at length. The use of special instrumentation such as the confocal microscope and laser microbeam workstation, is also discussed. The next section focuses on the preparation and staining of cells for fluorescence activated cell sorter (FACS) analysis and assays which measure various cell paramenters using FACS. This is followed by protocols detailing the fixation and staining of samples for electron microscopic evaluation of antibody labeling using colloidal gold. Applications combining immunocytochemistry with *in situ* hybridization are organized into a new section while the final section of the book focuses on additional applications relevant to the clinical laboratory. In particular, regulations and troubleshooting guidelines are discussed. Many of these special applications have never been presented within the context of a volume devoted to immunocytochemistry methods and protocols, and the desire is for researchers and clinicians considering an immunocytochemical approach to be as well informed about their options as possible.

I am deeply indebted to the authors of the various chapters in the first edition for their continued interest and suggestions for improving the volume. I appreciate all the authors' hard work, dedication, and willingness to share their expertise. The protocols presented here are routinely used in their laboratories as the extensive "Notes" they have provided will attest to. Once again, the high quality of the manuscripts made my job as an editor an easy one. I would also like to thank Dr. John Walker, *Methods in Molecular Biology* series editor, for his encouragement throughout the process of compiling this second edition. Final thanks go to Thomas Lanigan, Fran Lipton, and the staff of Humana Press for making this book a reality.

Lorette C. Javois, PhD

Contents

Contributors

ANDREA ABATI • *Cytopathology Section National Cancer Institute National Institutes of Health Bethesda, MD*

ROBERT R. AKSAMIT • *Food and Drug Administration Center for Biologics Evaluation & Research Division of Hematology Bethesda, MD*

GARY BRATTHAUER • *Division of Immunopathology Armed Forces Institute of Pathology Washington, D.C.*

JULIE BRENT • *In Vitro Technologies Baltimore, MD*

ROBERT CUNNINGHAM • *Department of Cellular Pathology Armed Forces Institute of Pathology Washington, D.C.*

EDNA ELFONT • *Histopathology Associates West Bloomfield, MI*

PATRICIA FETSCH • *Cytopathology Laboratory National Cancer Institute National Institutes of Health Bethesda, MD*

ALTON D. FLOYD • *Consultant in Biotechnology Edwardsburg, MI*

DENNIS FRISMAN • *Pathology Consultants Medical Group Torrence, CA*

LIANA HARVATH • *Food and Drug Administration Center for Biologics Evaluation & Research Division of Hematology Bethesda, MD*

VIRGINIA HEATWOLE • *R/M Nardone Associates Bethesda, MD*

GILBERT HERMAN • *Histopathology Associates West Bloomfield, MI*

ELAINE S. JAFFE • *Hematopathology Section Laboratory of Pathology National Cancer Institute National Institutes of Health Bethesda, MD*

MARIA CELIA JAMUR • *Department of Biology Sector of Biological Sciences Federal University of Parana Curitiba, Parana Brazil*

LORETTE JAVOIS • *Department of Biology The Catholic University of America Washington, DC*

LASZLO KRENACH • *A. Szent-Gyorgyi University of Medicine Szeged, Hungary*

TIBOR KRENACS • *A. Szent-Gyorgyi University of Medicine Szaged, Hungary*

UTE KENT • *University of Michigan Medical School Department of Pharmacology Ann Arbor, MI*

SU-YAU MAO • *Medimmune, Inc. Gaithersburg, MD*

MELISSA MELAN • *Biology Department Duquesne University Pittsburgh, PA*

MICHAEL MULLINS • *Biology Department The Catholic University of America Washington, DC*

ROLAND NARDONE • *Biology Department The Catholic University of America Washington, DC*

CONSTANCE OLIVER • *Department of the Navy Office of Naval Research - Arlington, VA*

P. SCOTT PINE • *Division of Applied Pharmacology & Research Center for Drug Evaluation & Research Food & Drug Administration Laurel, MD*

MARK RAFFELD • *National Institutes of Health National Cancer Institute Laboratory of Pathology Bethesda, MD*

DOUGLAS TERLE • *Applied Information Technology Food & Drug Administration Center for Biologics Evaluation & Research Bethesda, MD*

VISHAKA THAKER • *Department of Pathology In Situ Hybridization Unit National Cancer Institute National Institutes of Health Bethesda, MD*

MARK WILLINGHAM • *Department of Pathology & Laboratory Medicine Medical University of South Carolina Charleston, SC*

I

ANTIBODY PREPARATION

1

Overview of Antibody Use in Immunocytochemistry

Su-Yau Mao, Lorette C. Javois, and Ute M. Kent

1. Introduction

Immunocytochemistry, by definition, is the identification of a tissue constituent *in situ* by means of a specific antigen–antibody interaction where the antibody has been tagged with a visible label *(1)*. Cell staining is a powerful method to demonstrate both the presence and subcellular location of a particular molecule of interest *(2)*. Initial attempts to label antibodies with ordinary dyes were unsatisfactory because the label was not sufficiently visible under the microscope. A. H. Coons first introduced immunofluorescence in 1941, using specific antibodies labeled with a fluorescent dye to localize substances in tissues *(3)*. This technique was considered difficult, and its potential was not widely realized for nearly 20 yr. Early attempts focused on labeling the specific antibody itself with a fluorophore (*see* Chapter 6). The labeled antibody was then applied to the tissue section to identify the antigenic sites (direct method) *(3)* (*see* Chapter 15). Later, the more sensitive and versatile indirect method was introduced *(4)* (*see* Chapters 16–18). In this method, the specific antibody, bound to the antigen, was detected with a secondary reagent, usually another antibody that had been tagged with either a fluorophore or an enzyme.

Fluorochrome-labeled anti-immunoglobulin antibodies are now widely used in immunocytochemistry, flow cytometry (*see* Chapters 30–39), and hybridoma screening. The availability of fluorophores with different emission spectra has also made it possible to detect two or more antigens on the same cell or tissue section (*see* Chapter 14). Although fluorescent labeling offers sensitivity and high resolution, there are several disadvantages. First, it requires special instrumentation: a fluorescence microscope, a confocal microscope, or a flow cytometer. Second, background details are difficult to appreciate, and cellular autofluorescence can sometimes make the interpretation difficult. Finally,

From: *Methods in Molecular Biology, Vol. 115: Immunocytochemical Methods and Protocols*
Edited by: L. C. Javois © Humana Press Inc., Totowa, NJ

the preparations are not permanent. Nevertheless, the speed and simplicity of these methods have ensured that they remain popular, whereas advances in instrumentation have overcome many of the disadvantages (*see* Chapters 20 and 21).

Numerous attempts have been made to improve the methodology. The search for other labels that could be viewed with a standard light microscope resulted in widespread use of enzymes (*see* Chapters 23–27). Enzyme labels are detected by the addition of substrate at the end of the antigen–antibody reaction. The enzyme–substrate reactions yield intensely colored end products that can be viewed under a light microscope. Enzymatic labels are preferred by most researchers because they are less expensive, very sensitive, and can be used for permanent staining without special equipment requirements. Several enzymes are commonly used in immunocytochemistry, including peroxidase *(5)*, alkaline phosphatase *(6)*, and glucose oxidase *(7)* (*see* Chapter 23). Peroxidase catalyzes an enzymatic reaction with a very high turnover rate, offering good sensitivity within a short time. It is the enzyme of choice for immunocytochemistry. If two different enzymes are required, as in double-immuno enzymatic staining, alkaline phosphatase has generally been used as the second enzyme *(8)* (*see* Chapter 27). Alkaline phosphatase is relatively inexpensive, stable, and gives strong labeling with several substrates, thus offering a choice of differently colored reaction products. Glucose oxidase has also been used for double-immuno enzymatic labeling *(9)*. This enzyme has the advantage over peroxidase or alkaline phosphatase in that no endogenous enzyme activity exists in mammalian tissues. However, in practice, the endogenous enzyme activity of both peroxidase and alkaline phosphatase can easily be inhibited *(10)*.

If cellular localization of the antigen–antibody complex is not required, enzyme immunolabeling can be performed on cells adherent to a microtiter plate, and the color change resulting from the enzymatic reaction can be detected as a change in absorbance with an automatic plate reader (*see* Chapter 28). Biotinylation of antibodies and the use of the avidin–biotin complex has further extended the versatility and sensitivity of the enzymatic techniques (*see* Chapters 7 and 25–27). Most recently, the principles behind these techniques have been applied in combination with *in situ* hybridization techniques. Using nucleic acid–antibody complexes as probes, specific DNA or RNA sequences can be localized (*see* Chapters 46–49).

Other labels that have particular uses for electron microscopy are ferritin *(11)* and colloidal gold particles *(12,13)* (*see* Chapters 40–45). Gold particles are available in different sizes, therefore allowing simultaneous detection of several components on the same sample. Colloidal gold may also be detected with the light microscope following silver enhancement (*see* Chapter 29). In addition, radioactive labels have found some use in both light and electron microscopy *(14,15)*. The reasons for developing new labels are the continuing search for greater specificity and sensitivity of the reaction, together with the possibility of identifying two or more differently labeled antigens in the same preparation.

Immunocytochemical methods have become an integral part of the clinical laboratory, as well as the research setting (*see* Chapter 50). Clinically relevant specimens ranging from frozen sections and cell-touch preparations to whole-tissue samples are amenable to analysis (*see* Chapters 9–13). Panels of antibodies have been developed to aid in the differential diagnosis of tumors (*see* Chapter 51), and automated instrumentation has been designed to speed the handling of numerous specimens (*see* Chapter 52).

2. Sources of Antibodies

In institutions that are equipped with animal care facilities, polyclonal sera or ascites can be produced in house. Information on the generation of antibodies in animals can be found in several excellent references *(16–19)*. Alternatively, a number of service companies exist that can provide the investigator with sera and ascites, as well as help in the design of injection and harvesting protocols. Immune serum contains approx 10 mg/mL of immunoglobulins, 0.1–1 mg/mL of which comprise the antibody of interest. Therefore, polyclonal antibodies from sera of all sources should be purified by a combination of methods. Precipitation of immunoglobulins with ammonium sulfate is advisable, since this method removes the bulk of unwanted proteins and lipids, and reduces the sample volume (*see* Chapter 2). Additional purification can then be achieved by ion-exchange chromatography (*see* Chapter 3). If it is, however, necessary to obtain a specific antibody, the ammonium sulfate isolated crude immunoglobulins should be purified by affinity chromatography (*see* Chapter 4).

Monoclonal antibody generation has become a widely used technique and can be performed in most laboratories equipped with tissue culture facilities *(20,21)*. After an initial, labor-intensive investment involving spleen fusion followed by hybridoma selection, screening, and testing, these cells provide a nearly limitless supply of specific antibodies. In some instances, certain antibody-producing hybridomas have been deposited with the American Type Culture Collection (ATCC) and are available for a moderate fee. (In addition, under the auspices of the National Institute of Child Health and Human Development, a Development Studies Hybridoma Bank is maintained by the Department of Biological Sciences at the University of Iowa.) Ascites fluid contains approx 1–10 mg/mL of immunoglobulins. The majority of these antibodies (approx 90%) should be the desired monoclonal antibody. Ascites fluid can be purified by a combination of ammonium sulfate precipitation and ion-exchange chromatography, or by protein A or protein G affinity chromatography (*see* Chapter 5). For certain species and subtypes that bind poorly or not at all to protein A or protein G, ammonium sulfate precipitation followed by ion-exchange chromatography may be more suitable. Hybridoma culture supernatants contain 0.05–1 mg/mL of immunoglobulins, depending on whether or not the hybridomas are grown in the presence of calf serum. Antibodies from hybridoma culture supernatants may be most conveniently purified by affinity chromatography using either the specific antigen as a ligand or protein A/G. If the hybridoma culture supernatant contains fetal bovine serum, antigen affinity chromatography is preferred because of the presence of large quantities of bovine immunoglobulins. Protein A/G affinity purification will suffice for antibodies from hybridomas cultured in the absence of serum. Alternatively, these immunoglobulins may simply be concentrated by ammonium sulfate fractionation or ultrafiltration followed by dialysis (*see* Chapter 2).

Purified or semipurified antibodies are also commercially available from many sources. These are particularly useful if a certain technique requires the use of a species-specific secondary antibody. Several companies will also provide these antibodies already conjugated to reporter enzymes, fluorophores, avidin/biotin, or gold particles of various sizes.

3. Characteristics of a "Good" Antibody

The most desirable antibodies for immunocytochemical studies display high specificity and affinity for the antigen of interest and are produced in high titer. Immunoglobulins with these characteristics are preferred because they can be used at high dilution where false-positive reactions can be avoided. Under very dilute conditions, nonspecific antibody interactions can be minimized since these antibodies generally have lower affinities and will be less likely to bind. Also, nonspecific background staining owing to protein–protein interactions can be reduced, since the interacting molecule is diluted as well.

The affinity of an antibody is the strength of noncovalent binding of the immunoglobulin to a single site on the antigen molecule. These high-affinity antibodies are usually produced by the immunized animal in the later stages of the immune response where the antigen concentration becomes limiting. Affinities are expressed as affinity constants (K_a) and, for "good" antibodies, are generally in the range of 10^5–10^8 M^{-1} depending on the antigen. Antibody affinities can be determined by a number of methods *(22)*. The most reliable measurements are made by equilibrium dialysis. This technique is, however, best suited for antibodies raised to small soluble molecules that are freely diffusible across a dialysis membrane. Solution binding assays using radiolabeled immunoglobulins are generally performed to measure affinities for larger antigens. In some instances, avidity is used to describe the binding of the antibody–antigen interaction. Avidity refers to the binding of antibodies to multiple antigenic sites in serum and encompasses all the forces involved in the antibody–antigen interaction, including the serum pH and salt concentrations.

The titer of an antibody describes the immunoglobulin concentration in serum and is a measure of the highest dilution that will still give a visible antibody–antigen precipitation. Higher antibody titers are usually obtained after repeated antigen boosts. Antibody titers can be determined by double-diffusion assays in gels, enzyme-linked immunosorbent assays (ELISA), radioimmunoadsorbent assays (RIA), Western blotting, or other techniques *(17,22–24)*. These methods will detect the presence and also to some extent the specificity of a particular antibody, but will not ensure that the antibody is also suitable for immunocytochemistry *(25)*. For this reason, the antibody should be tested under the experimental conditions of fixing, embedding, and staining, and on the desired tissue to be used subsequently.

The power and accuracy of immunocytochemical techniques rely on the specificity of the antibody–antigen interaction. Undesirable or nonspecific staining can either be the result of the reagents used in the staining assay or crossreactivity of the immunoglobulin solution *(25)*. Background staining resulting from reagents can be overcome more easily by using purified reagents and optimizing conditions for tissue preparation and staining. Nonspecific binding can also be observed owing to ionic interactions with other proteins or organelles in the tissue preparation *(26)*. These interactions can be reduced by diluting the antibody and by increasing the salt concentration in the diluent and the washing solutions. In many instances, entire, sometimes semipure protein molecules, as well as conjugated or fusion proteins are used as immunogens. This leads to the production of a heterogenous antibody population with considerable crossreactivity to the contaminants. Therefore, these antibodies have to be purified by affinity chromatography before they can be used in immunocytochemical assays. The disadvantage of such purifications is that the most desirable immunoglobulins with the highest affinity will be bound the tightest and will be the most difficult to recover. Crossreactivities to the carrier protein to which the antigen has been conjugated or fused can be easily removed by affinity chromatography to the carrier. Increased antibody specificity can be obtained by using either synthetic peptides or protein fragments as antigens. Monoclonal antibodies are the most specific, since the isolation steps employed are designed to obtain a single clonal population of cells producing immunoglobulins against one antigenic site. Undesirable crossreactivities can, however, still occur if the antibody recognizes similar sites on related molecules or if the antigenic determinant is conserved in a family of proteins. Other potential sources of crossreactivity can be observed with tissues or cells containing F_c receptors that will bind the Fc region of primary or secondary immunoglobulins, in some cases with high affinity. These nonspecific sites have to be blocked first with normal serum or nonimmune immuno-globulins. If a secondary antibody is used for detection, the normal serum or immunoglobulin for blocking should be from the same species as the secondary antibody. Alternatively F(ab')$_2$ fragments can be used for detection.

4. Essential Controls for Specificity

As noted above, the specificity of the antibody–antigen reaction is critical for obtaining reliable, interpretable results. For this reason, the antibody has to be tested rigorously, and essential controls for antibody specificity should be included in any experimental design. A comprehensive discussion on antibody generation, specificity, and testing for immunocytochemical applications can be found in references *(27–29)* and, for specific applications, *see* Chapters 17, 50, and 51.

Initial specificity assays, such as Western blotting, immunoprecipitations, ELISAs, or RIAs, are performed with the purified antigen or a known positive cell extract. Specificity should also be demonstrated by preadsorbing the anti-body with the desired antigen, which should lead to loss of reactivity, whereas preadsorption with an irrelevant antigen should not diminish labeling. Alterna-tively, if the immunoreactive component is only partially purified from the tissue, detection of the desired component with the antibody should coincide with the presence of the molecule in fractions where the molecule of interest can be detected by its biochemical characteristics. These controls can be prob-lematic, however, since they require large amounts of purified or partially purified antigen. Controls in which a cell type completely lacks an antigen or into which an antigen's gene has been transfected into a negative cell type serve as better demonstrations of specificity.

A specific antibody should only stain the appropriate tissue, cell, or organelle. The use of either preimmune serum or an inappropriate primary antibody carried through the entire labeling assay serves as a negative control for the secondary antibody as well as the labeling procedure itself. Similarly, if the first antibody is omitted, no reaction due to inappropriate binding of the secondary antibody should occur. False positive reactions can be the result of background from fixed serum proteins within the tissue or faulty technique: inadequate washes, wrong antibody titers, overdigestion with protease, or arti-fact due to air drying. In clinical diagnoses, internal positive controls consist-ing of normal antigen-positive tissue adjacent to the tumor tissue are the most valuable since fixation is identical for both tissues.

References

1. VanNoorden, S. and Polak, J. M. (1983) Immunocytochemistry today: techniques and practice, in *Immunocytochemistry, Practical Applications in Pathology and*

Biology (Polak, J. M. and VanNoorden, S., eds.), Wright PSG, Bristol, England, pp. 11–42.

2. Sternberger, L. A. (1979) *Immunocytochemistry*, 2nd ed. Wiley, New York.

3. Coons, A. H., Creech, H. J., and Jones, R. N. (1941) Immunological properties of an antibody containing a fluorescent group. *Proc. Soc. Exp. Biol. Med.* **47,** 200–202.

4. Coons, A. H., Leduc, E. H., and Connolly, J. M. (1955) Studies on antibody production. I. A method for the histochemical demonstration of specific antibody and its application to a study of the hyperimmune rabbit. *J. Exp. Med.* **102,** 49–60.

5. Nakane, P. K. and Pierce, G. B., Jr. (1966) Enzyme-labeled antibodies: preparation and application for the localization of antigen. *J. Histochem. Cytochem.* **14,** 929–931.

6. Engvall, E. and Perlman, P. (1971) Enzyme-linked immunosorbent assay (ELISA). Quantitative assay of immunoglobulin G. *Immunocytochemistry* **8,** 871–874.

7. Massayeff, R. and Maillini, R. (1975) A sandwich method of enzyme immunoassay. Application to rat and human α-fetoprotein. *J. Immunol. Methods* **8,** 223–234.

8. Mason, D. Y. and Woolston, R. E. (1982) Double immunoenzymatic labeling, in *Techniques in Immunocytochemistry*, vol. 1 (Bullock, G. and Petrusz, P., eds.), Academic, London, pp. 135–152.

9. Campbell, G. T. and Bhatnagar, A. S. (1976) Simultaneous visualization by light microscopy of two pituitary hormones in a single tissue section using a combination of indirect immunohistochemical methods. *J. Histochem. Cytochem.* **24,** 448–452.

10. Mason, D. Y., Abdulaziz, Z, Falini, B., and Stein, H. (1983) Double immunoenzymatic labeling, in *Immunocytochemistry, Practical Applications in Pathology and Biology* (Polak, J. M. and VanNoorden, S., eds.), Wright PSG, Bristol, England, pp. 113–128.

11. Singer, S. J. (1959) Preparation of an electron-dense antibody conjugate. *Nature* **183,** 1523–1524.

12. Faulk, W. P. and Taylor, G. M. (1971) An immunocolloid method for the electron microscope. *Immunochemistry* **8,** 1081–1083.

13. Roth, J., Bendagan, M., and Orci, L. (1978) Ultrastructural localization of intracellular antigens by use of Protein-A gold complex. *J. Histochem. Cytochem.* **26,** 1074–1081.

14. Larsson, L.-I. and Schwartz, T. W. (1977) Radioimmunocytochemistry—a novel immunocytochemical principle. *J. Histochem. Cytochem.* **25,** 1140–1146.

15. Cuello, A. C., Priestley, J. V., and Milstein, C. (1982) Immunocytochemistry with internally labeled monoclonal antibodies. *Proc. Natl. Acad. Sci. USA* **78,** 665–669.

16. Livingston, D. M. (1974) Immunoaffinity chromatography of proteins. *Methods Enzymol.* **34,** 723–731.

17. Clausen, J. (1981) Immunochemical techniques for the identification and estimation of macromolecules, in *Laboratory Techniques in Biochemistry and Molecular Biology,* vol. 1, pt. 3 (Work, T. S. and Work, E., eds.), Elsevier, Amsterdam, pp. 52–155.

18. Brown, R. K. (1967) Immunological techniques (general). *Methods Enzymol.* **11,** 917–927.

19. Van Regenmortel, M. H. V., Briand, J. P., Muller, S., and Plaué, S. (1988) Immunization with peptides. Synthetic peptides as antigens, in *Laboratory Techniques in Biochemistry and Molecular Biology,* vol. 19 (Burdon, R. H. and van Knippenberg, P. H., eds.), Elsevier, Amsterdam, pp. 131–158.

20. Kohler, G. and Milstein, C. (1975) Continuous cultures of fused cells secreting antibody of predefined specificity. *Nature* **256,** 495–497.

21. Galfre G. and Milstein, C. (1981) Preparation of monoclonal antibodies: strategies and procedures. *Methods Enzymol.* **73,** 3–46.

22. Nisonoff, A. (1984) Specificities, affinities, and reaction rates of antihapten antibodies, in *Introduction to Molecular Immunology.* Sinauer, Sunderland, MA, pp. 29–43.

23. Oudin, J. (1980) Immunochemical analysis by antigen–antibody precipitation in gels. *Methods Enzymol.* **70,** 166–198.

24. VanVunakis, H. (1980) Radioimmunoassays: an overview. *Methods Enzymol.* **70,** 201–209.

25. Vandesande, F. (1979) A critical review of immunocytochemical methods for light microscopy. *J. Neurosci. Methods* **1,** 3–23.

26. Grube, D. (1980) Immunoreactivities of gastrin (G) cells. II. Nonspecific binding of immunoglobulins to G-cells by ionic interactions. *Histochemistry* **66,** 149–167.

27. DeMey, J. and Moeremans, M. (1986) Raising and testing polyclonal antibodies for immunocytochemistry, in *Immunocytochemistry: Modern Methods and Applications* (Polak, J. M. and VanNoorden, S., eds.), Wright, Bristol, England, pp. 3–12.

28. Ritter, M. A. (1986) Raising and testing monoclonal antibodies for immunocytochemistry, in *Immunocytochemistry: Modern Methods and Applications* (Polak, J. M. and VanNoorden, S., eds.), Wright, Bristol, England, pp. 13–25.

29. VanNoorden, S. (1986) Tissue preparation and immunostaining techniques for light microscopy, in *Immunocytochemistry: Modern Methods and Applications* (Polak, J. M. and VanNoorden, S., eds.), Wright, Bristol, England, pp. 26–53.

2

Purification of Antibodies
Using Ammonium Sulfate Fractionation
or Gel Filtration

Ute M. Kent

1. Introduction

In this chapter, two commonly used techniques that are utilized in many immunoglobulin purification schemes are described. The first procedure, ammonium sulfate fractionation, is generally employed as the initial step in the isolation of crude antibodies from serum or ascitic fluid *(1–5)*. Ammonium sulfate precipitation, in many instances, is still the method of choice because it offers a number of advantages. Ammonium sulfate fractionation provides a rapid and inexpensive method for concentrating large starting volumes. "Salting out" of polypeptides occurs at high salt concentrations where the salt competes with the polar side chains of the protein for ion pairing with the water molecules, and where the salt reduces the effective volume of solvent. As expected from these observations, the amount of ammonium sulfate required to precipitate a given protein will depend mainly on the surface charge, the surface distribution of polar side chains, and the size of the polypeptide, as well as the pH and temperature of the solution. Immunoglobulins precipitate at 40–50% ammonium sulfate saturation depending somewhat on the species and subclass *(3)*. The desired saturation is brought about either by addition of solid ammonium sulfate or by addition of a saturated solution. Although the use of solid salt reduces the final volume, this method has a number of disadvantages. Prolonged stirring, required to solubilize the salt, can lead to denaturation of proteins in the solution at the surface/air interface *(6)*. Localized high concentrations of the ammonium sulfate salt may cause unwanted proteins to precipitate. Since ammonium sulfate is slightly acidic in solution, the pH of the protein solution requires constant monitoring and adjustment if solid salt is added.

From: *Methods in Molecular Biology, Vol. 115: Immunocytochemical Methods and Protocols*
Edited by: L. C. Javois © Humana Press Inc., Totowa, NJ

Therefore, it is advisable to add a buffered solution of saturated ammonium sulfate. A saturated ammonium sulfate solution is considered to be 100%, and for most antibody purification purposes, serum or ascites are mixed with an equal volume of saturated ammonium sulfate to give a 50% solution. Tables for determining amounts of solid or saturated solution to be added to achieve a desired percentage of saturation or molarity can be found in most biochemical handbooks *(7)*. The density of a saturated ammonium sulfate solution at 20°C is 1.235 g/cm^3 *(4)*. This is sufficiently low to allow removal of precipitated proteins by centrifugation. Ammonium sulfate has been found to stabilize proteins in solution by raising the midpoint temperature at which proteins can be unfolded *(8)*. This effect is thought to be the result of the interaction of the salt with the structure of water. Precipitated immunoglobulins can therefore be solubilized in a minimal volume of buffer and stored for extended periods without significant loss of bindability or proteolytic degradation. Complete precipitation occurs within 3–8 h at 4°C. The precipitate is then collected by centrifugation, solubilized in an appropriate volume of buffer for storage at –80°C, or dialyzed to remove residual salt prior to further purification. Although fractionation with ammonium sulfate provides a convenient method for substantial enrichment of immunoglobulins, it should not be used as a single-step purification, since the precipitated material still contains considerable quantities of contaminating proteins. Additional procedures for further purification of antibodies are discussed in Chapters 3–5. In most instances, residual high concentrations of salt interfere with subsequent purification methods or further use of the antibody. Ammonium sulfate can easily be removed by dialysis of the protein solution against large volumes of the desired buffer. Although dialysis is still a very common method of salt removal, it is somewhat time-consuming. An alternative method for removal of small molecules from proteins is gel-filtration or gel-permeation chromatography. Gel filtration is a general, simple, and gentle method for fractionating molecules according to their size. Excellent reviews on gel-permeation chromatography theory and principles can be found in **refs. *9–11***. Successful resolution in gel filtration depends mainly on the inclusion and exclusion range of the stationary matrix, the column dimensions, and the size of the sample applied. The matrix should be compatible with the buffers of choice, exhibit good flow characteristics, and not interact significantly with the proteins in the sample. The eluting buffer should, therefore, contain a certain concentration of salt, usually 50–150 m*M*, to minimize nonspecific protein–matrix interactions. Agarose-based gel-filtration matrices like the Sephadex G series, Sepharose CL, or Superose (Pharmacia-LKB, Piscataway, NJ) have been widely used since they provide all of these desired characteristics. Superose 6 is extremely useful since it has a large separation range for molecules of 5×10^3 to 5×10^6 Dalton.

The pore size of the matrix should be chosen according to the particular application. For simple desalting, buffer exchange, or for the removal of small reaction byproducts (*see* Chapter 6), the matrix should retain the small molecules within the total column volume, whereas the proteins of interest should elute in the excluded or void volume. In general, the excluded volume represents about one-third of the total column volume. The major disadvantage of gel-filtration chromatography is the limited sample size that can be applied at one time. The volume of the sample is critical for optimal separation and should not exceed 1–10% of the total column volume. For good resolution of complex protein mixtures that chromatograph within the included volume of the column, the sample size should not exceed 1%, whereas for desalting procedures, the sample volume may approach 5–10% of the total column volume. For this reason, it is usually necessary to include a concentration step prior to gel filtration.

2. Materials

1. Serum or ascites (100 mL).
2. BBS (200 m*M* sodium borate, 160 m*M* sodium chloride): Dissolve 247.3 g of boric acid, 187 g of NaCl and 75 mL of 10 *M* NaOH in 4 L of water. Check the pH of the solution and adjust with 10 *M* NaOH to pH 8.0. Add water to bring the solution to a final volume of 20 L (*see* **Note 1**).
3. Saturated ammonium sulfate (enzyme grade): Dissolve 800 g of ammonium sulfate in 1 L of hot BBS. Filter the solution through Whatman no. 1 paper and cool to room temperature. Confirm that the pH of the solution is 8.0 with a strip of narrow-range pH paper. Cool the saturated ammonium sulfate solution and store at 4°C (*see* **Note 2**).
4. Whatman no. 1 filter paper.
5. pH paper.
6. 200 m*M* Sodium bicarbonate, 5 m*M* EDTA.
7. Millipore- or HPLC-quality water (*see* **Note 3**).
8. 20% Ethanol and 20% ethanol (HPLC grade).
9. Dialysis tubing: Spectrapor, mol-wt cut of 3–10,000 Dalton.
10. 10- to 50-mL Round-bottom polycarbonate centrifuge tubes.
11. 100-mL Graduated cylinder.
12. 10-mL Pipets.
13. 4-L Beaker.
14. Protein concentrator, 50 mL (Amicon, Beverly, MA or Pharmacia).
15. Ultrafiltration membranes, 43 mm, PM30 (Amicon).
16. N_2 tank.
17. Superose 6 column (Pharmacia).
18. FPLC system (Pharmacia, P-500 pumps, Frac-100 fraction collector, HR flow cell, UV-1 flow-through monitor, V-7 valve).
19. 0.22-μm Millex-GV syringe filters (Millipore, Bedford, MA).

20. Buffer filtration device (either a glass filtration unit, fitted with a 0.45-μm membrane, connected to a side-arm flask or a tissue culture sterilization filter unit).
21. Centrifuge equipped with a rotor that will accommodate 50-mL round-bottom tubes and can be operated at 10,000g.
22. Vacuum source or a bath sonicator to degas buffers.
23. Spectrophotometer and UV-light-compatible cuvets.

3. Methods

3.1. Ammonium Sulfate Precipitation and Dialysis

1. Pipet 25 mL of serum or ascitic fluid into each of four 50-mL polycarbonate tubes. Centrifuge at 10,000g for 30 min at 4°C to remove any large aggregates (*see* **Note 4**).
2. Carefully decant the supernatant into a 100-mL graduated cylinder, and adjust the volume to 96 mL with BBS (*see* **Note 5**).
3. Pipet 16 mL of the clarified serum or ascites into each of six clean 50-mL polycarbonate tubes. Add 16 mL of cold, saturated ammonium sulfate solution and stir gently with a pipet (*see* **Note 6**).
4. Let the solutions stand on ice for 3 h (*see* **Note 7**).
5. Centrifuge at 10,000g for 30 min at 4°C.
6. Carefully decant the supernatant (*see* **Note 8**).
7. Dissolve the pellet in a minimal volume (approx 30 mL) of cold BBS (*see* **Note 9**).
8. Prepare the dialysis tubing. Cut the dry tubing into strips of manageable lengths (approx 30–40 cm) (*see* **Note 10**).
9. Add the cut tubing to the sodium bicarbonate/EDTA solution and heat to 90°C for 30 min. Stir the tubing periodically with a polished glass rod (*see* **Note 11**).
10. Rinse the tubing well with several changes of deionized water. Store tubing in 20% ethanol at 4°C until needed.
11. Remove the required number of strips of dialysis membrane and rinse them well with water to remove all traces of ethanol.
12. Tie a knot into one end of the tubing and check for leakage (*see* **Note 12**).
13. Fill the tubing to approx one-half of its capacity with the crude immunoglobulin solution from **step 9** (*see* **Note 13**), and close the tubing with a knot or a dialysis tubing clip.
14. Place the filled bag into a 1-L graduated cylinder filled with cold BBS. Dialyze with stirring against at least four to five changes of buffer for a minimum of 4 h each time.

3.2. Protein Concentration and Storage

1. Remove the dialyzed protein solution and estimate the amount of protein recovered (*see* **Note 14**). Dilute 100 μL of the dialyzed protein solution with 900 μL of BBS. Using BBS as a blank, read the absorbance of the diluted solution at 280 nm. A 1-mg/mL solution of protein consisting mainly of immunoglobulins will have an absorbance of approx 1.4 if read in a cuvet with a 1-cm path length. Therefore,

divide the measured absorbance reading by 1.4 to arrive at a concentration estimate in mg/mL for the 10-fold diluted sample (*see* **Note 15**).

2. Assemble the protein concentration apparatus according to the manufacturer's instructions or *see* **ref. *12*** (*see* **Note 16**).

3. Concentrate the immunoglobulin solution to approx 10 mg/mL under N_2 on ice with gentle stirring.

4. Estimate the final protein content as in **step 1** above and store the antibody solution at 4°C for short-term storage (weeks) or at –80°C for long-term storage (months to years) (*see* **Note 17**).

3.3. Gel Filtration by Fast Protein Liquid Chromatography (FPLC)

1. Filter BBS through a 0.45-μm filtration membrane. Degas the buffer by applying vacuum for 30 min or by sonicating in a bath sonicator for 5 min.

2. Connect the Superose 6 column to the FPLC system (*see* **Note 18**), and equilibrate the column with 50 mL of BBS at a flow rate of 0.5 mL/min. Check the manufacturer's recommendations for optimal operating back pressures.

3. Filter the protein sample through a 0.22-μm syringe filter, and inject the sample onto the column (*see* **Note 19**).

4. Elute with BBS at 0.5 mL/min and monitor the effluent at 280 nm. Collect 0.5- to 1-mL fractions.

5. The monomeric immunoglobulins will elute after about 30 min (*see* **Note 20**).

6. Collect the IgG-containing fractions and determine the protein concentration by reading the absorbance at 280 nm (*see* **Note 21**).

7. Wash the column with 50 mL of BBS. For short-term storage (days) the column can be stored in BBS (*see* **Note 22**). For long-term storage, wash the column with 75–80 mL of water, followed by 50 mL of 20% ethanol (HPLC grade). Disconnect the column and cap the ends to prevent the matrix from drying out.

4. Notes

1. BBS is a good buffer for storing antibodies because of its bacteriostatic qualities. Care must, however, be taken when adding antibodies in BBS to living cells so that the final volume of BBS added does not exceed 10%.

2. The pH of saturated ammonium sulfate can be checked either directly with narrow-range pH paper or after 10-fold dilution with a pH meter. Excess ammonium sulfate should precipitate out in the cold. The solution above the ammonium salt is considered to be 100% saturated.

3. The quality of water used in chromatography and antibody purification is important for long-term antibody integrity, as well as column and equipment performance and longevity. The water should also have a low-UV absorbance in order not to interfere with the detection of the desired protein and be free of particulate material, which can clog the columns and tubing. Therefore, Millipore- or HPLC-grade water is preferable. Alternatively, glass-distilled, filtered water can be used. Glass-distilled water does, however, sometimes contain dissolved organic material that can lead to a high baseline and interfere with protein detection.

4. In general, serum should be heat-inactivated by heating at 56°C for 15 min to inactivate complement components prior to ammonium sulfate fractionation. Ascitic fluid should first be filtered through a cushion of glass wool.

5. All steps should be performed in a cold room or on ice to avoid denaturation of proteins or proteolysis.

6. A number of references indicate that ammonium sulfate should be added gradually while stirring on ice. The main reason for this suggestion is to reduce the possibility of local high concentrations of the saturated salt, which can lead to precipitation of undesirable proteins. This is generally only of major concern when trying to precipitate a particular enzyme at a very defined concentration of salt. For immunoglobulin purification, this need not be considered, since antibodies comprise the major fraction of protein in serum or ascites. When pipeting protein solutions, try to avoid bubble formation since this can lead to denaturation of proteins.

7. Since ammonium sulfate fractionation is a crude procedure for antibody purification, this step may also be extended from 3 h to overnight for convenience.

8. If a cleaner precipitate is required, the pellet can be redissolved and reprecipitated at this step.

9. Dislodge the pellet from the sides of the tube with a pipet and gently resuspend the precipitate by pipeting up and down without creating bubbles. The precipitate may be solubilized more easily after letting the dislodged pellet sit in buffer on ice for 30 min.

10. Dialysis tubing is treated with glycerol and preservatives that need to be removed prior to use. Handle the membrane with gloves to avoid introduction of proteolytic enzymes and to reduce punctures.

11. There should only be enough tubing in the beaker to allow free movement of the tubing when stirred. Do not boil the membrane, since this can change the pore size. Do not let the tubing dry out at any time after this step.

12. For additional safety, a second successive knot should be tied at the end of the tubing. Alternatively, the ends of the tubing can be closed with dialysis tubing clips. Test the tubing for leaks by filling it with water, pinching the ends closed, and applying slight pressure to the bag.

13. Leave enough space in the dialysis bag so that the volume can double during dialysis.

14. After dialysis, the protein solution will still be somewhat opalescent. Any precipitated material containing mainly denatured proteins should be removed by centrifugation.

15. Since ammonium sulfate fractionation will also cause precipitation of other proteins, antibody concentrations obtained from absorbance measurements at 280 nm are only estimates. Alternatively, a sample of the dialyzed solution can be resolved on a SDS-polyacrylamide gel alongside a series of known concentrations of IgG. Staining the gel with Coomassie blue can then be used to estimate the amount of immunoglobulin obtained and can also give an estimate of purity.

16. The ultrafiltration membrane is treated with glycerol and preservatives that need to be removed prior to use. Float the membrane, shiny side down, on water for a few hours. Rinse the membrane and insert it into the ultrafiltration apparatus with the shiny side up. The membrane can be stored in 20% ethanol and reused.

17. To obtain an accurate absorbance reading within the linear range, the sample may need to be diluted more than 10-fold. The absorbance of the diluted sample should not be >1.5. The protein concentration in mg/mL is obtained by dividing the absorbance of the diluted sample by 1.4 and multiplying by the dilution factor. Aliquot the appropriate quantities that may be required for later use or subsequent purification steps. Optimal concentrations for storage are between 1 and 10 mg/mL, depending on the antibody. Avoid repeated freezing and thawing of protein solutions, since this denatures the polypeptides.

18. Other systems with similar components can also be used, provided they can be operated at flow rates that will be compatible with the column-operating pressures. For some systems, additional column fittings may be required to facilitate connection of the Superose 6 column. If the purpose of the gel-filtration step is to exchange buffers, then the column should be equilibrated and eluted with the buffer that the sample is to be exchanged into. Optimal separation of sample components can be achieved with a sample volume of 200 µL. For desalting or buffer exchange, a sample volume of up to 2 mL can be used.

19. Avoid drawing bubbles into the syringe. If injected onto the column, these bubbles will be detected by the UV monitor as spurious peaks.

20. In general, a threefold dilution of the injected sample volume is to be expected.

21. If necessary, the antibodies can be concentrated after this step. This can be conveniently accomplished using Centricon centrifuge concentrators (Amicon).

22. In general, chromatography columns should not be left connected to pumps or to the UV monitors in salt solutions. Always include a wash step with water to remove any salt from the system. It is preferable to store the columns disconnected in 20% ethanol and to rinse the entire FPLC system, including pumps, tubing, and UV flow cell with water, followed by 20% ethanol. Keep a record of the column performance, and use it to determine when filter changes or column-cleaning steps are required.

References

1. Manil, L., Motte, P., Pernas, P., Troalen, F. Bohuon, C., and Bellet, D. (1986) Evaluation of protocols for purification of mouse monoclonal antibodies. *J. Immunol. Methods* **90,** 25–37.

2. Holowka, D. and Metzger, H. (1982) Further characterization of the beta-component of the receptor for immunoglobulin E. *Mol. Immunol.* **19,** 219–227.

3. Harlow, E. and Lane, D. (1988) Storing and purifying antibodies, in *Antibodies. A Laboratory Manual.* Cold Spring Harbor Laboratory, Cold Spring Harbor, NY, Chapter 8.

4. England, S. and Seifter, S. (1990) Precipitation techniques. *Methods Enzymol.* **182,** 285–296.

5. Jaton, J. C., Brandt, D. Ch., and Vassalli, P. (1979) The isolation and characterization of immunoglobulins, antibodies, and their constituent polypeptide chains, in *Immunological Methods*, vol. 1 (Lefkovits, I. and Pernis, B., eds.), Academic, New York, pp. 43–67.
6. Dixon M. and Webb, E. C. (1979) Enzyme techniques, in *Enzymes*. Academic, New York, pp. 11–13.
7. Suelter, C. H. (1985) Purification of an enzyme, in *A Practical Guide to Enzymology*, Chapter 3. Wiley, New York, pp. 78–84.
8. von Hippel, P. H. and Wong, K.-Y. (1964) Neutral salts: The generality of their effects on the stability of macromolecular conformations. *Science* **145,** 577–580.
9. *Gel Filtration–Theory and Practice.* (1984) Pharmacia Fine Chemicals, Rahms i Lund, Uppsala, Sweden.
10. Stellwagen, E. (1990) Gel filtration. *Methods Enzymol.* **182,** 317–328.
11. Harris, D. A. (1992) Size-exclusion high-performance liquid chromatography of proteins, in *Methods in Molecular Biology, vol. 11: Practical Protein Chromatography* (Kenney, A. and Fowell, S., eds.), Humana, Clifton, NJ, pp. 223–236.
12. Cooper, T. G. (1977) Protein purification, in *The Tools of Biochemistry,* Chapter 10. Wiley, New York, pp. 383–385.

3

Purification of Antibodies Using Ion-Exchange Chromatography

Ute M. Kent

1. Introduction

Ion-exchange chromatography is a rapid and inexpensive procedure employed to purify antibodies partially from different sources and species *(1–4)*. It is a particularly useful tool for isolating antibodies that either do not bind or that bind only weakly to protein A (e.g., mouse IgG_1) *(3)*. This purification method should not be used alone to obtain purified immunoglobulins from crude starting material, but should either be preceded by ammonium sulfate fractionation (*see* Chapter 2) or followed by affinity chromatography (*see* Chapter 4). The principles and theory of ion-exchange chromatography are discussed in detail by Himmelhoch *(5)*, and in reference *(6)*. Ion-exchange chromatography separates proteins according to their surface charge. Therefore, this separation is dependent on the p*I* of the protein of interest, the pH and salt concentration of the buffer, and on the charge of the stationary ion-exchange matrix. Proteins are reversibly bound to a charged matrix of beaded cellulose, agarose, dextran, or polystyrene. This interaction can be disrupted by eluting with increasing ionic strength or a change in pH. An ion-exchange matrix should be stable, have good flow characteristic, and not interact nonspecifically with proteins. The most commonly used matices with these qualities are the weak carboxymethyl cation-exchangers Cellex CM and CM Sephacel or strong sulfopropyl (SP) exchangers (Bio-Rad, Hercules, CA; Pharmacia-LKB, Piscataway, NJ), and the weak diethylaminoethyl anion exchangers Cellex D and DEAE-Sephacel, or strong quaternary aminoethyl (QAE) exchangers (Bio-Rad, Pharmacia). A protein will have a net positive charge below its p*I* and bind to a cation-exchanger, whereas above its p*I*, it will

From: *Methods in Molecular Biology, Vol. 115: Immunocytochemical Methods and Protocols*
Edited by: L. C. Javois © Humana Press Inc., Totowa, NJ

have a net negative charge and bind to an anion-exchange resin *(6)*. For optimal binding and elution, the pH of the equilibration buffer should be one pH unit above the p*I* of the protein of interest for cation-exchange and one pH unit below the p*I* for anion-exchange chromatography. Antibodies can be purified by either method, but are most frequently isolated by ion-exchange chromatography with DEAE resins using either a batch or column procedure *(1,7,8)*. Since antibodies have a net neutral charge at a pH near neutrality, two purification techniques can be employed. If the pH of the antibody solution is maintained at pH 6.5–7.0, the immunoglobulins will not be retained on the column and will elute first *(8,9)*. The disadvantage is that the trailing edge of the immunoglobulin peak is usually contaminated with other proteins. Alternatively, the immunoglobulins can be bound to the stationary matrix by ionic interactions near pH 8.0 and then eluted with a gradient of increasing ionic strength *(4)*. Monoclonal antibodies from ascites have also been successfully purified by ion-exchange using a Mono Q exchanger (Pharmacia) *(4,10,11)*. The Mono Q matrix is composed of a stable polymer for fast, high resolution. The Mono Q matrix contains quaternary amino groups ($-CH_2-N^+[CH_3]_3$) and belongs to the strong ion exchangers that allow separations to be carried out at pH ranges of 3.0–11.0. It can also be used to bind molecules in the presence of moderate concentrations of salts. This is particularly useful for some immunoglobulins that require a certain concentration of ionic strength for solubility. The exchanger has an ionic capacity of 300 μmol/mL or approx 20–50 mg of protein/mL of gel, and is stable to denaturants and organic solvents.

2. Materials

1. 15 mL Mouse hybridoma tissue-culture supernatant (approx 0.5 mg/mL).
2. Millipore- or HPLC-quality water (*see* Chapter 2; **Note 3**).
3. Buffer A: 20 m*M* triethanolamine, pH 7.7.
4. Buffer B: 20 m*M* triethanolamine, pH 7.7, 350 m*M* NaCl.
5. 2 *M* Sodium chloride.
6. 2 *M* Sodium hydroxide.
7. 20% Ethanol (HPLC grade).
8. Mono Q (HR5/5) (Pharmacia-LKB).
9. FPLC components (two P 500 pumps, V7 injection valve, gradient controller, UV-1 detector, Frac-100 fraction collector, 50 mL Superloop [Pharmacia], dual-channel chart recorder, or similar components).
10. 0.22-μm Millex-GV Syringe filter (Millipore, Bedford, MA).
11. 20-mL Disposable syringe.
12. Glass-filtration device, or a 500-mL filter-sterilization flask with a 0.45-μm membrane.
13. 0.45-μm Membranes (Millipore).

3. Methods

3.1. Mono Q Ion-Exchange Chromatography by Fast Protein Liquid Chromatography (FPLC)

3.1.1. Sample Application and Elution

1. Dialyze the tissue-culture supernatant against 500 mL buffer A for 4 h or overnight at 4°C (*see* **Note 1**).
2. Remove any precipitated proteins by centrifugation at 10,000*g* for 30 min at 4°C.
3. Filter the sample through a 0.22-μm syringe filter.
4. Filter all buffers or solutions to be used in the chromatography steps through a 0.45-μm filter (*see* **Note 2** and Chapter 2, gel filtration), and equilibrate the Mono Q column with 10 mL buffer A at a flow rate of 1 mL/min.
5. Apply the sample with a 50-mL Superloop in buffer A (*see* **Notes 3** and **4**).
6. Elute the mouse immunoglobulins with a gradient of 0% buffer B to 100% buffer B in 25 min at a flow rate of 1 mL/min.
7. Collect 1-mL fractions and monitor the effluent at 280 nm.
8. The major peak, containing the desired IgG, should elute near 150–180 m*M* NaCl (*see* **Note 5**).

3.1.2. Column Regeneration and Storage

1. Disconnect the column and reconnect it in reverse (*see* **Note 6**).
2. Inject 1 mL of filtered 2 *M* sodium chloride and wash with 10 mL of buffer B at 0.2 mL/min. Inject 1 mL of filtered 2 *M* sodium hydroxide.
3. Wash with 20 mL of Millipore-quality water, and re-equilibrate the column in equilibration buffer if another run is to be performed.
4. For storage, the column should be equilibrated with 10 mL 20% ethanol after the water wash in **step 3**.

4. Notes

1. If the starting material is ascitic fluid or serum, then the sample should first be partially purified by ammonium sulfate fractionation followed by extensive dialysis or gel filtration (*see* Chapter 2).
2. All buffers and solutions used for chromatography should be prepared with high-quality water, filtered, and degassed. Careful attention to this will result in decreased buffer backgrounds or spurious peaks owing to contaminants or air bubbles. Particles in the buffers can shorten the column life and plug the column or tubing.
3. For some samples, the buffer pH or gradient conditions may need to be adjusted for optimal binding and separation. It is advisable to test unknown samples by first injecting only 100–200 μg of protein. If no binding occurs, raise the pH of the starting buffer by 0.5-U increments.
4. Although the theoretical capacity for the column is higher, the recommended quantity of protein that can be loaded is 25 mg.

5. The flanking fractions of the main IgG peak may contain small quantities of contaminating proteins. Each fraction should be analyzed by SDS-polyacrylamide gel electrophoresis before the desired fractions are pooled (*see* Chapter 2, **Note 15**).
6. Reversing the column flow results in a more efficient removal of proteins and other impurities. Since the majority of these molecules are likely bound at the highest concentration at the top of the matrix, washing the column in reverse also ensures that these molecules are not washed through the entire matrix. The column should be disconnected from the UV monitor during the initial washing steps, so large protein aggregates do not block the narrow-flow cell tubing.

References

1. Sampson, I. A., Hodgen, A. M., and Arthur, I. H. (1984) The separation of IgM from human serum by FPLC. *J. Immunol. Methods* **69**, 9–15.
2. James, K. and Stanworth, D. R. (1964) Studies on the chromatography of human serum proteins on diethylaminoethyl(DEAE)-cellulose. (I) The effect of the chemical and physical nature of the exchanger. *J. Chromatog.* **15**, 324–335.
3. Manil, L., Motte, P., Pernas, P., Troalen, F., Bohuon, C., and Bellet, D. (1986) Evaluation of protocols for purification of mouse monoclonal antibodies. Yield and purity in two-dimensional gel electrophoresis. *J. Immunol. Methods* **90**, 25–37.
4. Clezardin, P., McGregor, J. L., Manach, M., Boukerche, H., and Dechavanne, M. (1985) One-step procedure for the rapid isolation of mouse monoclonal antibodies and their antigen binding fragments by fast protein liquid chromatography on a mono Q anion-exchange column. *J. Chromatogr.* **319**, 67–77.
5. Himmelhoch, S. R. (1971) Chromatography of proteins on ion-exchange adsorbents. *Methods Enzym.* **22**, 273–286.
6. *FPLC Ion Exchange and Chromatofocusing—Principles and Methods.* (1985) Pharmacia-LKB, Offsetcenter, Uppsala, Sweden.
7. Jaton, J.-C., Brandt, D. Ch., and Vassalli, P. (1979) The isolation and characterization of immunoglobulins, antibodies, and their constituent polypeptide chains, in *Immunological Methods*, vol. 1 (Lefkovits, I. and Pernis, B., eds.), Academic, New York, pp. 45,46.
8. Webb, A. J. (1972) A 30 min preparative method for isolation of IgG from human serum. *Vox Sang.* **23**, 279–290.
9. Phillips T. M. (1992) *Analytical Techniques in Immunochemistry.* Marcel Dekker, New York, pp. 22–39.
10. Burchiel S. W., Billman, J. R., and Alber, T. R. (1984) Rapid and efficient purification of mouse monoclonal antibodies from ascites fluid using high performance liquid chromatography. *J. Immunol. Methods* **69**, 33–42.
11. Tasaka, K., Kobayashi, M., Tanaka, T., and Inagaki, C. (1984) Rapid purification of monoclonal antibody in ascites by high performance ion exchange column chromatography for diminishing non-specific staining. *Acta Histochem. Cytochem.* **17**, 283–286.

4

Purification of Antibodies Using Affinity Chromatography

Ute M. Kent

1. Introduction

The effectiveness of affinity chromatography relies on the ability of a molecule in solution to recognize specifically an immobilized ligand *(1–3)*. This type of separation, unlike other chromatographic methods, uses the intrinsic biological activity of a molecule to bind to a substrate, hapten, or antigen. Principles of matrix selection, gel preparation, and coupling of ligands have been reviewed extensively by Ostrove *(2)*. Antibody affinity chromatography has been employed to isolate antigen-specific antibodies (antibodies raised against a particular protein), hapten-specific antibodies (antipeptide antibodies, antiphosphotyrosine antibodies, anti-TNP antibodies), or species-specific immunoglobulins, or to separate crossreacting immunoglobulins from the antibody of interest *(3–7)*.

Several types of affinity matrices are commercially available. The most common matrix for coupling of molecules is CNBr-activated Sepharose (Pharmacia-LKB, Piscataway, NJ) *(8)*. It is ideally suited for affinity chromatography for several reasons. Sepharose exhibits little nonspecific protein adsorption, is stable over a wide pH range, and can be used with denaturants or detergents. Because of its large pore size (exclusion limit of 2×10^7), the matrix has a high capacity and, therefore, allows for internal ligand attachment. Covalent coupling of ligands to the activated Sepharose occurs spontaneously at pH 8.0–9.0 through the unprotonated primary amino groups of the ligand. One disadvantage of this matrix, however, is that the isourea linkage formed between the activated matrix and the ligand is not completely stable, and will hydrolyze with time. This does not pose a significant problem when large pro-

From: *Methods in Molecular Biology, Vol. 115: Immunocytochemical Methods and Protocols*
Edited by: L. C. Javois © Humana Press Inc., Totowa, NJ

teins like immunoglobulins are used as an affinity ligand, since the protein is usually bound by several attachments. Another disadvantage of this type of affinity matrix is that the ligand is attached directly to the stationary matrix without an intervening spacer arm. This can lead to stearic hindrance in some applications, as in hapten affinity chromatography. In this type of isolation method, the ligand is generally only bound by a single attachment, and therefore, the linkage should also be more stable. For these instances, matrices containing different chemical coupling groups attached by spacer arms have been developed. Affi-gel 10 (Bio-Rad, Hercules, CA) provides an example of such a matrix composed of crosslinked agarose to which a neutral 10-atom spacer arm has been coupled via a stable ether bond. The reactive N-hydroxysuccinamide groups can react spontaneously with primary amino groups forming a stable amide linkage.

2. Materials

1. CNBr-activated Sepharose 4B (Pharmacia).
2. 10 mM HCl.
3. Sintered glass funnel (coarse, 50 mL).
4. Millipore-quality or distilled water (*see* Chapter 2, **Note 3**).
5. 5 mg Rat immunoglobulin (or other desired species).
6. Coupling buffer A: 100 mM NaHCO$_3$, pH 8.0, 500 mM sodium chloride.
7. 200 mM Glycine, pH 8.0.
8. 100 mM Sodium acetate buffer, pH 4.0, 500 mM sodium chloride.
9. Capped polycarbonate tubes (15 and 50 mL).
10. BBS: 200 mM boric acid, 160 mM sodium chloride, pH 8.0 (for preparation *see* Chapter 2).
11. Poly Prep chromatography columns, 0.8 × 4 cm (Bio-Rad).
12. Rabbit antirat IgG (from approx 20 mL rabbit serum).
13. 100 mM Glycine, pH 3.0.
14. 1M Tris-HCl, pH 8.0.
15. Peristaltic pump.
16. Affi-gel 10 resin (Bio-Rad).
17. Isopropanol (cold).
18. Coupling buffer B: 100 mM HEPES, pH 7.5.
19. 150 mM Phosphotyramine in coupling buffer B (*see* **refs. 6** and **9**).
20. Phosphate-buffered saline (PBS), pH 7.4: 1.7 mM potassium phosphate monobasic, 5 mM sodium phosphate dibasic, and 150 mM sodium chloride.
21. Ammonium sulfate-precipitated antiphosphotyrosine antibodies (*see* Chapter 2; hybridomas are available from ATCC).
22. Elution buffer: PBS, pH 7.4, and 10 mM p-nitrophenyl phosphate.
23. pH paper.
24. 0.02% Sodium azide in PBS (w/v).
25. Spectrophotometer and quartz or UV-compatible plastic cuvets.

3. Methods

3.1. Affinity Chromatography with CNBr-Activated Sepharose

3.1.1. Resin Preparation and Coupling

1. Weigh out 1 g CNBr-activated Sepharose 4B and sprinkle it over 20 mL 10 mM HCl (*see* **Note 1**).
2. Wash the swollen gel on a 50-mL coarse sintered glass funnel with 4 × 50 mL 10 mM HCl by repeatedly suspending the matrix in the HCl solution followed by draining with vacuum suction (*see* **Note 2**).
3. Suspend 5 mg of rat immunoglobulin in 5 mL of coupling buffer A.
4. Add this suspension to the gel and mix end over end in a capped 15-mL polycarbonate tube overnight at 4°C (*see* **Note 3**).
5. Pour the matrix into a sintered glass funnel and drain the gel. Reserve the eluate to estimate how much antibody has been coupled (*see* **Note 4**).
6. Wash the matrix with 100 mL of coupling buffer A to remove any unbound ligand.
7. Suspend the matrix in 45 mL 200 mM glycine, pH 8.0, and tumble end over end in a 50-mL capped tube at 4°C overnight to block any unreacted groups.
8. Drain and wash the gel with three cycles of alternating pH. First, suspend the drained gel in 50 mL 100 mM sodium acetate, pH 4.0, and 500 mM NaCl. Drain with vacuum suction and wash with 50 mL coupling buffer A. Drain and repeat the alternating pH washes twice.
9. Suspend the gel in 20 mL of BBS. The affinity matrix is now ready for use in column chromatography or batch adsorption.

3.1.2. Sample Application and Elution

1. Pack the matrix in a Poly Prep column (Bio-Rad) (*see* **Note 5**).
2. Attach the column outlet to a peristaltic pump, and wash the column with 5 mL BBS at 0.5 mL/min.
3. Drain most of the BBS, leaving approx 0.5 mL on top of the gel bed.
4. Apply 15 mL of rabbit antirat IgG to the column, and circulate the solution through the matrix at 0.2 mL/min for 3 h at 4°C.
5. Drain the column as in **step 3** and save the eluate (*see* **Note 6**).
6. Wash the matrix with approx 10 column volumes of BBS until the absorbance of the eluate is <0.02 at 280 nm compared to the column buffer.
7. Remove the bound antibody with 5 mL of 100 mM glycine, pH 3.0, at 0.5 mL/min.
8. Collect 1-mL fractions into tubes containing 500 μL 1 M Tris-HCl, pH 8.0.
9. Pool the immunoglobulin-containing samples and concentrate as necessary (*see* **Note 7**).

3.1.3. Column Regeneration and Storage

1. Neutralize the column matrix immediately by washing with 20 mL of BBS. Ensure that the column is neutralized by checking the effluent with pH paper.
2. Store the column closed and capped in BBS at 4°C.

3. If the top of the column becomes dirty, remove a few millimeters of the discolored gel from the top of the matrix. The column should then be washed with several cycles of alternating pH. This is accomplished by first washing the column with 3 column volumes of coupling buffer A, followed by 3 column volumes of 100 m*M* glycine, pH 3.0. Repeat this cycle several times and re-equilibrate the column with BBS.

3.2. Antihapten Affinity Chromatography with Affi-gel 10

3.2.1. Resin Preparation and Ligand Coupling

1. Transfer sufficient Affi-gel 10 slurry to give a 3-mL packed gel to a 50-mL coarse sintered glass funnel (*see* **Note 8**).
2. Drain and wash the matrix with 10 mL of cold isopropanol.
3. Wash the matrix with 10 mL cold water.
4. Suspend the gel cake in 10 mL coupling buffer B containing 150 m*M* phosphotyramine (*see* **Note 9**).
5. Tumble the matrix end over end in a 15-mL capped polypropylene tube overnight at 4°C.
6. Drain and wash the matrix with a minimum of 10 column volumes of coupling buffer B or until the absorbance at 270 nm is <0.02.
7. Equilibrate the matrix with PBS, pH 7.4, and pack the matrix into an Poly Prep disposable column.

3.2.2. Sample Application and Elution

1. Dialyze the antiphosphotyrosine immunoglobulins against PBS, pH 7.4, overnight at 4°C (*see* **Note 10**).
2. Remove any precipitates by centrifugation at 10,000*g* for 30 min at 4°C.
3. Connect the column outlet to a peristaltic pump and wash the column with 5–10 mL of PBS, pH 7.4, at 0.5 mL/min.
4. Drain most of the buffer, leaving approx 0.5 mL on top of the gel bed (*see* **Note 11**).
5. Apply the dialyzed immunoglobulin sample in a volume of approx 15 mL to the column. Circulate the sample through the column for 3 h at 0.2 mL/min (*see* **Note 12**).
6. Drain the column as in **step 4** and save the eluent (*see* **Note 13**).
7. Wash the column with a minimum of 10 column volumes PBS, pH 7.4, or until the absorbance at 280 nm is <0.02.
8. Drain the column, leaving 100 µL of buffer on top of the bed. Elute the bound antiphosphotyrosine antibodies with PBS, pH 7.4, containing 10 m*M* *p*-nitrophenol (elution buffer). Apply one column volume of the elution buffer to the column and elute until the elution buffer just reaches the column outlet. Stop the flow and incubate the column in elution buffer for 30 min at 4°C.
9. Apply a second column volume of elution buffer, drain, and save the first volume of eluent. The second volume may also be collected after 30 min incubation by applying one column volume of PBS and draining the second volume of eluent.
10. Combine all elution buffer fractions and dialyze against several changes of PBS until the majority of the hapten has been removed (*see* **Note 14** and Chapter 2 for concentration procedure).

3.2.3. Column Regeneration and Storage

1. Wash the matrix with 10 column volumes of PBS, pH 7.4.
2. Wash the matrix either with two column volumes of 100 mM glycine, pH 3.0, or 100 mM acetic acid.
3. Equilibrate the column with PBS, pH 7.4. Confirm that the column pH has been neutralized by testing the effluent with pH paper.
4. Store the column stoppered in PBS, pH 7.4, containing 0.02% sodium azide at 4°C.

4. Notes

1. The gel will swell instantly. One gram of dry gel will yield approx 3.5 mL of hydrated matrix.
2. The dry matrix contains additives that need to be removed by these washing steps.
3. Avoid mechanical stirring of the gel, since this cannot only damage the gel matrix, but can also lead to denaturation of the immunoglobulins. A convenient way to keep the matrix gently suspended is by placing the tube on a nutator or serum incubator.
4. The amount of protein that is bound to the column can be estimated by subtracting the quantity of IgG that is eluted. This is only an estimate, but generally sufficient for antibody purification purposes. Continue with the subsequent steps if no more than 20% of the applied protein concentration is found in the eluate.
5. Poly Prep columns are convenient since they are unbreakable, disposable, can be capped easily and securely at both ends, and have graduation markings for measuring column volumes. After the column has been packed, it should be stored at 4°C. Do not let the column warm up again or dry out, since this will introduce air bubbles which can cause protein denaturation. All subsequent purification steps should be performed at 4°C.
6. The column capacity for any given antibody solution may not be the same and has to be determined empirically for each batch. Therefore, this column eluate and the first two column volumes of wash should be saved, since they may still contain some of the desired antibodies. The titer of the eluate can be tested, or the eluate can be reapplied to the gel at the end of the first purification.
7. The samples containing the highest absorbance at 280 nm should be pooled. Any precipitated antibodies can be removed by centrifugation at 10,000g for 30 min at 4°C. The immunoglobulins can then be concentrated and stored as described in Chapter 2.
8. The following wash steps should be completed in the cold as quickly as possible, since the N-hydroxysuccinamide reactive groups of Affi-gel 10 will undergo gradual hydrolysis in aqueous solutions at neutral pH. Washing should be accomplished in <20 min.
9. Phosphotyramine can be conveniently monitored because of its absorbance at 280 nm. The synthesis of phosphotyramine can be performed as described in **refs. 6** and **9**.
10. Ascites or serum should not be applied directly, but should be first purified by ammonium sulfate fractionation.

11. Be careful not to let the column dry out. If the top of the column becomes dry, the matrix will need to be resuspended in equilibration buffer by gentle stirring with a glass rod. If the trapped bubbles cannot be removed in this manner, the column has to be repacked.

12. Circulate at low flow rates to allow sufficient interaction time between the antibody and the hapten. Generally three complete passes through the column are adequate for most purposes.

13. Save the eluent for the same purpose as in **Note 6**.

14. *p*-Nitrophenylphosphate is slightly yellow at neutral pH, and its removal can be conveniently monitored by its absorbance at 350 nm. Alternatively phenylphosphate or phosphotyrosine may also be used to elute the antibodies.

References

1. *Affinity Chromatography—Principles and Methods.* (1983) Pharmacia-LKB, Ljungfoerefagen AB, Oerebro, Sweden.

2. Ostrove, S. (1990) Affinity chromatography: general methods. *Methods Enzymol.* **182,** 357–379.

3. Kenney, A. C. (1992) Ion-exchange chromatography of proteins, in *Methods in Molecular Biology, vol. 11: Practical Protein Chromatography* (Kenney, A. and Fowell, S., eds.), Humana, Totowa, NJ, pp. 249–258.

4. Conklyn, M. J., Kadin, S. B., and Showell,H. J. (1990) Inhibition of IgE-mediated *N*-acetylglucosaminidase and serotonin release from rat basophilic leukemia cells (RBL-2H3) by Tenidap: A novel anti- inflammatory agent. *Int. Arch. Allerg. Appl. Immunol.* **91,** 369–373.

5. Glenney, J. R., Jr., Zokas, L., and Kamps, M. P. (1988) Monoclonal antibodies to phosphotyrosine. *J. Immunol. Methods* **109,** 277–285.

6. Ross, A. H., Baltimore, D., and Eisen, H. N. (1981) Phosphotyrosine-containing proteins isolated by affinity chromatography with antibodies to synthetic hapten. *Nature* **294,** 654–656.

7. Wofsy, L. and Burr, B. (1969) The use of affinity chromatography for the specific purification of antibodies and antigens. *J. Immunol.* **103,** 380–382.

8. Kenney, A., Goulding, L., and Hill, C. (1988) The design, preparation and use of immunopurification reagents, in *Methods in Molecular Biology, vol. 3: New Protein Techniques* (Walker, J. M., ed.), Humana, Clifton, NJ, pp. 99–110.

9. Rithberg, P. G., Harris, T. J. R., Nomoto, A., and Wimmer, E. (1978) O^{4-} (5'Uridylyl) tyrosine is the bond between the genome-linked protein and the RNA of poliovirus. *Proc. Natl. Acad. Sci. USA* **75,** 4868–4872.

5

Purification of Antibodies
Using Protein A-Sepharose and FPLC

Ute M. Kent

1. Introduction

Protein A chromatography is a type of affinity chromatography that relies on the specific interaction of protein A with the Fc region of immunoglobulins from a number of species *(1)*. Protein A is a 42,000-Dalton polypeptide originally isolated from the cell walls of *Staphylococcus aureus* *(2)*. Because of its extended shape, protein A does not, however, run true to its actual size on SDS-polyacrylamide gels. Protein A is well characterized, and for a detailed review, *see* Langone *(3)* and **ref. 1**. Protein A has been expressed in recombinant form, and its crystal structure has been solved *(4)*. The affinity of protein A for Fc regions is very high ($K_d = 10^{-7}$). The molecule contains four binding sites, but only two Fc domains can be bound at any one time. Since the antibody combining site is left free, protein A, when covalently coupled to stationary supports like agarose or Sepharose beads, provides an excellent reagent for isolating immune complexes or immunoglobulins from a crude solution *(5,6)*. Protein A has also been useful for separating Fc fragments from Fab fragments after proteolytic digestion. The advantages of protein A are mainly its stability and specificity. Protein A is stable over a wide pH range (pH 2.0–11.0) and under most denaturing conditions commonly used in chromatography (e.g., 4 M urea, 6 M guanidinium hydrochloride) *(7)*. After neutralization or removal of the denaturant, protein A is able to refold and regain its ability to bind Fc regions. Although protein A has a high affinity for Fc regions of antibodies, this binding can be reversed by lowering the pH or by using denaturing agents *(8)*. It has been observed that the affinity of protein A for immunoglobulins from different species and subclasses is not the same *(9)*. In general, polyclonal antibodies of human, pig, rabbit, or guinea pig origin bind to protein A very well, whereas

From: *Methods in Molecular Biology, Vol. 115: Immunocytochemical Methods and Protocols*
Edited by: L. C. Javois © Humana Press Inc., Totowa, NJ

mouse antibodies bind only moderately well. However, human monoclonal antibodies of the subtype IgG_3 and mouse subtype IgG_1 are only minimally bound. (A compete listing of immunoglobulin affinities for protein A can be found in **refs. 5** and **10–12**). For these subtypes, as well as antibodies from rat, goat, sheep, and chickens, another bacterial cell-wall polypeptide, protein G, is more suitable. Protein G is a 30,000- to 35,000-Dalton polypeptide originally isolated from *Streptococci (13)*. Like protein A, it also binds the Fc region of immunoglobulins. Initially, a major drawback in the use of protein G was that it also contained a binding site for albumin. Molecular cloning and genetic manipulations have now succeeded in generating a protein G molecule that does not bind albumin, therefore making it the ideal substitute for protein A in many instances. Protein A/G bound to agarose is available from many sources with similar capacities. Prepacked columns as well as entire kits, including matrices and buffers, are also available (e.g., Pharmacia-LKB, Piscataway, NJ; Pierce, Rockford, IL). The following purification protocol using protein A Sepharose is suitable for isolating mouse antibodies from approx 500 mL of serum-free hybridoma tissue culture supernatants of subclass IgG2a or IgG2b. For other subclasses or species, some matrix or buffer modifications may be required *(6)*.

2. Materials

1. Protein A-Sepharose 4 Fast Flow (Pharmacia).
2. Phosphate-buffered saline: 1.7 mM potassium phosphate monobasic, 5 mM sodium phosphate dibasic, pH 7.4, 150 mM sodium chloride.
3. 100 mM Acetic acid.
4. HR 5/10 column (Pharmacia).
5. 100 mM Sodium phosphate, pH 8.0.
6. 100 mM Glycine, pH 3.0.
7. Millipore-quality or HPLC-grade water.
8. Filtration device with 0.45-μm membrane (Millipore, Bedford, MA).
9. 0.22-μm Millex-GV Syringe filters, low protein binding (Millipore).
10. 1 M Tris-HCl, pH 8.0; 100 mM Tris-HCl, pH 8.0.
11. 20% Ethanol, HPLC grade.
12. Hybridoma culture supernatant (serum-free).
13. Spectrophotometer and quartz or UV-compatible plastic cuvets.

3. Methods

3.1. Column Preparation

1. Remove 12 mL of protein A-Sepharose and allow it to settle.
2. Aspirate the storage solution.
3. Suspend the matrix in 20 mL PBS and collect the protein A-Sepharose by centrifugation at 500g for 5 min. Aspirate the supernatant. Repeat this wash step three times with 20 mL PBS.

4. Resuspend the matrix in 20 mL of 100 mM acetic acid and rock gently for 10 min at 4°C.

5. Remove the acid and wash the matrix three times in filtered 100 mM sodium phosphate, pH 8.0 (*see* **Note 1**).

6. Degas the matrix under vacuum and pack the HR 5/10 column with the washed matrix.

7. Equilibrate the column with 40 mL 100 mM sodium phosphate and check for proper column packing. The matrix should be free of particles, air bubbles, or cracks.

8. Apply 20 mL of filtered 100 mM glycine, pH 3.0, followed by 20 mL of 100 mM sodium phosphate, pH 8.0 (*see* **Note 2**).

3.2. Protein A-Sepharose Chromatography by Fast Protein Liquid Chromatography (FPLC)

1. Filter all buffers through a 0.45-µm filter and degas the solutions under vacuum or by sonication.

2. Dialyze the hybridoma culture supernatant against 100 mM sodium phosphate, pH 8.0 (*see* **Note 3**).

3. Filter the dialyzed culture supernatant through a 0.22-µm filter.

4. Equilibrate the column with 20 mL of 100 mM sodium phosphate, pH 8.0.

5. Apply the hybridoma culture supernatant to the protein A column and monitor the eluent for protein at 280 nm. Wash the column with 30–40 mL of phosphate buffer, pH 8.0 (*see* **Note 4**).

6. Elute the bound immunoglobulins with 20 mL of 100 mM glycine, pH 3.0 (*see* **Note 5**).

7. Collect 1-mL fractions into tubes containing 500 µL of 1 M Tris-HCl, pH 8.0, and mix (*see* **Note 6**).

8. Determine the concentration of the immunoglobulin-containing fractions by measuring the absorbance at 280 nm. A 1-mg/mL solution will have an absorbance of 1.4 in a cuvet with a 1-cm path length (*see* **Note 7**).

3.3. Column Regeneration and Storage

1. Neutralize the column immediately by washing with 20–30 mL of PBS, pH 7.4. Check the pH of the column effluent with pH paper to ensure that the pH is back to neutrality.

2. For short-term storage (days), the column can be equilibrated with PBS containing 0.02% sodium azide.

3. For long-term storage (weeks to months), the column should be washed with 20–30 mL of water followed by 10 mL of 20% ethanol (*see* **Note 8**).

4. Notes

1. It is recommended that these washing steps be accomplished batch-wise prior to packing the column to avoid clogging the column filters or frits with fine particles or removed protein A.

2. It is good practice when using any column for the first time to perform one or two mock purifications. This ensures that the column as well as the matrix are compatible with the

buffers one plans to use during purification. This is also a good time to check and record the normal operating pressures of a blank run to monitor column performance over time. Increasing back pressures during subsequent purifications are signs of trouble, and indicate a dirty or clogged column that needs to be cleaned or replaced.

3. Serum, ascites, or hybridoma culture supernatant can, after filtration, be applied to the column directly. The crude antibody solution should then be diluted with 1/10 volume of 1 *M* Tris-HCl, pH 8.0, and chromatography should be performed with 100 m*M* Tris-HCl, pH 8.0. It is, however, advisable first to precipitate the immunoglobulins with ammonium sulfate (*see* Chapter 2). This not only reduces the sample volume, but more importantly removes lipids, particularly from serum and ascitic fluid, extending the life of the column.

4. After the breakthrough (unbound proteins) has cleared the column, the column should be washed with five additional column volumes prior to low-pH elution.

5. The Pharmacia Frac-100 fraction collector comes equipped with a tray that can be filled with ice. Alternatively, the tubes containing the eluted and neutralized immunoglobulins can be removed and placed on ice immediately.

6. Mix the eluted antibody solution with the 1 *M* Tris-HCl, pH 8.0, gently. Immunoglobulins are very stable and the majority will renature after the pH is raised. Nevertheless, some denaturation will occur and the aggregated antibodies should be removed by centrifugation.

7. Protein A chromatography yields immunoglobulins in very concentrated form. Therefore, the absorbance of the solution should not be measured directly. Dilute the sample into PBS, and take a reading using PBS as a blank. The optical density of the diluted sample should not be above 1–1.5 to fall within the range of linearity.

8. Protein A columns can be used many times. It is not recommended to use the same column for purification of different antibodies because of possible crosscontamination. Should this, however, become necessary, the column has to be washed with several column volumes of alternating pH (PBS followed by 100 m*M* glycine, pH 3.0, or 100 m*M* acetic acid). This step should also include a denaturant wash with 2 *M* urea.

References

1. Goodswaard, J., van der Dank, J. A., Noardizij, A., van Dam, R. H., and Vaerman, J.-P. (1978) Protein A reactivity of various mammalian immunoglobulins. *Scan. J. Immunol.* **8**, 21–28.

2. Forsgrem, A. and Sjoquist, J. (1966) "Protein A" form *S. aureus I*. Pseudo-immune reaction with human gamma-globulin. *J. Immunol.* **97**, 822–827.

3. Langone, J. J. (1982) Applications of immobilized Protein A in immunochemical techniques. *J. Immunol. Methods* **55**, 277–296.

4. Deisenhofer, T. (1981) Crystallographic refinement and atomic models of a human Fc fragment and its complex with fragment B of Protein A from *Staphylococcus aureus* at 2.9- and 2.8-Å resolution. *Biochemistry* **20**, 2361–2370.

5. Lindmark R., Thoren-Talling, K., and Sjoquist, J. (1983) Binding of immunoglobulins to Protein A and immunoglobulin levels in mammalian sera. *J. Immunol. Methods* **62**, 1–14.

6. Kristiansen, T. (1974) Studies on bloodgroup substances. V. Bloodgroup substance A coupled to agarose as an immunosorbent. *Biochim. Biophys. Acta* **362,** 567–574.

7. *Affinity Chromatography–Principles and Methods.* (1983) Pharmacia-LKB, Ljungfoeretagen, Oerebro AB, Sweden.

8. Bywater, R., Eriksson, G.-B., and Ottosson, T. (1983) Desorption of immunoglobulins from Protein A-Sepharose Cl-4B under mild conditions. *J. Immunol. Methods* **64,** 1–6.

9. Ey, P. L., Prowse, S. J., and Jemkin, C. R. (1978) Isolation of pure IgG1, IgG2a and IgG2b immunoglobulins from mouse serum using Protein A-Sepharose. *Immunochemistry* **15,** 429–436.

10. Kruger, N. J. and Hammond, J. B. W. (1988) Purification of immunoglobulins using protein A-Sepharose, in *Methods in Molecular Biology, vol. 3: New Protein Techniques* (Walker, J. M., ed), Humana, Clifton, NJ, pp. 363–371.

11. Akerstrom, B., Brodin, T., Reis, K., and Bjock, L. (1985) Protein G: A powerful tool for binding and detection of monoclonal and polyclonal antibodies. *J. Immunol.* **135,** 2589–2592.

12. Harlow, E. and Lane, D. (1988) Reagents, in *Antibodies: A Laboratory Manual.* Cold Spring Harbor Laboratory, Cold Spring Harbor, NY, Chapter 15.

13. Bjorck, L. and Kronvall, G. (1984) Purification and some properties of *Streptococcal* Protein G, a novel IgG-binding reagent. *J. Immunol.* **133,** 969–974.

6

Conjugation of Fluorochromes to Antibodies

Su-Yau Mao

1. Introduction

The use of specific antibodies labeled with a fluorescent dye to localize sub-stances in tissues was first devised by A. H. Coons and his associates. At first, the specific antibody itself was labeled and applied to the tissue section to identify the antigenic sites (direct method) *(1)*. Later, the more sensitive and versatile indirect method *(2)* was introduced. The primary, unlabeled, antibody is applied to the tissue section, and the excess is washed off with buffer. A second, labeled antibody from another species, raised against the IgG of the animal donating the first antibody, is then applied. The primary antigenic site is thus revealed. A major advantage of the indirect method is the enhanced sensitivity. In addition, a labeled secondary antibody can be used to locate any number of primary antibodies raised in the same animal species without the necessity of labeling each primary antibody.

Four fluorochromes are commonly used; fluorescein, rhodamine, Texas red, and phycoerythrin (*see* Chapter 14). They differ in optical properties, such as the intensity and spectral range of their absorption and fluorescence. Choice of fluorochrome depends on the particular application. For maximal sensitivity in the binding assays, fluorescein is the fluorochrome of choice because of its high quantum yield. If the ligand is to be used in conjunction with fluorescence microscopy, rhodamine coupling is advised, since it has superior sensitivity in most microscopes and less photobleaching than fluorescein. Texas red *(3)* is a red dye with a spectrum that minimally overlaps with that of fluorescein; there-fore, these two dyes are suitable for multicolor applications. Phycoerythrin is a 240-kDa, highly soluble fluorescent protein derived from cyanobacteria and eukaryotic algae. Its conjugates are among the most sensitive fluorescent probes available *(4)* and are frequently used in flow cytometry and immunoassays *(5)*.

From: *Methods in Molecular Biology, Vol. 115: Immunocytochemical Methods and Protocols*
Edited by: L. C. Javois © Humana Press Inc., Totowa, NJ

Thiols and amines are the only two groups commonly found in biomolecules that can be reliably modified in aqueous solution. Although the thiol group is the easiest functional group to modify with high selectivity, amines are common targets for modifying proteins. Virtually all proteins have lysine residues, and most have a free amino terminus. The ε-amino group of lysine is moderately basic and reactive with acylating reagents. The concentration of the free-base form of aliphatic amines below pH 8.0 is very low. Thus, the kinetics of acylation reactions of amines by isothiocyanates, succinimidyl esters, and other reagents is strongly pH-dependent. Although amine acylation reactions should usually be carried out above pH 8.5, the acylation reagents degrade in the presence of water, with the rate increasing as the pH increases. Therefore, a pH of 8.5–9.5 is usually optimal for modifying lysines.

Where possible, the antibodies used for labeling should be pure (*see* Chapters 2–5). Affinity-purified, fluorochrome-labeled antibodies demonstrate less background and nonspecific fluorescence than fluorescent antiserum or immunoglobulin fractions. The labeling procedures for the isothiocyanate derivatives of fluorescein and sulfonyl chloride derivatives of rhodamine are given below *(6)*. The major problem encountered is either over- or undercoupling, but the level of conjugation can be determined by simple absorbance readings.

2. Materials

1. IgG.
2. Borate buffered saline (BBS): 0.2 M boric acid, 160 mM NaCl, pH 8.0.
3. Fluorescein isothiocyanate (FITC) or Lissamine rhodamine B sulfonyl chloride (RBSC).
4. Sodium carbonate buffer: 1.0 M NaHCO$_3$-Na$_2$CO$_3$ buffer, pH 9.5, prepared by titrating 1.0 M NaHCO$_3$ with 1.0M Na$_2$CO$_3$ until the pH reaches 9.5.
5. Absolute ethanol (200 proof) or anhydrous dimethylformamide (DMF).
6. Sephadex G-25 column.
7. Whatman DE-52 column.
8. 10 mM Sodium phosphate buffer, pH 8.0.
9. 0.02% Sodium azide.
10. UV spectrophotometer.

3. Methods

3.1. Coupling of Fluorochrome to IgG

1. Prior to coupling, prepare a gel-filtration column to separate the labeled antibody from the free fluorochrome after the completion of the reaction. The size of the column should be 10 bed volumes/sample volume (*see* **Note 1**).
2. Equilibrate the column in phosphate buffer. Allow the column to run until the buffer level drops just below the top of bed resin. Stop the flow of the column by using a valve at the bottom of the column.

3. Prepare an IgG solution of at least 3 mg/mL in BBS, and add 0.2 vol of sodium carbonate buffer to IgG solution to bring the pH to 9.0. If antibodies have been stored in sodium azide, the azide must be removed prior to conjugation by extensive dialysis (*see* **Note 2**).

4. Prepare a fresh solution of fluorescein isothiocyanate at 5 mg/mL in ethanol or RBSC at 10 mg/mL in DMF immediately before use (*see* **Note 3**).

5. Add FITC at a 10-fold molar excess over IgG (about 25 µg of FITC/mg IgG). Mix well and incubate at room temperature for 30 min with gentle shaking. Add RBSC at a 5-fold molar excess over IgG (about 20 µg of RBSC/mg IgG), and incubate at 4°C for 1 h.

6. Carefully layer the reaction mixture on the top of the column. Open the valve to the column, and allow the antibody solution to flow into the column until it just enters the bed resin. Carefully add phosphate buffer to the top of the column. The conjugated antibody elutes in the excluded volume (about one-third of the total bed volume).

7. Store the conjugate at 4°C in the presence of 0.02% sodium azide (final concentration) in a light-proof container. The conjugate can also be stored in aliquots at −20°C after it has been snap-frozen on dry ice. Do not refreeze the conjugate once thawed.

3.2. Calculation of Protein Concentration and Fluorochrome-to-Protein Ratio

1. Read the absorbance at 280 and 493 nm. The protein concentration is given by **Eq. 1**, where 1.4 is the optical density for 1 mg/mL of IgG (corrected to 1-cm path length).

$$\text{Fluorescein-conjugated IgG conc. (Fl IgG conc.) (mg/mL)} = (A_{280\,nm} - 0.35 \times A_{493\,nm})/1.4 \tag{1}$$

The molar ratio (F/P) can then be calculated, based on a molar extinction coefficient of 73,000 for the fluorescein group, by **Eq. 2** (*see* **Notes 4** and **5**).

$$\text{F/P} = (A_{493\,nm}/73,000) \times (150,000/\text{Fl IgG conc.}) \tag{2}$$

2. For rhodamine-labeled antibody, read the absorbance at 280 and 575 nm. The protein concentration is given by **Eq. 3**.

$$\text{Rhodamine-conjugated IgG conc. (Rho IgG conc.) (mg/mL)} = (A_{280\,nm} - 0.32 \times A_{575\,nm})/1.4 \tag{3}$$

The molar ratio (F/P) is calculated by **Eq. 4**.

$$\text{F/P} = (A_{575\,nm}/73,000) \times (150,000/\text{Rho IgG conc.}) \tag{4}$$

4. Notes

1. Sephadex G-25 resin is the recommended gel for the majority of desalting applications. It combines good rigidity, for easy handling and good flow characteristics, with adequate resolving power for desalting molecules down to about 5000 Dalton mol wt. If the volume of the reaction mixture is <1 mL, a prepacked

disposable Sephadex G-25 column (PD-10 column from Pharmacia, Piscataway, NJ) can be used conveniently.

2. When choosing a buffer for conjugation of fluorochromes, avoid those containing amines (e.g., Tris, azide, glycine, and ammonia), which can compete with the ligand.

3. Both sulfonyl chloride and isothiocyanate will hydrolyze in aqueous conditions; therefore, the solutions should be made freshly for each labeling reaction. Absolute ethanol or dimethyl formamide (best grade available, stored in the presence of molecular sieve to remove water) should be used to dissolve the reagent. The hydrolysis reaction is more pronounced in dilute protein solution and can be minimized by using a more concentrated protein solution. Caution: DMSO should not be used with sulfonyl chlorides, because it reacts with them.

4. An F/P ratio of two to five is optimal, since ratios below this yield low signals, whereas higher ratios show high background. If the F/P ratios are too low, repeat the coupling reaction using fresh fluorochrome solution. The IgG solution needs to be concentrated prior to reconjugation (e.g., Centricon-30 microconcentrator from Amicon Co., Beverly, MA, can be used to concentrate the IgG solution).

5. If the F/P ratios are too high, either repeat the labeling with appropriate changes or purify the labeled antibodies further on a Whatman DE-52 column (diethylaminoethyl microgranular preswollen cellulose, 1-mL packed column/1–2 mg of IgG). DE-52 chromatography removes denatured IgG aggregates and allows the selection of the fraction of the conjugate with optimal modification. Equilibrate and load the column with 10 mM phosphate buffer, pH 8.0. Wash the column with equilibrating buffer and elute with the same buffer containing 100 mM NaCl (first) and 250 mM NaCl (last). Measure the F/P ratios of each fraction, and select the appropriate fractions.

References

1. Coons, A. H., Creech, H. J., and Jones, R. N. (1941) Immunological properties of an antibody containing a fluorescent group. *Proc. Soc. Exp. Biol. Med.* **47**, 200–202.

2. Coons, A. H., Leduc, E. H., and Connolly, J. M. (1955) Studies on antibody production. I. A method for the histochemical demonstration of specific antibody and its application to a study of the hyperimmune rabbit. *J. Exp. Med.* **102**, 49–60.

3. Titus, J. A., Haugland, R., Sharrow, S. O., and Segal, D. M. (1982) Texas Red, a hydrophilic, red-emitting fluorophore for use with fluorescein in dual parameter flow microfluorometric and fluorescence microscopic studies. *J. Immunol. Methods* **50**, 193–204.

4. Oi, V. T., Glazer, A. N., and Stryer, L. (1982) Fluorescent phycobiliprotein conjugates for analyses of cells and molecules. *J. Cell Biol.* **93**, 981–986.

5. Bochner, B. S., McKelvey, A. A., Schleimer, R. P., Hildreth, J. E., and MacGlashan, D. W., Jr. (1989) Flow cytometric methods for the analysis of human basophil surface antigens and viability. *J. Immunol. Methods* **125**, 265–271.

6. Schreiber, A. B. and Haimovich, J. (1983) Quantitative fluorometric assay for detection and characterization of Fc receptors. *Methods Enzymol.* **93**, 147–155.

7

Biotinylation of Antibodies

Su-Yau Mao

1. Introduction

The high affinity and specificity of the avidin–biotin interaction permit diverse applications in immunology, histochemistry, *in situ* hybridizations, affinity chromatography, and many other areas (*1*) (*see* Chapters 25 and 26). It was first exploited in immunocytochemical applications in the mid-1970s (*2,3*), and has since been commonly used to localize antigens in cells and tissues. In this technique, a biotinylated primary or secondary antibody is first applied to the sample, and the detection is accomplished by using labeled avidin. Avidin with a variety of labels are available commercially, including fluorescent, enzyme, iodine, ferritin, or gold labels.

Both avidin and its bacterial counterpart, streptavidin, are standard reagents for histochemical procedures. Avidin is a 66-kDa, positively charged glycoprotein with an isoelectric point of about 10.5 (*4*). The positively charged residues and the oligosaccharide component of avidin can interact nonspecifically with negatively charged cell surfaces and nucleic acids, sometimes causing background problems in histochemical and cytometric applications. Avidin is, however, inexpensive and one of the most commonly used reagents for these applications. On the other hand, streptavidin, a 60-kDa nonglycosylated protein with a near-neutral isoelectric point, exhibits less nonspecific binding than avidin. Both avidin and streptavidin bind four biotin equivalents per molecule with high affinity (K_a is about $10^{14}\ M^{-1}$) and low reversibility, thus permitting numerous combinations of avidin, biotin, and antibody. It is possible to create a widely branching complex and build up high amounts of label over the original antigenic site to increase the sensitivity. One such technique was developed by Hsu et al. (*5*).

Labeling antibodies by covalent coupling of a biotinyl group is simple and normally does not have any adverse effect on the antibody (*6*). Most bio-

From: *Methods in Molecular Biology, Vol. 115: Immunocytochemical Methods and Protocols*
Edited by: L. C. Javois © Humana Press Inc., Totowa, NJ

tinylations are performed using a succinimide ester of biotin. The reagent reacts with primary amines of the lysine residues or the amino terminus on the antibody to form amide bonds. The protocol described below *(7,8)* uses a water-soluble analog of *N*-hydroxylsuccinimide biotin (Pierce, Rockford, IL), which can be dissolved directly in the reaction buffer. Biotin-coupled antibodies are stable under normal storage conditions.

2. Materials
1. IgG.
2. Sulfosuccinimidobiotin: 2 mg/mL in sodium borate buffer (*see* **Note 1**).
3. Sodium borate buffer: 0.2 *M* boric acid, 160 m*M* NaCl, pH 8.5.
4. Dialysis tubing.
5. Phosphate-buffered saline (PBS): 10 m*M* sodium phosphate, 150 m*M* NaCl, pH 7.4.
6. UV spectrophotometer.
7. 0.02% Sodium azide.

3. Methods
1. Prepare an IgG solution of at least 3 mg/mL in sodium borate buffer. If the antibodies have been stored in sodium azide, the azide must be removed prior to conjugation. Dialyze extensively against the borate buffer (*see* **Note 2**).
2. Add the sulfosuccinimidobiotin at a 30-fold molar excess over IgG (about 90 µg/mg IgG). Mix well, and incubate at room temperature for 30 min with gentle shaking (*see* **Note 3**).
3. Dialyze extensively against several changes (>10^6 vol) of phosphate-buffered saline or other desired buffer to remove uncoupled biotin.
4. Centrifuge dialysate (8000*g*, 10 min, 4°C) to remove any precipitate formed during dialysis.
5. Determine the IgG concentration by measuring the absorbance at 280 nm (absorbance of 1 mg/mL = 1.4, measured using cuvet with 1-cm path length).
6. Store the conjugate at 4°C in the presence of 0.02% sodium azide (final concentration). The conjugate can be stored in aliquots at –20°C for very long periods if previously snap-frozen on dry ice (*see* **Notes 4** and **5**).

4. Notes
1. The sulfonated esters will hydrolyze in aqueous conditions. Therefore, the solutions should be made freshly before each use. The hydrolysis reaction is more pronounced in dilute protein solutions and can be minimized by increasing protein concentration.
2. When choosing a buffer for biotinylations, avoid those containing amines (e.g., Tris, azide, glycine, and ammonia), which can compete with the ligand. Phosphate buffers may result in the "salting out" of the biotin reagents containing sulfo-*N*-hydroxysuccinimide moieties.
3. Alternatively, the water-insoluble biotinylated succinimide ester can be first dissolved in fresh distilled dimethylsulfoxide (10 mg/mL), and then added to the IgG solution.

4. Many biotinylated succinimide esters are now available. Most of these variations alter the size of the spacer arm between the succinimide coupling group and the biotin. The additional spacers could facilitate avidin binding and, thus, may be critical for some applications *(9)*.

5. If a free amino group forms a portion of the protein that is essential for activity (e.g., the antigen-combining site for antibody), biotinylation with the succinimide ester will lower or destroy the activity of the protein, and other methods of labeling should be tried. Biotin hydrazide has been used to modify the carbohydrate moieties of antibodies *(10,11)*. Other alternatives are the thiol-reactive biotin maleimide *(12)* or biotin iodoacetamide *(13)*.

References

1. Roffman, E., Meromsky, L., Ben-Hur, H., Bayer, E. A., and Wilchek, M. (1986) Selective labeling of functional groups on membrane proteins or glycoproteins using reactive biotin derivatives and ^{125}I-streptavidin. *Biochem. Biophys. Res. Comm.* **136**, 80–85.
2. Becker, J. M. and Wilchek, M. (1972) Inactivation by avidin of biotin-modified bacteriophage. *Biochim. Biophys. Acta* **264**, 165–170.
3. Heitzmann, H. and Richards, F. M. (1974) Use of the biotin-avidin complex for specific staining of biological membranes in electron microscopy. *Proc. Natl. Acad. Sci. USA* **71**, 3537–3541.
4. Green, N. M. (1975) Avidin. *Adv. Protein Chem.* **29**, 85–133.
5. Hsu, S.-M., Raine, L., and Fanger, H. (1981) Use of avidin–biotin–peroxidase complex (ABC) in immunoperoxidase techniques. *J. Histochem. Cytochem.* **29**, 577–580.
6. Guesdon, J. L., Ternynck, T., and Avrameas, S. (1979) The use of avidin–biotin interaction in immuno-enzymatic techniques. *J. Histochem. Cytochem.* **27**, 1131–1139.
7. Lee, W. T. and Conrad, D. H. (1984) The murine lymphocyte receptor for IgE. II. Characterization of the multivalent nature of the B lymphocyte receptor for IgE. *J. Exp. Med.* **159**, 1790–1795.
8. LaRochelle, W. J. and Froehner, S. C. (1986) Determination of the tissue distributions and relative concentrations of the postsynaptic 43-kDa protein and the acetylcholine receptor in Torpedo. *J. Biol. Chem.* **261**, 5270–5274.
9. Suter, M. and Butler, J. E. (1986) The immunochemistry of sandwich ELISAs. II. A novel system prevents the denaturation of capture antibodies. *Immunol. Lett.* **13**, 313–316.
10. O'Shannessy, D. J., Dobersen, M. J., and Quarles, R. H. (1984) A novel procedure for labeling immunoglobulins by conjugation to oligosaccharide moieties. *Immunol. Lett.* **8**, 273–277.
11. O'Shannessy, D. J. and Quarles, R. H. (1987) Labeling of the oligosaccharide moieties of immunoglobulins. *J. Immunol. Methods* **99**, 153–161.
12. Bayer, E. A., Zalis, M. G., and Wilchek, M. (1985) 3-(*N*-Maleimido-propionyl)-biocytin: a versatile thiol-specific biotinylating reagent. *Anal. Biochem.* **149**, 529–536.
13. Sutoh, K., Yamamoto, K., and Wakabayashi, T. (1984) Electron microscopic visualization of the SH1 thiol of myosin by the use of an avidin-biotin system. *J. Mol. Biol.* **178**, 323–339.

II

TISSUE PREPARATION
FOR LIGHT MICROSCOPIC ANALYSES

8

Overview of Cell Fixatives
and Cell Membrane Permeants

Melissa A. Melan

1. Introduction

The localization of proteins and carbohydrates within cells and tissues with specific antibodies has long been proven to be a valuable technique. Immuno-localization procedures allow one to detect not only well-characterized cellular structures but also provide information about newly characterized proteins and carbohydrates. This chapter will review some of the advantages and drawbacks of common chemical fixation and permeabilization methods used for immuno-localization at the level of light microscopy.

It is usually impossible to perform immunohistochemical microscopic studies with living specimens. Antibodies are large molecules that need to be microinjected into living cells if internal structures or proteins are to be local-ized. Microinjection, however, is not practical if the antigen to be localized resides within an organelle. In the case of surface antigens, antibody solutions can be applied to the outer surface of living cells, but long incubation with antibodies can result in the internalization of antibody-protein complexes through endocytosis. Antibodies can also nonspecifically bind to cell surfaces by their F_c regions and lead to false results (1). Specific proteins can be studied with the technique of fluorescent analog cytochemistry, which allows fluoro-chrome-derivatized proteins to be functionally traced in living cells (2). Although a powerful technique, it usually has limited applications and is often not prac-tical for most laboratories or routine procedures (1). Large amounts of protein need to be isolated and conjugated to a fluorochrome in a way that does not disturb its functional properties (2). Therefore, most laboratories rely on chemical fixation and permeabilization of cells before treatment with antibody solutions to determine the spatial distribution of antigens within tissues and cells (3).

From: Methods in Molecular Biology, Vol. 115: Immunocytochemical Methods and Protocols
Edited by: L. C. Javois © Humana Press Inc., Totowa, NJ

It is helpful to know the chemistry of fixatives in order to understand their action and avoid artifacts *(4)*. Most commonly studied antigens are either proteins or carbohydrates. Many of these molecules are soluble in aqueous solutions and need to be fixed in place in cells. Insoluble antigens also need to be structurally preserved *(1)*. All chemical fixatives will cause chemical and conformational changes in the protein structure of cells with lesser changes noted for carbohydrate antigens *(5)*. Secondary and tertiary structures of proteins are the most important for eliciting antigenicity and chemical fixatives usually disturb these conformations *(3)*.

Under ideal conditions, fixation of tissues and cells should minimally change cellular structure and chemical composition *(4)*. For immunohistochemical analysis, the fixative should preserve cellular structure as well as prevent the loss and/or migration of antigens. Unfortunately, those fixation methods that minimally affect antigenic epitopes are generally not the best for preserving morphology. And conversely, the methods that best preserve morphology are those most disruptive to antigenic sites *(4)*. The loss of antigenicity increases with the fixative concentration and the duration of fixation *(3)*. One should therefore realize that there is no "perfect" fixative and strike a balance between preserving cellular structure and maintaining the antigenicity of the epitopes of interest *(3)*.

There are two basic types of fixatives: coagulants and crosslinking agents. Although coagulant agents tend to induce artifacts (extraction, tissue shrinkage, granular/reticular cytoplasm), they have been found to be effective in light microscopy, particularly for large-molecular-weight antigens and polymerized structural proteins *(1,3–5)*. Because crosslinking fixatives act by forming chemical bonds, severe conformational changes of proteins can occur as a result of the modification of reactive groups *(1)*. Crosslinking fixatives can also cause artifacts either by linking low-molecular-weight antigens to larger structural proteins or by causing steric blockage of antibody access to the antigenic epitopes *(1)*. Protein and carbohydrate antigens can also be lost through extraction during fixation procedures caused by the solubility of most antigens in aqueous solutions *(6)*.

Antibodies are large molecules. Immunoglobulin G molecules have an arm-to-arm distance of 146Å *(7)*, and F_{ab} fragments have dimensions of 30 X 40 X 50Å *(8)*. Molecules of this size cannot diffuse into and out of cells. Fixatives, particularly crosslinking fixatives, act on membrane proteins and reduce the overall membrane permeability *(3,9)*. Therefore, one needs to "open" plasma membranes and organelle membranes in order to allow antibody access to intracellular and intraorganellar antigens. Solvents, saponins, and nonionic detergents are the most commonly used reagents for membrane permeabilization. It should be noted, however, that plasma membranes and some organelle membranes often have different properties and therefore may need different conditions to affect permeabilization *(10)*.

2. Commonly Used Fixatives

2.1. Solvents

Solvents such as alcohols and acetone are strong coagulant fixatives. They act by displacing water, breaking hydrogen bonds, and thus disrupting the tertiary structure of proteins *(11)*. With these reagents, soluble proteins are precipitated *(11)*, but neither carbohydrates nor nucleic acids are fixed in place and are removed by subsequent washing *(5)*. Lipids in both the membranes and cytoplasm are solubilized and extracted *(5)*. The displacement of water causes cellular shrinkage and destroys most organelles within the cell *(7,10)*. At low temperatures (0 to –20°C) ethanol precipitates proteins without denaturing them and has been found to be practical for fixing large-molecular-weight antigens, such as assembled cytoskeletal proteins *(1,11,12)*. Lower molecular-weight antigens (<100 kDa) are generally extracted *(1,6)*. These fixatives, because of their extreme disruptive effects on cellular organelles, are useful only for light microscopy and not for electron microscopy *(1)*.

2.2. Formaldehyde/Paraformaldehyde

Formaldehyde is a colorless gas that is soluble in water *(3)*. Commercial aqueous preparations of formalin contain 37–40% w/w solubilized gas. They also contain formic acid (<0.05%) and 10–15% methanol, which is added to prevent the polymerization of formaldehyde into paraformaldehyde *(3,11)*. Methanol and formic acid make these solutions an unacceptable fixative for fine structures *(9)*. Paraformaldehyde is a polymerized form of formaldehyde that dissociates at 60°C and neutral pH. Freshly prepared solutions of paraformaldehyde are preferred for most immunochemical procedures because they provide a fixative free of extraneous additives and are usually the conservative fixatives of choice when beginning the development of a fixation protocol *(3,5)*.

Although it is the simplest aldehyde, the chemistry of formaldehyde reactions with proteins is quite complex *(11)*. Formaldehyde crosslinks proteins by addition to amino, amido, guanidino, thiol, phenolic, imidazolyl, and indolyl groups and forms hemiacetal derivatives *(3)*. If the hemiactetal addition products are in close proximity to other proteins, they react by condensation to form chemically stable methylene bridges that crosslink the proteins *(11)*. Formaldehyde addition reactions are readily reversible by washing with water or alcohol *(5,11)*. Prolonged washing of tissues can, in some cases, restore antigenicity to fixed proteins *(1)*. The maximum levels of protein crosslinkages occur in the pH range of 7.5–8.0 *(3)*. At lower pH, primary amino groups are unreactive *(1)*, and thus crosslinking reactions are not favored. Addition of bicarbonate to formaldehyde is reported to minimize extraction of proteins from tissues, presumably by raising intracellular pH and increasing protein crosslinkages (3).

Formaldehyde prevents the extraction of glycogen but does not preserve soluble polysaccharides. Acid mucopolysaccharides are also not preserved unless they are bound to proteins (3). Formaldehyde is a good fixative for lipids, particularly if 1–2 mM Ca^{2+} or Mg^{2+} are included in the fixative vehicle (4,5,11). Membrane fixation is improved by reducing lipid extraction (4). It is also thought that fixation with formaldehyde lowers the solubility of membrane phospholipids in water (11).

2.3. Glutaraldehyde

Glutaraldehyde (glutaric acid dialdehyde) is a fixative that is very effective in preserving fine structure (3). Glutaraldehyde fixation is usually paired with osmium tetroxide postfixation to provide excellent cytological preservation for electron microscopy. The postfixation steps, however, severely lower the antigenicity of proteins for immunodetection, either through protein cleavage, oxidation, or conformational changes induced by osmium tetroxide (1,9). Fixation with glutaraldehyde alone also results in lowered protein antigenicity (3,9).

Because it is a dialdehyde, glutaraldehyde acts to crosslink proteins by means of its two aldehyde groups. The chemical reactions of glutaraldehyde with proteins are not well understood and are the topic of much debate. Current opinion holds that glutaraldehyde is most reactive with the ε-amino groups of lysine (3). Glutaraldehyde acts to rapidly crosslink proteins and thus renders them insoluble (3). Intramolecular crosslinkages tend to predominate over intermolecular bonding, and major conformational changes occur because of the disruption α-helical structures. These types of protein shape changes can lead to lowered immunoreactivity by blocking or masking reactive epitopes (3,9). The reactions of glutaraldehyde with carbohydrates are also not well understood. The fixative most likely reacts with polyhydroxyl compounds to form polymers in mucopolysaccharides (3). Fixation with glutaraldehyde is essentially irreversible (1).

Highly purified solutions of glutaraldehyde are used by enzyme cytochemists but seem to seldom be used by immunocytochemists (1). Impurities in glutaraldehyde, cyanide, and arsenic in some commercial preparations can greatly contribute to reducing protein antigenicity (13). Undefined "impurities" that absorb at 235 nm can form with prolonged exposure to air (4). The purity of glutaraldehyde solutions can be checked by determining the $A_{235/280}$ ratio of the solution. A value of >0.2 is generally associated with impure solutions. Therefore, fresh solutions of glutaraldehyde prepared from stocks packaged under inert gas are usually the best. Charcoal absorption or distillation of the solutions are also options for purification (3,4).

Incomplete fixation with glutaraldehyde can cause serious artifacts in immunocytochemistry resulting from unreacted free aldehyde groups. These reactive groups, if left unchanged, can react with antibodies and either inactivate or nonspecifically link them to proteins. The free aldehyde groups must be irreversibly blocked either through reduction with sodium borohydride *(14)* or by blocking with phenylhydrazine *(11)* ethanolamine, or lysine *(1)*. Unreacted free aldehyde groups of the fixative can also cause high background autofluorescence in fluorescence procedures *(14)*.

Heating of tissue samples during fixation with glutaraldehyde has been shown to produce superior morphological preservation for electron microscopy *(15)*. It is presumed that heating glutaraldehyde increases the formation of its more penetrable and reactive free monomeric form. Samples heated to 40°C for a period of 10–60 s show morphological preservation equal to or better than that obtained with fixation for 1–2 h at room temperature. Sample heating can be achieved using either microwave irradiation, convective, or resistive methods *(15)*. Although the preservation of immunoreactive sites is not addressed in these studies, this technique or variations thereof hold promise for immunolocalizations because the duration of fixation is drastically reduced. In addition, minimal heat denaturation effects would be expected, particularly for tissues from mammals, as the samples are heated to only 40°C.

2.4. N-Hydroxylsuccinimide Esters

Homobifunctional *N*-hydroxylsuccinimide esters have recently been found to be excellent fixatives for the preservation of cytoskeletal elements in immunofluorescence and scanning electron microscopy *(16)*. These esters act at physiological pH to crosslink proteins by forming stable inter- and intramolecular covalent linkages, primarily with the ε-amino groups of lysine residues. The crosslinkers are available with different spacer-arm lengths, the longer of which can react with more distantly spaced amino groups. *N*-hydroxylsuccinimide esters with a spacer-arm length of 12Å, a "medium" length, have been shown to produce the most favorable results for both soluble and polymerized proteins *(16)*. These agents are particularly useful in immunofluorescence microscopy, as they do not produce any detectable background autofluorescence.

2.5. Other Chemical Fixatives

Picric acid (trinitrophenol) and trinitroresorcinol, when added to fixative solutions, give greater fine structural preservation of cells *(11,12)*. These compounds cause coagulation of proteins by forming salts with positively charged groups of proteins *(11)*. The protein precipitates that form retain their antigenicity *(3)*. Picric acid or trinitroresorcinol are most often added to formaldehyde solutions, and fixative solutions of glutaraldehyde that contain these compounds have been shown to be effective for immunoelectron microscopic techniques *(13)*. These fixatives work particularly well to preserve membrane structure *(4)*.

Carbodiimide crosslinking fixatives have also been tested as fixative agents. These compounds act by crosslinking carboxyl groups to amine groups through amide bond linkages *(1)*. The carboxyl and amine reactive groups must be in close proximity, however, for crosslinkage to occur *(17)*. They have been found to be inferior to aldehydes in preserving cellular structure, particularly for electron microscopic procedures. They may, however, preserve some antigenic sites that are destroyed by aldehyde fixatives *(1,17)*. Carbodiimides show promise for fluorescence techniques in that they do not cause significant background fluorescence *(17)*.

3. Commonly Used Permeabilization Agents

3.1. Solvents

Alcohols and acetone are the simplest kinds of membrane permeabilization agents. They act by dissolving membrane lipids, thus rendering the membrane permeable to antibodies *(5)*. Because of their coagulant effects on proteins, these solvents can be used as a "one-step" fixative and permeant *(18)*.

3.2. Saponins and Lysolecithin

Saponins are natural compounds derived from plants. Saponins are generally the best permeant choice for routine cytoplasmic antigen localizations *(10)*. They act on membranes by interacting with cholesterol, plant sterols, phospholipids, and proteins. Saponin treatment is thought to break the associations between cholesterol and phospholipids, causing the formation of 120–150 Å membrane openings resulting from small losses of cholesterol *(19)*. Some of these membrane openings are transient in nature, and some are permanent, with most being formed 10–20 s after treatment with the agent *(19)*. Because of the transient nature of some of these membrane openings, it is usually recommended that saponin be included in all solutions throughout antibody treatments.

Lysolecithins act by dissolving cholesterol and cause massive losses of the sterol from membranes *(19)*. Lysolecithins have been shown to cause the formation of openings 300–400 Å in diameter in erythrocyte plasma membranes *(20)*. Unlike saponins, lysolecithin membrane openings are permanent.

3.3. Nonionic Detergents

Polyoxyethylene nonionic detergents (Triton X-100, Nonidet P-40, Tween-20, Brij 35, etc.) are used most often in immunochemical techniques because they generally do not denature proteins. Detergents act by intercalating into phospholipid bilayers, solubilizing lipids, and integral membrane proteins, thereby disrupting the membrane *(21)*. Hydrophobic proteins become enveloped in detergent and are easily washed away. For some antigens, particularly those localized within mitochondria or the nucleus, nonionic detergent treatments are required, because the membranes of these organelles do not contain large amounts of cholesterol and are not rendered permeable by saponins *(10)*.

Treatment of cells with nonionic detergents, while considered mild, is not without hazards; it can never be assumed that a detergent will not affect protein structure. Detergents with long hydrocarbon chains can denature some proteins *(21)*. High levels of oxidizing impurities capable of reacting with sulfhydryl groups in proteins have been found in commercial preparations of Triton X-100 and Brij 35 *(22)*. These oxidizing compounds can lead to loss of antigenicity and to high background autofluorescence in some fluorescence procedures. Some integral membrane proteins may be removed from the plasma membrane and organelle membranes by detergent treatments even after the cells have been fixed *(10)*. Such extraction of hydrophobic proteins can artifactually suggest lack of reactivity with antibodies.

4. Special Considerations for Plant Cells

The presence of a cellulosic cell wall and vacuoles constitute the major structural differences between plant and animal cells. These two structures allow the plant cell to maintain a high internal turgor pressure, a factor that needs to be considered when choosing an osmotically compatible fixative vehicle. Fixative solutions, nevertheless, are generally the same for plant and animal tissues *(3)*. The cytoplasm of animal cells has a higher protein concentration per unit volume than plant cells, and thus crosslinking fixatives give good results with animal cells *(3)*. Meristematic plant cells with dense cytoplasm and small vacuoles show better fixation with these agents than mature cells *(3)*.

Plant plasma membranes and tonoplasts (vacuole membranes) are particularly sensitive to fixation conditions, and fixation of the vacuoles can be quite problematic. If the fixation is inadequate, then the tonoplast can rupture and release hydrolytic enzymes into the cytoplasm *(3)*. Often fixatives at higher concentrations are used for plant cells in order to compensate for dilution by the vacuolar contents. The air in intercellular spaces can also hinder the penetration of fixatives into plant tissues; however, putting the tissue under vacuum during fixation assists fixative penetration. It is usually best to apply a slight vacuum, as stronger vacuums have been found to cause structural damage, most often resulting in separation of the plasma membrane from the cell wall *(3)*.

Although some report that antibodies are able to fully penetrate cell walls, immunofluorescence micrographs of cell-wall protein localizations suggest that this is not the case *(23)*. Antibody solutions infiltrated under vacuum into stem tissues do not appear to reach the inner surfaces of the cell walls, whereas antibodies applied to cut surfaces of a stem clearly do [*see* **Fig. 3** in **ref.** *23*]. In order to facilitate the penetration of antibodies into plant cells, the cell walls need to either be "opened" or removed. This is most often accomplished by digestion with the enzymes cellulase and/or pectinase *(24)*.

5. Choosing a Fixation/Permeabilization Protocol

Many factors need to be considered when choosing and/or developing a fixation protocol for antigen localization. The fixative regime will depend in large part on the antigen being studied *(1)*. Some fixation methods can be epitope-specific, in that a particular antigen may or may not react with different antibodies *(1)*. This can be either a disadvantage or an advantage in that the same tissue may not react with different antibodies, or one can test different stocks of antibodies prepared against a particular antigen in order to find one that reacts with tissues fixed by a "preferred" method. The choice of fixative vehicle is important because of its osmotic and ionic effects on cellular structures, particularly organelles *(9)*. Extraction of soluble proteins has been found to be a particular problem with hyperosmolar fixatives, and isotonic or hypotonic solutions are generally preferred *(1)*. Fixative vehicles, as with the fixatives themselves, usually need to be adjusted for the particular cell type or tissue under investigation *(9)*.

Melan and Sluder *(6)* showed that localizations of soluble proteins in cells can differ markedly from in vivo distributions depending on the fixation/permeabilization regime chosen. **Figure 1** (reproduced from *6*) shows an

---

Fig. 1. *(opposite page)* Distribution of FITC-conjugated BSA in various fibroblast cell lines under different fixation/permeabilization regimes. (**A–D**) Protein distribution in living cells: (A) PtK$_1$, (B) CHO, (C) 3T3, and (D) HeLa cells. The protein is excluded from the nuclei of all cells. (**E–H**) Protein distribution in cells extracted for 10 min with 0.1% Triton X-100 before fixation for 30 min with 3.7% formaldehyde: (E) PtK$_1$, (F) CHO, (G) 3T3, and (H) HeLa cells. Nuclear fluorescence is seen in (E) PtK$_1$ and (G) 3T3 cells. (**I–L**) Protein distribution in cells extracted for 10 min with 1% Triton X-100 before fixation for 30 min with 3.7% formaldehyde: (I) PtK$_1$, (J) CHO, (K) 3T3, and (L) HeLa cells. No fluorescence is detected in the cells with the exception of some nuclear fluorescence seen in (L) HeLa cells. (**M–P**) Protein distribution in cells fixed for 30 min with 3.7% paraformaldehyde before permeabilization for 10 min with 0.1% Triton X-100. Fluorescence is seen primarily in the cytoplasm with the exception that nuclear fluorescence is seen in (M) PtK$_1$ and (N) CHO cells. (**Q–T**) Protein distributions in cells fixed for 5 min with 90% methanol, 50 m*M* EGTA at –20°C: (Q) PtK$_1$, (R) CHO, (S) 3T3, and (T) HeLa cells. All cells show an overall low fluorescence, fibrous-textured cytoplasmic fluorescence, and bright staining at the periphery of the nucleus. 10 mm per scale division (black bar). (Reproduced with permission from **ref.** *6*.)

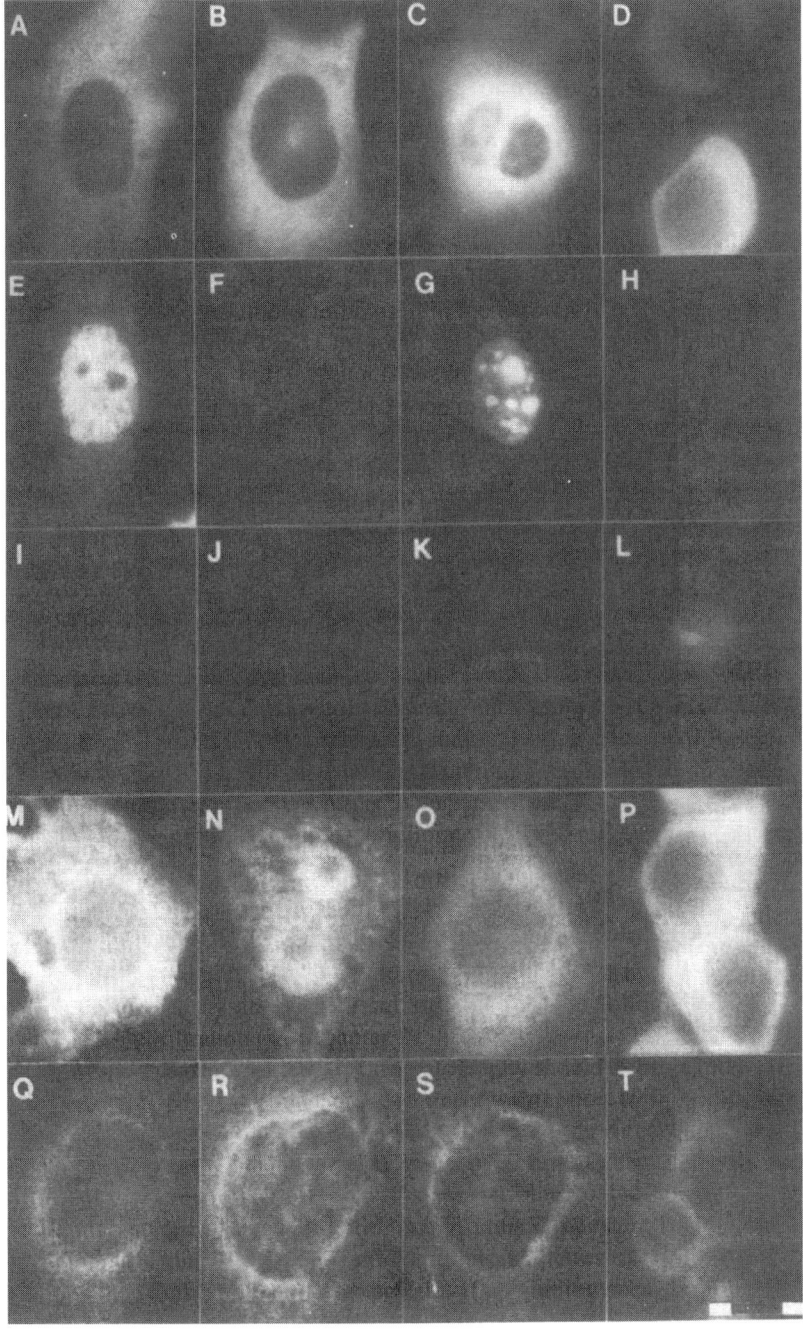

Fig. 1.

array of artifactual localizations obtained with soluble fluorescein-labeled bovine serum albumin (FITC-BSA) in various fibroblast cell lines. Although the FITC-BSA was evenly distributed in the cytoplasm and excluded from the nuclei in living cells (**Fig. 1A–D**), different fixation and permeabilization regimes led to striking relocation and extraction artifacts. A low level of global redistribution of extracted soluble proteins to all cells, presumably by transfer through washing solutions, was also noted *(6)*.

This study cautions that unless the protein and carbohydrate composition of a cellular structure is thoroughly known, immunolocalizations in that structure could be suspect *(4,6)*. Often the intracellular distribution of an antigen is not known beforehand, and these kinds of artifacts can lead to uncertainty in the results from immunolocalizations. It is recommended that the location of antigens be determined by several methods (i.e., various fixatives and permeabilization agents) before strong conclusions are drawn. Optimal fixation, therefore, requires a systematic evaluation of reagents and procedures *(3)*. It is generally best to begin with an established procedure and then modify the procedure as needed while keeping in mind the potential for artifactual localizations *(3)*.

Acknowledgments

I thank Drs. Greenfield Sluder and Dean P. Whittier for critical evaluation and helpful comments on the original manuscript. I also thank Mr. Frederick J. Miller and Mr. Michael Rodriguez for photographic assistance. **Fig. 1** is reproduced with permission from the Company of Biologists, Ltd.

References

1. Larsson, L. I. (1988) *Immunocytochemistry: Theory and Practice.* CRC, Boca Raton, FL.
2. Wang, Y. L. (1989) Fluorescent analog cytochemistry: tracing functional protein components in living cells, in *Methods in Cell Biology, vol. 29* (Taylor, D. L. and Wang, Y. L., eds.), Academic, San Diego, CA, pp. 1–12.
3. Hyat, M. A. (1981) *Fixation for Electron Microscopy.* Academic, New York.
4. Bowers, B. and M. Maser. (1988) Artifacts in fixation for transmission electron microscopy, in *Artifacts in Biological Electron Microscopy* (Crang, R. F. E. and Klomparens, K. L., eds.), Plenum, New York, pp. 13–42.
5. Humason, G. L. (1967) *Animal Tissue Techniques.* W. H. Freeman, San Francisco.
6. Melan, M. A. and Sluder, G. (1992) Redistribution and differential extraction of soluble proteins in permeabilized cultured cells: implications for immunofluorescence microscopy. *J. Cell Sci.* **101,** 731–743.
7. Huber, R., Deisenhofer, J., Coleman, P. M., Matsushima, M., and Palm, W. (1976) Crystallographic structure studies of an IgG molecule and an F_c fragment. *Nature* **264,** 415–420.
8. Poljak, R. J., Amzel, L. M., Avey, H. P., Becka, L. N., and Nissonoff, A. (1972) Structure of Fab' New at 6Å resolution. *Nature New Biol.* **235,** 137–140.

9. Glauert, A. M. (1974) Fixation, dehydration and embedding of biological specimens, in *Practical Methods in Electron Microscopy, vol. 3* (Glauert, A. M., ed.), North-Holland, Amsterdam, pp. 1–201.

10. Goldenthal, K. L., Hedman, K., Chen, J. W., August, J. T., and Willingham, M. C. (1985) Postfixation detergent treatment for immunofluorescence suppresses localization of some integral membrane proteins. *J. Histochem. Cytochem.* **33,** 813–820.

11. Kiernan, J. A. (1981) *Histological and Histochemical Methods. Theory and Practice.* Pergamon, Oxford, UK.

12. Ito, S. and Karnovsky, M. J. (1968) Formaldehyde-glutaraldehyde fixatives containing trinitro compounds. *J. Cell Biol.* **39,** 168a.

13. Newman, G. R., Jasani, B., and Williams, E. D. (1983) A simple post-embedding system for the rapid demonstration of tissue antigens under the electron microscope. *Histochem. J.* **15,** 543–555.

14. Weber, K., Rathke, P. C., and Osborn, M. (1978) Cytoplasmic microtubular images in glutaraldehyde-fixed tissue culture cells viewed by electron microscopy and by immunofluorescence microscopy. *Proc. Natl. Acad. Sci. USA* **75,** 1820–1824.

15. Leonard, J. B. and Shepardson, S. P. (1994) A comparison of heating modes in rapid fixation techniques for electron microscopy. *J. Histochem. Cytochem.* **42,** 383–391.

16. Safiejko-Mroczka, B. and Bell, P. B. (1996) Bifunctional protein cross-linking reagents improve labeling of cytoskeletal proteins for qualitative and quantitative fluorescence microscopy. *J. Histochem. Cytochem.* **44,** 641–656.

17. Kendall, P. A., Polak, J. M., and Pearse, A. G. E. (1971) Carbodiimide fixation for immunohistochemistry: observations on the fixation of polypeptide hormones. *Experientia* **27,** 1104–1106.

18. Harris, P., Osborn, M., and Weber, K. (1980) Distribution of tubulin containing structures in the egg of the sea urchin *Strongylocentrotus purpuratus* from fertilization through first cleavage. *J. Cell Biol.* **84,** 668–679.

19. Assa, Y., Shany, S., Gestetner, B., Tencer, Y., Birk, Y., and Bondi, A. (1973) Interaction of alfalfa saponins with components of the erythrocyte membrane in hemolysis. *Biochim. Biophys. Acta* **307,** 83-91.

20. Seeman, P. (1967) Transient holes in the erythrocyte membrane during hypotonic hemolysis and stable holes in the membrane after lysis by saponin and lysolecithin. *J. Cell Biol.* **32,** 55-70.

21. Neugebauer, J. (1988) *A Guide to the Properties and Uses of Detergents in Biology and Biochemistry.* Calbiochem-Novabiochem Corp., La Jolla, CA.

22. Ashani, Y. and Catravas, G. N. (1980) Highly reactive impurities in Triton X-100 and Brij 35: Partial characterization and removal. *Anal. Biochem.* **109,** 55–62.

23. Melan, M. A. and Cosgrove, D. J. (1988) Evidence against the involvement of ionically bound cell wall proteins in pea epicotyl growth. *Plant Physiol.* **86,** 469–474.

24. Wick, S. M., Seagull, R. W., Osborn, M., Weber, K., and Gunning, B. E. S. (1981) Immunofluorescence microscopy of organized microtubule arrays in structurally stabilized meristematic plant cells. *J. Cell Biol.* **89,** 685–690.

9

Preparation of Frozen Sections for Analysis*

Gary L. Bratthauer

1. Introduction

Fresh tissue must be preserved in some manner before analysis because the substances to be tested in the tissue are often labile and cannot withstand analytical procedures without first being preserved. The best method for antigen preservation in immunocytochemical analysis is freezing. Freezing is a very suitable means for preserving antigens that lose their immunoreactivity *(1)*. Fresh tissue, after being obtained, is quick-frozen in a flask of liquid nitrogen for a few seconds depending on the size of the tissue. The rapid introduction of ultracold temperatures prevents soluble materials from degrading and reinforces the structural components, steadfastly holding them in place. This method of preservation does not specifically alter the tissue in any way other than to cause some labile or low-concentration solutes to degrade slightly or lyophilize. In performing assays on frozen specimens, there is always a bit of denaturation as the sections are being prepared, because the cut section thaws in order to adhere to the glass slide. There is also the risk of lyophilization of important components as well as the threat of freeze/thaw conditions occurring in the freezer after the specimens are preserved. However, this method of producing frozen specimens for analysis by antibody is the closest one can come to in vivo conditions, as no chemical changes have been forced on the tissue. Instead, the tissue is bathed in the cold liquid form of the gas until frozen. The quickness with which this occurs is the reason that this process is so good for preservation. Sometimes it is desirable to speed up the process even

*The opinions or assertions contained herein are the private views of the author and are not to be construed as official or as reflecting the views of the Department of the Army or the Department of Defense.

From: *Methods in Molecular Biology, Vol. 115: Immunocytochemical Methods and Protocols*
Edited by: L. C. Javois © Humana Press Inc., Totowa, NJ

further, and an alcohol is used to directly bathe the specimen, which can penetrate the tissue more quickly than can liquid nitrogen alone.

2. Materials
2.1. Cutting Sections

1. Gloves.
2. Goggles.
3. Liquid nitrogen, in tank ($-196°C$).
4. Dewar flask, styrofoam, or some such insulated thermos.
5. Mounting block.
6. Medium such as Tissue-Tek OCT media (Baxter Diagnostics, McGaw Park, IL) or gum tragacanth (Fisher Scientific, Pittsburgh, PA) 7% in H_2O.
7. Large forceps.
8. Cryostat microtome.
9. Clean glass slides.
10. Freezer, $-70°C$.
11. Staining racks and dishes.
12. Stirring block with stir bars.
13. Zip-lock bags and slide boxes.
14. Tin foil.
15. Isopentane.
16. $60–80°C$ Oven.

2.2. Immunostaining of Frozen Sections

1. 0.1% poly-L-lysine solution (Sigma, St. Louis, MO).
2. Fixative solution of 95% ethanol, 5% glacial acetic acid, which should be prepared in advance and kept cold ($4°C$).
3. Phosphate buffered saline (PBS): 0.01 M sodium phosphate, 0.89% sodium chloride, pH 7.40 ± 0.05.
4. Endogenous enzyme blocking solution, 1.5% hydrogen peroxide (H_2O_2) in PBS prepared from a dilution of 30% H_2O_2.
5. Serum blocking solution, 10% animal serum in PBS (the species of serum should match the detecting system antibody, *see* **Note 1**).

3. Method
3.1. Preserving Tissue in Liquid Nitrogen

1. Fill a dewar flask half full with liquid nitrogen (*see* **Note 2**).
2. Apply some OCT embedding medium or 7% suspension of gum tragacanth compound in tap water to the end of a mounting block. Attach long forceps to the other end.
3. Immerse in liquid nitrogen for 2 s to adhere the compound to the block.
4. Obtain the fresh specimen and set it into the compound on the end of the mounting block. Make sure the tissue is in the desired orientation (*see* **Note 3**).

5. Plunge the sample directly into the liquid nitrogen immediately after securing it to the block (*see* **Note 4**).
6. Time to freeze is dependent on the size of the specimen, but for most applications it should only be 10–15 s (*see* **Note 4**).
7. Remove frozen sample and place in ultralow freezer in zip lock bag to prevent lyophilization (*see* **Note 5**).

3.2. Sectioning Frozen Tissue

3.2.1. Preparing Coated Slides

1. Place clean glass slides in a staining rack.
2. Immerse the slides for 30 min in a large staining dish containing 1:10 dilution of 0.1% poly-L-lysine solution in deionized water (*see* **Note 6**).
3. Remove the slides and oven dry for 1 h at 60°C.

3.2.2. Preparing the Sections

1. Fasten the mounting block to the block holder in the cryostat.
2. Align the knife to touch the surface of the tissue.
3. Set the thickness to 4 μ.
4. Begin cutting slowly until a choice section clings to the knife.
5. Touch a poly-L-lysine-coated slide to the section so that it binds to the surface of the glass. Allow it to air-dry (*see* **Notes 6** and **7**).
6. Store cut sections at –70°C in slide boxes.

3.3. Preparation of Section for Immunostaining

1. Sections in slide rack should be rapidly removed from the freezer and placed in a staining dish with the ethanol/acetic acid fixative solution for 2–10 min depending on tissue and antigen assayed (*see* **Note 7**).
2. Rinse the fixed sections with three changes of tap water for 5 min each to remove the fixative solution.
3. Block endogenous peroxidase and peroxidase-like activity by incubation in 1.5% H_2O_2 in PBS endogenous enzyme blocking solution for 15 min with constant stirring (*see* **Note 8**).
4. Rinse in water with three changes for 5 min each to remove all of the hydrogen peroxide.
5. Block charged sites on tissue surface with incubation in the 10% serum in PBS serum blocking solution overnight at 4°C (*see* **Note 1**).

4. Notes

1. The tissue cells exist in an electrically charged environment. To prevent antibodies from binding because of excess charges on the tissue surface, a proteinaceous solution is used to bind to these sites in advance of the antibody incubations. A 10% normal animal serum in PBS is used, obtained from the same species as that providing the secondary or detecting system antibody. The charged protein mol-

ecules of the serum will bind to the charged areas on the section, preventing the antibody reagents from binding to these areas nonspecifically. The use of other species' sera may cause the antibodies to adhere nonspecifically or cross-react. A problem that sometimes occurs when dealing with frozen or fresh material is that the tissue, being less denatured, is often more prone to protein binding to functionally preserved receptors *(2)*. In this instance, the use of serum may cause a problem by functionally reacting with the tissue and creating "exogenous antigen." This is especially true when assaying for a substance present in high concentrations in the serum. When applying serum to the section, the desired antigen is also being applied and may bind to areas that would be unavailable in a paraffin-embedded section. If the antiserum has a broad spectrum of reactivity and can detect the antigen in various species, the antigen could be identified where it has bound inadvertently. If this is a potential concern, charged sites on frozen sections may be blocked with the use of 2% bovine serum albumin. An overnight incubation at 4°C is best for the complete removal of available charged sites for nonspecific binding, but a 2-h room temperature incubation will suffice.

2. Be extra careful when handling the liquid nitrogen as it can cause serious injury. Wear goggles and gloves.

3. It is important to work quickly because the tissue will start to deteriorate the moment it is obtained. As always, when handling fresh tissue, it is imperative to wear gloves and protective clothing. If desired, specimens may be wrapped in tin foil and immersed into the liquid nitrogen directly. This way, the sample may be refrozen at a later date with proper orientation on an embedding-medium-coated mounting block.

4. It is more detrimental to stop the freezing too soon than to let it continue too long, so add a few seconds to the actual freezing time. It may be necessary to quick-freeze with a minimum of artifact. This may be accomplished by the use of isopentane cooled by immersion into liquid nitrogen and used as the freezing medium. This liquid-nitrogen-supercooled isopentane will preserve more rapidly and thus more efficiently. The alcohol will penetrate the tissue and allow the cold temperatures to freeze the tissue more quickly; however, isopentane will freeze with too much liquid nitrogen exposure.

    a. Place the isopentane in a Pyrex™ beaker or large test tube that is small enough to be put in the dewar flask. Add the isopentane and surround the beaker with liquid nitrogen.

    b. After 2 min, the temperature should be –140°C. Plunge the specimen in the isopentane for 10 s.

    Rapid freezing is the best preservation tool. The use of liquid nitrogen and cooled isopentane are for the purposes of cooling quickly. The tissue should be relatively small in size to allow for rapid and thorough penetration of the cold-temperature chemicals.

5. To store a block for later cutting, place in an airtight container, preferably one that is not too much bigger than the block. Small Zip-lock bags are good for this. Place into an ultralow-temperature freezer (–70°C) as soon as possible. Avoid freezers with automatic freeze–thaw cycles such as "frost-free" types.

6. The sections should be cut as thin as possible on glass slides that are coated to prevent the tissue from coming off later in the process. Immunocytochemical procedures require the sections to be in buffers and solutions for hours or days, and they can float off the slides during that time if not affixed properly. Some slide coatings like poly-L-lysine can create background staining with certain techniques. Experimentation can determine the slide coating that provides clean backgrounds with the technique of choice. In addition to the use of poly-L-lysine–coated glass slides, slides may also be coated with gelatin as follows:

   a. Prepare solution of 0.5% gelatin in deionized H_2O (heat to dissolve).
   b. Immerse the slides for 15 min.
   c. Allow to air-dry for 24 h.

   Also, slides may be coated with Silane in the same manner as poly-L-lysine, or the use of Plus slides (Fisher Scientific) can control the loss of tissue sections common with these techniques.

7. Optimum time of fixation can be experimentally determined. The use of alternate fixatives may also be employed to identify specific compounds. One of the reasons for using a frozen preparation is to examine the tissue or cells in as close to the in vivo state as possible. Also, there are many antigens that cannot be evaluated in fixed tissues *(3)*. Therefore, when analyzing these sections, it is unwise to fix them in an efficiently crosslinking fixative such as glutaraldehyde or formalin. However, alcohol, while a relatively mild fixative, still can cause enough distortion to warrant the use of even milder agents. The classic is acetone, which is frequently recommended.

   a. The sections are fixed for 10 min in cold acetone.
   b. The sections are allowed to air-dry.

   A large number of antisera are being generated against acid-precipitated proteins, and the presence of acid denaturation in the tissue preparations is often required for antibody recognition. This is the reason the ethanol/acetic acid fixation protocol is featured in this chapter. It should be stated though, that for sections to be the closest to in vivo conditions, simple acetone fixation is preferred. Other possible fixatives may include methanol, Bouin's fluid, or formalin—for the best morphologic preservation and only if the antigen is known to survive aldehyde crosslinking *(4)*. Formalin fixation, if attempted, should be brief, 30 s to 1 min. The fixative used is important because frozen sections inherently sacrifice some morphology for the improved protein viability, and one does not want to negate that advantage with the choice of the wrong fixative. If a particularly labile antigen is to be detected, the sections can be immediately fixed after adhering to the glass slide. Sections fixed immediately can be stored in PBS at 4°C for up to 5 d *(5)*.

8. The enzyme or the chromogen detection system determines whether any endogenous material must first be destroyed. If a peroxidase marker molecule is to be used, endogenous peroxidase or peroxidase-like activity should be blocked. Because these preparations are more fragile than a fixed embedded sample, endogenous enzyme is inactivated with a weaker blocking solution than would

otherwise be used. The standard endogenous oxidation blocking solution when using peroxidase enzymes is a solution of 3% hydrogen peroxide in methanol. For frozen sections, though, PBS is substituted for the same reason that formalin is avoided. The compound of interest should not be subjected to any more denaturing agents than is necessary *(6)*. Also, the amount of H_2O_2 can be reduced; a 1.5% solution is usually sufficient. Sometimes even 1.5% H_2O_2 will cause visible bubbles of gas to develop, and if this is too extensive, it may succeed in lifting the section off the slide. Careful monitoring and gentle tapping of the container will help to prevent this. Wear gloves and exercise caution when handling the 30% H_2O_2 since it is caustic and can cause burns.

References

1. Cuello, A. C. (1983) *Immunocytochemistry*, Wiley, New York, p. 4.
2. Ditzel, H., Erb, K., Nielsen, B., Borup-Christensen, P., and Jensenius, J. (1990) A method for blocking antigen-independent binding of human IgM to frozen tissue sections when screening human hybridoma antibodies. *J. Immunol. Meth.* **133**, 245–251.
3. Nadji, M. and Morales, A. (1984) Immunoperoxidase: pt. II. Practical applications. *Lab. Med.* **15**, 33–37.
4. Tse, J. and Goldfarb, S. (1988) Immunohistochemical demonstration of estrophilin in mouse tissues using a biotinylated monoclonal antibody. *J. Histochem. Cytochem.* **36**, 1527–1531.
5. Miller, R. (1991) Immunohistochemistry in the community practice of pathology: pt. I. *Lab. Med.* **22**, 457–464.
6. Van Bogart, L. (1985) Present status of estrogen-receptor immunohistochemistry. *Acta Histochem.* **76**, 29–35.

10

Processing of Cytological Specimens*

Gary L. Bratthauer

1. Introduction

Immunocytochemistry can be a valuable tool for the determination of cellular contents from individual cell suspensions. Samples that can be analyzed include blood smears, aspirates, and swabs from any cellular site. Each sample is treated differently, yet all the methods are interchangeable. There is no one way to prepare these types of cell samples for immunocytochemical analysis. This chapter will deal with the most common forms of cell sample, the swab, aspirate, smear, and touch preps. Blood can be analyzed as a smear, but it presents more of a problem because of the concentration of red blood cells. These cells have an oxidative type function, and when using a peroxidase-based detection system, it can greatly interfere with the test. Concentrated cellular suspensions that exist in a low-viscosity medium make good candidates for smear preparations. Dilute cell suspensions existing in a dilute medium are best suited for the preparation of cytospins through cytocentrifugation. Cell suspensions that exist in a high-viscosity medium, are best suited to be tested as swab preparations (1). The constant among these preparations is that the whole cell is present on the slide surface. For any intercellular reaction to take place, immunoglobulin must first traverse the cell membrane that is intact in these preparations. Reactions taking place in the nucleus can be more difficult, and the extracellular fluids can create unique obstacles in the performance of immunocytochemistry.

In the event that smears or aspirates cannot be adequately performed, or the sample is too small for extra studies, touch preparations can provide a means

---

*The opinions or assertions contained herein are the private views of the author and are not to be construed as official or as reflecting the views of the Department of the Army or the Department of Defense.

From: Methods in Molecular Biology, Vol. 115: Immunocytochemical Methods and Protocols
Edited by: L. C. Javois © Humana Press Inc., Totowa, NJ

of quick cell identification or examination *(2)*. Touch preparations, or imprints, enable the examination of the whole cell apart from the tissue aspect. The cells are obtained by touching a wet tissue with a glass slide. Cells adhere to the glass in roughly the same orientation as they exist on the surface of the lesion touched. Fixing them in place enables one to examine the cells for a rapid investigation without having to freeze and cut through a tissue block. If cytological information is needed, touch preparations can be obtained easily from the surface of otherwise large tissue fragments. Cells that are acquired in this manner can provide information about such things as membrane receptors and some cell adhesion molecules in the absence of the tissue structural components. Sometimes tissue structural elements can interfere with the ability to identify a substance associated with a particular cell. Cell touch preparations made at the time of specimen removal can then be used.

2. Materials

2.1. Preparation of Cell Smears or Touch Preps

1. Gloves.
2. Clean glass slides.
3. Beveled edge slide.
4. Small glass transfer pipet.
5. Sterile swabs.
6. 0.1% poly-L-lysine solution (Sigma, St. Louis, MO).
7. Staining racks and dishes.
8. Stirring block with stir bars.
9. Cyto Prep spray fixer (Fisher Scientific, Pittsburgh, PA).
10. Methanol.
11. 10% neutral buffered formalin (NBF).
12. Gauze.
13. Coplin jar.
14. Forceps.

2.2. Preparation of Cytospin Slides

1. Cytocentrifuge.
2. Slide holder apparatus for cytocentrifuge.
3. Sample chambers for cytocentrifuge.
4. Filter cards for use with cytocentrifuge.
5. Fixative solution of 95% ethanol, 5% glacial acetic acid, which should be prepared in advance and kept cold (4°C).

2.3. Immunostaining of Cytological Specimens

1. Phosphate buffered saline (PBS): 0.01 M sodium phosphate, 0.89% sodium chloride, pH 7.40 ±0.05.

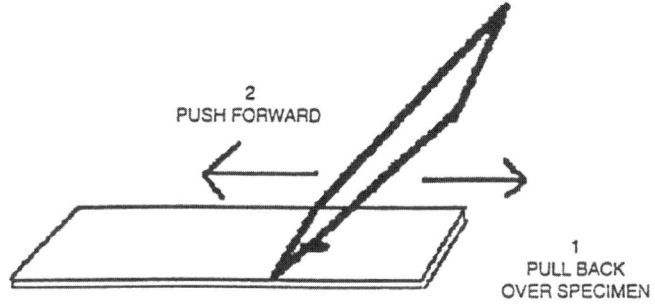

2
PUSH FORWARD

1
PULL BACK
OVER SPECIMEN

Fig. 1. Preparing a smear.

2. 0.25% Triton X-100 and 5% dimethylsulfoxide (DMSO) in PBS.
3. Endogenous enzyme blocking solution: 1.5% hydrogen peroxide (H_2O_2) in PBS prepared from a dilution of 30% H_2O_2.
4. Serum blocking solution: 10% animal serum in PBS (the species of serum should match the detecting system antibody, *see* **Note 1**).
5. Acetone, refrigerated.

3. Methods
3.1. Preparation of Coated Slides

1. Position clean glass slides in a staining rack.
2. Immerse the slides for 30 min in a large staining dish containing a 1:10 dilution of 0.1% poly-L-lysine solution in deionized water.
3. Remove the slides and oven dry for 1 h at 60°C.

3.2. Preparation of Cytology Smears
3.2.1. Cell Film Preparation

1. Add a drop of cell material (blood, cell suspension, and so on) to the end of the coated glass slide (*see* **Note 2**).
2. Hold the beveled edge slide at a 45° angle to the plane of the coated slide and gently touching the surface of the slide, back the edge over the drop of cells so they spread within the 45° angle, the width of the slide (*see* **Fig. 1**).

3. Slide the beveled edge slide toward the other end of the preparation slide (in the direction of the 135° angle) with a rapid uniform motion (*see* **Note 3**).
4. Allow to air dry for 30 min.
5. Immerse in methanol for 30 s to fix the cells.
6. Rinse slides three times with deionized water.
7. Incubate the slides in a solution of 0.25% Triton X-100, 5% DMSO in PBS for 10 min to permeabilize the membranes (*see* **Note 4**).
8. Rinse three times in deionized water for 5 min each to remove detergents.
9. Block endogenous peroxidase and peroxidase-like activity by incubation in 3% H_2O_2 in methanol solution for 45 min with constant stirring (*see* **Note 5**).
10. Rinse three times with deionized water for 5 min each to remove hydrogen peroxide and methanol.
11. Block charged sites in the cell preparation with an incubation in the 10% animal serum in PBS overnight at 4°C (*see* **Note 1**).

3.2.2. Swabbed Slide Preparation

1. Obtain sample on sterile swab.
2. Smear the sample onto the glass slide, using the majority of the surface area to distribute the specimen (*see* **Note 2**).
3. Spray fix the material with Cyto Prep.
4. Allow to air-dry.
5. Postfix the slides in 10% neutral buffered formalin for 30 s.
6. Rinse three times with deionized water for 5 min each to remove fixative.
7. Incubate the slides in a solution of 0.25% Triton X-100, 5% DMSO in PBS for 10 min to permeabilize the membranes (*see* **Note 4**).
8. Rinse three times in deionized water for 5 min each to remove detergents.
9. Incubate in 3% H_2O_2 methanol for 45 min with constant stirring (*see* **Note 5**).
10. Rinse three times with deionized water for 5 min each to remove hydrogen peroxide and methanol.
11. Incubate in 10% normal serum of secondary species overnight at 4°C (*see* **Note 1**).

3.3. Preparation of Cytospin Specimens

1. Position slides in slide holders with filter cards and sample chambers and attach to cytocentrifuge rotor.
2. Prepare cell suspension of 500 cells/mm³ (µL) with PBS (*see* **Note 6**).
3. Add 0.1 mL cell suspension to chamber and centrifuge at 1000 rpm for 5 min (*see* **Note 7**).
4. Remove slide and immediately dip in 95% ethanol, 5% glacial acetic acid fixative for 2 min (*see* **Notes 8** and **9**).
5. Rinse three times with deionized water for 5 min each to remove fixative.
6. Incubate the slides in a solution of 0.25% Triton X-100, 5% DMSO in PBS for 10 min to permeabilize the membranes (*see* **Note 4**).
7. Rinse three times in deionized water for 5 min each to remove detergents.
8. Incubate in 1.5–3% H_2O_2 in methanol for 30 min (*see* **Note 5**).

9. Rinse three times with deionized water for 5 min each to remove hydrogen peroxide and methanol.
10. Incubate in 10% normal serum of secondary species overnight at 4°C (*see* **Note 1**).

3.4. Preparation of Touch Prep Specimens

1. Take the refrigerated acetone and add to a Coplin jar.
2. Position the excised tissue directly above the slides, best side facing the slide, using forceps (*see* **Note 10**).
3. Align the slide horizontally on flat surface (*see* **Note 10**).
4. Gently touch the selected exposed tissue surface down onto the slide (*see* **Note 11**).
5. Apply slight pressure then remove the tissue after a few seconds (*see* **Note 11**).
6. With a minimum of disturbance, immerse slide into the cold acetone solution (*see* **Note 8**).
7. Remove it after 10 min and air-dry.
8. Rehydrate the imprints with three changes of deionized water for 5 min each.
9. Incubate the slides in a solution of 0.25% Triton X-100, 5% DMSO in PBS for 10 min to permeabilize the membranes (*see* **Note 4**).
10. Rinse three times in deionized water for 5 min each to remove detergents.
11. Block endogenous peroxidase and peroxidase-like activity by incubation in 1.5–3% H_2O_2/PBS solution for 15 min with constant stirring (*see* **Note 5**).
12. Rinse in water with three changes for 5 min each to remove all of the hydrogen peroxide.
13. Block charged sites on tissue surface with incubation in the 10% serum in PBS serum blocking solution overnight at 4°C (*see* **Note 1**).

4. Notes

1. The cells exist in a sometimes mucoid, often electrostatic extracellular fluid. To prevent antibodies from binding as a result of excess surface charges or mucus, a proteinaceous solution is allowed to bind to these sites in advance of the antibody incubations. The charged protein molecules of the serum will bind to the charged sites or stick to the mucoid fluid, preventing the antibody reagents from binding to these areas nonspecifically. A 10% normal animal serum in PBS is used, from the same species as that providing the secondary or detecting system antibody. Since these preparations are cellular and not related to tissue, the extent to which serum proteins are used to block nonspecific charged adherence of antibodies is variable. Sometimes no protein is needed at all, while at other times, the use of a 2% bovine serum albumin solution will be needed. As in the case of frozen sections, there is also a danger of exogenous antigen addition when dealing with whole cells on slides. How the preparation is made depends to a large extent on what antigen is being studied. This is especially true when assaying for a compound present in high concentrations in the serum. When applying serum to the section, the compound of interest is also being applied and may bind to areas that would be unavailable in a paraffin-embedded section. If the antiserum has a broad spectrum of reactivity and can detect the antigen in various species, the antigen

could be identified where it has bound inadvertently. An overnight incubation at 4°C is best for the complete removal of available charged sites for nonspecific binding, but a 2-h room temperature incubation will suffice.

2. As always, when handling fresh blood or body fluids, it is imperative to wear gloves and protective clothing.

3. In the preparation of smears, it is necessary to get a uniform thin film to avoid cells bunching up in layers. Start with a smaller sample amount if this is occurring. The cells will spread out over the surface of the slide and form a film with a feathered edge, if done properly.

4. Swabs that are heavily mucoid, such as gynecological specimens, present a problem in physically locating the antibody near the antigen. These specimens need some additional treatments in the form of cell membrane disruption with DMSO or detergents to ensure intracytologic components will be available for testing *(3)*. Also, the intact nature of cell touch preps may require detergent incubation. It is important to completely dissolve the Triton X-100 in the PBS before adding the DMSO. While each of the procedures outlined here calls for the pretreatment step with Triton X-100 and DMSO, it should be pointed out that this may not be necessary in all cases. The accessibility of any one antigen for any one antibody is because of many factors, including location, structure, and concentration of the antigen to be identified. The use of detergents and other solutions to gain access to an antigen are only necessary if the antigen is difficult to detect. Obviously, the fewer steps taken to analyze the cells, the better, so as not to cause undue manipulation. These protocols are intended as guidelines and as templates with which to examine individual systems, subject to experimentation.

5. If a peroxidase marker molecule is to be used for immunostaining, endogenous peroxidase or peroxidase-like activity should be blocked. Because of the large amount of red blood cells in blood preparations, some remaining endogenous enzyme and enzyme-like activity may occur if using a peroxidase system. Even though the standard 3% solution is recommended, it may be necessary to increase the amount of hydrogen peroxide or increase the time of incubation. The amount of red blood cells present may cause a visible bubbling action on the surface of the slide. This is not detrimental, but it should be monitored in case the oxidation is violent enough to remove cells from the slide. In the case of cytospins, though, sometimes the cell suspension is free of endogenous enzyme or enzyme-like material, and the amount of H_2O_2 needed can be reduced. Wear gloves and exercise caution when handling the 30% H_2O_2, as it is caustic and can cause burns. Touch preps often require slightly different handling. Because these preparations are more fragile than a fixed embedded sample and do not share the support of a surrounding mucoid environment, endogenous enzyme is inactivated with a weaker blocking solution than would otherwise be used. The standard endogenous oxidation blocking solution when using peroxidase enzymes is a solution of 3% hydrogen peroxide in methanol. For these fresh cells, though, PBS is substituted for the same reason that formalin is avoided. The compound of interest should not be subjected to any more denaturing agents than is necessary. Also, the amount of H_2O_2 can be reduced; a 1.5% solution is usually sufficient.

6. Preparations that are too dilute for cytocentrifugation may be dropped with a pipet, using the location of the filter card as a guide, and allowed to dry *(4)*.

7. It is important to get just the right amount of speed for cytocentrifugation, as too much speed will flatten the cell, and too little will not allow the cells to adequately bind to the slide.

8. For touch preps, the slide should be immersed in the acetone as soon as possible, but the cells need a moment to adhere to the plane of the glass. Slowly dip the slide into the acetone, as a violent action at this point could wash off some of the cells. As is the case with frozen sections, fixation is a matter of choice. In this instance, the use of acetone is preferred because the cells are still whole, and the membranes require disruption in order for the contents to be accessible for later analysis. However, an ethanol/acid fixative (95% ethanol, 5% glacial acetic acid) is perfectly acceptable if desired. For the other cytology specimens, experimentation within individual systems is necessary. There is no right or wrong way to make these preparations. There is a fine balance struck between the need for good morphology and the need for antigen preservation. Alcohol, whether ethanol or methanol, is a mild fixative that preserves protein epitopes for antibody recognition. Yet, with some whole cell preparations, a short postfixation (fixing a second time) in an aldehyde crosslinking fixative sometimes stabilizes antigenic determinants better. This method can only be used in cases where antigen is not destroyed by formaldehyde crosslinking *(5)*. Other possible fixatives to try may include methanol or Bouin's fluid. Some have had success with methanol and acetone mixed in a 1:1 ratio *(6)*. A large number of antisera are being generated against acid-precipitated proteins, and the presence of acid denaturation in the tissue preparations is often required for antibody recognition. This is the reason the ethanol/acetic acid fixation protocol is featured in this chapter. It should be stated, though, that for cells to be the closest to in vivo conditions, simple acetone fixation is best. Formalin fixation, if attempted, should be brief, 30 s to 1 min.

9. In preparing cells for cytospin slides, there are times when immediate fixation is necessary. For these instances, the cells can be fixed before centrifugation. Starting at **step 2** for the cytocentrifuged specimen (**Subheading 3.3.**), continue with:
   a. Prepare cell suspension by diluting to 1000 cells/μL with PBS.
   b. Add 0.25 mL of cell suspension to 0.25 mL of 95% ethanol, 5% glacial acetic acid and mix immediately.
   c. Incubate for 2 min and add to sample chambers.
   d. Centrifuge at 2000 rpm for 5 min.
   e. Continue with the above procedure from **step 4**.
      Caution: If the cells are fixed before centrifugation they may bunch up and require more vigorous mixing prior to centrifugation.

10. The purpose for the preparation of these types of slides is to examine the cells of a tissue quickly, with no need for tissue preparation and cutting. It is important to work swiftly, as the tissue will start to deteriorate the moment it is obtained. Therefore, it is important to have all the preparations ready for the rapid han-

dling of the excised tissue. As soon as a surface of the tissue is decided on, it should immediately be touched to a waiting slide, oriented properly and ready to be fixed.

11. Excess tissue fluid may be absorbed with a little gauze, but avoid touching any wet cells. Also, gentle pressure is required, but if too forceful, some cells may be destroyed.

References

1. Kobayashi, T., Ueda, M., Araki, H., Toyoda, K., Ohmori, K., and Sawaragi, I. (1987) Immunocytochemical demonstration of chlamydia infection in the urogenital tracts. *Diag. Cytopathol.* **3,** 303–306.
2. Masood, S. (1989) Use of monoclonal antibody for assessment of estrogen receptor content in fine-needle aspiration biopsy specimen from patients with breast cancer. *Arch. Pathol. Lab. Med.* **113,** 26–30.
3. Li, C., Lazcano-Villareal, O., Pierre, R., and Yam, L. (1987) Immunocytochemical identification of cells in serous effusions. *Am. J. Clin. Pathol.* **88,** 696–706.
4. Janssens, P., Kornaat, N., Tieleman, R., Monnens, L., and Willems, J. (1992) Localizing the site of hematuria by immunocytochemical staining of erythrocytes in urine. *Clin. Chem.* **38,** 216–222.
5. Tse, J. and Goldfarb, S. (1988) Immunohistochemical demonstration of estrophilin in mouse tissues using a biotinylated monoclonal antibody. *J. Histochem. Cytochem.* **36,** 1527–1531.
6. Bein, G., Bitsch, A., Hoyer, J., and Kirchner, H. (1991) The detection of human cytomegalovirus immediate early antigen in peripheral blood leucocytes. *J. Immunol. Meth.* **137,** 175–180.

11

Processing of Tissue-Culture Cells*

Gary L. Bratthauer

1. Introduction

In research, a technique that has increased in importance is the analysis of the living cell as seen in cell culture. It is within the special boundaries of cell culture that conditions can be manipulated in order to examine living cells and to better understand their behavior *(1)*. The fact that transformed or malignant cells grow very nicely in rather simple conditions of chemical formulation, make these cells ideal laboratories in which to study cellular processes. Immunocytochemistry as applied to cell culture, has allowed experiments on living cells to be examined through the detection of products produced as a result of those experiments *(2)*. Also, the immunocytochemical analysis of cell culture can be used to examine viral infections before obvious cytopathic effect *(3)*. It becomes important therefore, to prepare these cells adequately in order to fully understand the implications of the various experiments or inoculations. There are many ways of doing so, with and without cell removal from the culture flask. Once the cells are ready to be evaluated by immunocytochemical means, there are several preparative procedures available, each effective for a slightly different set of conditions. Cells are grown in culture to confluence, or suspension, in the case of nonadherent cells. The techniques for doing so are beyond the scope of this text; most cell culture protocols are readily available in the literature. When the cells have grown confluent to the point of study, they can be removed from culture by digestion and centrifugation, or allowed to remain adherent and assayed directly. They may be obtained in solution

*The opinions or assertions contained herein are the private views of the author and are not to be construed as official or as reflecting the views of the Department of the Army or the Department of Defense.

From: *Methods in Molecular Biology, Vol. 115: Immunocytochemical Methods and Protocols*
Edited by: L. C. Javois © Humana Press Inc., Totowa, NJ

and treated the same as cells from aspirates, or they can be concentrated and prepared in a cell block as is described for whole tissue in Chapter 12.

2. Materials
2.1. Culture Slide or Block Preparation

1. Gloves.
2. Lab-tek slides (Nunc, Naperville, IL).
3. Cell scraper.
4. 15-mL Polystyrene conical centrifuge tube.
5. 50-mL Polypropylene centrifuge tubes.
6. Vortex.
7. Centrifuge (3000 rpm).
8. 3-mL syringe with 22-gauge needle.
9. Forceps.
10. Trypsin 0.1% in culture medium.

2.2. Immunostaining of Specimens

1. Fixative solution: 95% ethanol, 5% glacial acetic acid.
2. Phosphate buffered saline (PBS): 0.01 M sodium phosphate, 0.89% sodium chloride, pH 7.40 ±0.05).
3. 0.25% Triton X-100 and 5% Dimethylsulfoxide (DMSO).
4. Endogenous enzyme blocking solution: 1.5% hydrogen peroxide (H_2O_2) in PBS prepared from a dilution of 30% H_2O_2.
5. Serum blocking solution: 10% animal serum in PBS (the species of serum should match the detecting system antibody; see **Note 1**).
6. 10% neutral buffered formalin (NBF).

3. Methods
3.1. Preparation of Cell-Culture Slides (Method 1)

This method is only appropriate for cells that adhere to the flask.

1. Transfer 200 μL of cell culture to the wells of a Lab-tek slide chosen to facilitate the experiment (see **Notes 2** and **3**).
2. Allow cells to grow to confluence with the addition of fresh media.
3. Wash the cells thoroughly with five changes of PBS for 2 min each (see **Note 4**).
4. Drain the PBS and add fixative directly to the cells. Fix cells for 3 min (see **Note 5**).
5. Wash away excess fixative with five changes of water for 2 min each.
6. Incubate the slides in the solution of 0.25% Triton X-100, 5% DMSO in PBS for 10 min to permeabilize the membranes (see **Note 6**).
7. Rinse three times in deionized water for 5 min each to remove detergents.
8. Block the endogenous enzyme or enzyme-like activity, if necessary, by incubation in 1.5% H_2O_2 in PBS solution for 15 min (see **Note 7**).
9. Rinse three times in deionized water for 5 min each to remove detergents.

10. Incubate in the 10% animal serum in PBS overnight at 4°C. The charged protein molecules of the serum will bind to available sites, preventing the antibody reagents from binding to these areas nonspecifically (*see* **Note 1**).

3.2. Preparation of Cell Culture Cell Blocks (Method 2)

If the desire in examining cells in culture is to relate the findings to solid tissues, then the investigator may opt to test the cultured cells as though they were a solid tissue by embedding and sectioning them.

1. Grow cells in flask to confluence or to a heavy suspension if nonadherent (*see* **Note 2**).
2. Remove confluent cells by the addition of 0.1% trypsin in media solution for 2 min, followed by extensive media washing and decanting to a 50-mL centrifuge tube. Cells grown in suspension are merely decanted to a 50-mL centrifuge tube (*see* **Note 8**).
3. Centrifuge at 800*g* for 5 min to remove cells from the medium (*see* **Note 9**).
4. Gently decant the medium from the cells and add 20 mL of PBS, vortexing slowly.
5. Centrifuge at 800g for 5 min to remove cells from the medium.
6. Gently decant the medium from the cells, and add 20 mL of PBS, vortexing slowly.
7. Centrifuge at 800g for 5 min to remove cells from the medium.
8. Gently decant the PBS from the cells and add 10 mL of PBS to the tube and vortex.
9. After suspension is thoroughly mixed, decant to a 15-mL polystyrene centrifuge tube.
10. Centrifuge at 1500g for 10 min to pellet the cells.
11. Decant the PBS and carefully add 4 mL of 10% neutral buffered formalin down the side of the tube, overlaying the pellet without causing turbulence (*see* **Note 10**).
12. Incubate overnight at 4°C.
13. Decant the formalin, and gently overlay pellet with 5 mL water.
14. Using the syringe and needle, gently enter the water down the side of the tube and draw 2 mL into the syringe.
15. Carefully undermine the pellet with the needle along the wall of the tube and face the bevel toward the plastic face on the inside of the tube.
16. Rapidly express the syringe, and the cells will dislodge as a solid pellet. The undermining action may have to be repeated several times to dislodge the pellet, depending on the type of cells and the force of centrifugation. Also, even though some of the pellet might break off and be lost, this should not affect the bulk of the cells, which ought to still be in the form of the pellet (*see* **Note 11**).
17. Add 5 mL of water slowly and swirl the pellet in the liquid, being careful not to disrupt the cells.
18. As the pellet is in motion within the tube, quickly pour water and pellet in wide-mouth glass container that is easily accessible with forceps.
19. When pellet has settled, decant the water and add enough 70% ethanol to cover the pellet.
20. The pellet is now ready for processing in an automated tissue processor. If one is not available, the pellet can be embedded in paraffin by hand, gradually replacing the ethanol with xylene and then gradually replacing the xylene with paraffin.

21. Once the pellet is embedded in paraffin, it can be treated as any tissue specimen would be treated (*see* Chapter 12).

4. Notes

1. The cells exist in a formulated medium that contains serum. If washing is incomplete or if the medium is a little viscous, nonspecific antibody binding may occur. To prevent antibodies from binding nonspecifically, a proteinaceous solution is used to bind to these sites in advance of the antibody incubations. A 10% normal animal serum in PBS is used, from the same species as that providing the secondary or detecting system antibody. The use of other species' sera may cause the antibodies to adhere nonspecifically or crossreact. A problem that sometimes occurs when dealing with fresh material is that the cells, being less denatured, are often more prone to protein binding to functionally preserved receptors (*4*). In this instance, the use of serum may cause a problem by functionally reacting with the cells and creating "exogenous antigen." This is especially true when assaying for a substance present in high concentrations in the serum. When applying serum to the cells, the antigen of interest is also being applied and may bind to areas that would be unavailable in a paraffin-embedded section. If the antiserum has a broad spectrum of reactivity and can detect the antigen in various species, the antigen could be identified where it has bound inadvertently. If this is a potential concern, charged sites on cell culture cells may be blocked with the use of 5% bovine serum albumin in PBS. An overnight incubation at 4°C is best for the complete removal of available sites for nonspecific binding, but a 2-h room temperature incubation will suffice.
2. Care must be used in handling live cells in culture because potential biohazards exist. Personal precautions should be followed, including the wearing of gloves and protective clothing.
3. Choice of slide design is often dictated by the experiment. Some slides have four wells, some have eight, some are glass, and some are plastic. Glass is recommended simply because the slide becomes more versatile, acetone can be used as a fixative, and finished slides can be dehydrated in ethanol and cleared in xylene. Slides with individual culture wells make the examination of more than one cell area possible on the same slide and are good for the study of different conditions. If these slides are not available, a cell-culture flask can be used by cutting the bottom out and assaying as if it were a slide. Different conditions can be tested by using a cutout template of thickened cardboard and confining solutions to prescribed areas on the flask bottom. When assaying individual wells differently, it is necessary to keep the plastic well cover in place. However, when the specific individual treatments are finished, all of the wells can be assayed together as an entire slide. In order to do this, the rubber gasket separating the chambers should be removed using forceps. This may be difficult depending on the fixation used and the step in the protocol. Usually, the use of alcohol softens the rubber sufficiently to allow for easy removal. This must be done anyway prior to coverslipping. Simply pull on one end of the gasket and tear the whole piece off. If

the gaskets are left in place, there will be a bit of background associated with the edge where the gasket was positioned. Since these culture slides must be sterile and are sold in sealed packages, the cells are not fixed to the surface with any special compound. This makes these preparations a bit more fragile and generally prohibits the use of enzymatic digestion protocols.

4. One problem unique to these slides is the closed system in place in the cell culture well. As these cells grow, their metabolites are shed into the surrounding medium. If a particular analyte is in high-enough concentrations to be contained in the spent medium, it could also be left behind attached to some cells because of inadequate washing. It is very important to wash the spent medium away with many PBS washes before fixation. This way, no extracellular constituent will cause aberrant reactions resulting from its remaining on the cell inappropriately.

5. The assay of cells grown on slides becomes very much like the assay of cells described in Chapters 9 and 10. The style and use of fixative is up to the investigator. The limitations when using plastic surfaces involve the use of acetone or xylene in the preparations. Acetone would have deleterious effects, so 95% ethanol, 5% glacial acetic acid can be used as a mild yet competent fixative. When using a multiwell type of slide, different conditions of fixation can easily be tested, confined to separate wells on the slide. Generally though, with the advantage of freshly fixed cells to work with, the mildest fixative that preserves adequately will be best for an immunocytochemical study.

6. Depending on the desired antigen to be identified and its location within the cell, the cells could be subjected to a detergent incubation *(5)*. A pretreatment of DMSO, Triton X-100 in PBS can be employed for better immunoglobulin penetration. It is important to completely dissolve the Triton X-100 in the PBS before adding the DMSO.

7. Depending on the cells being analyzed, the removal of endogenous enzyme prior to immunostaining may not be necessary. With some cell lines, there are no endogenous enzyme or enzyme-like substances present, and this step would be unnecessary. If that is the case, it is best to avoid the additional treatment. Cells grown and tested like this provide a system that is close to what might be seen in vivo. Therefore, the fewest manipulations necessary to allow the detection of the desired analyte, the better the accuracy of that detection. If a peroxidase marker molecule is to be used, and the cells have either the enzyme or a high level of oxidative function, the endogenous peroxidase or peroxidase-like activity should be blocked with 1.5% hydrogen peroxide in PBS. The enzyme or the chromogen detection system determines whether any endogenous material must first be destroyed. Depending on the type of cells assayed, the concentration and duration of the hydrogen peroxide step may be altered. If necessary, only that concentration and incubation time needed to effectively quench endogenous enzyme is advised. Because these preparations are more fragile than a fixed, embedded sample, endogenous enzyme is inactivated with a weaker blocking solution than would otherwise be used. The standard endogenous oxidation blocking solution when using peroxidase enzymes is a solution of 3% hydrogen peroxide in methanol. For cell culture slides, though, PBS is substituted for the same reason that

formalin is avoided. The compound of interest should not be subjected to any more denaturing agents then is necessary *(6)*. Also, the amount of H_2O_2 is reduced; a 1.5% solution is usually sufficient. Sometimes even 1.5% H_2O_2 will cause visible bubbles of gas to develop and, if too intense, may succeed in lifting cells off the slide. Careful monitoring and gentle tapping of the container will help to prevent this from happening. Wear gloves and exercise caution when handling the 30% H_2O_2 as it is caustic and can cause burns.

8. Avoid over trypsinizing, as too much will not only destroy some cells, but may alter the antigen in those remaining. If trypsin presents a problem, the cells could be scraped off with a cell scraper.
9. Proper centrifugation speed is necessary to avoid crushing the cells together.
10. The fixative used is again a matter of preference. However, the main reason for preparing this type of specimen is usually to compare fresh cells to the cells found in tissue specimens. Therefore, it is recommended that formalin fixation be used, since this is the fixative of choice for most tissue protocols, and the crosslinking that develops helps to hold the cells together as a pellet. Specimens can be fixed in 95% ethanol/5% glacial acetic acid but will be more fragile in pellet form.
11. Technique is the major factor in successfully extricating a pellet from the tube. Sometimes it is necessary to practice with a few samples. The nice part about cell culture is that there are unlimited numbers of cells to study. It is wise to set up several tubes in hope of obtaining a decent pellet to embed. Sometimes the cells are in short supply or such that more fortification is needed for the pellet to stick together. In these instances, following **Subheading 3.2., step 10**, an equal volume of 2% gelatin (warmed to liquid) is added to the pellet and the pellet resuspended. Sample is placed at −70°C to speed up gelatin solidification. The sample is then fixed as in **step 11**. This provides a matrix in which the cells will adhere. It can be easier to dislodge a more intact pellet this way.

References

1. Carone, F., Nakamura, S., Schumacher, B., Punyarit, P., and Bauer, K. (1989) Cyst-derived cells do not exhibit accelerated growth or features of transformed cells in vitro. *Kid. Int.* **35**, 1351–1357.
2. Silverman, T., Rein, A., Orrison, B., Langloss, J., Bratthauer, G., Miyazaki, J., and Ozato, K. (1988) Establishment of cell lines from somite stage mouse embryos and expression of major histocompatibility class I genes in these cells. *J. Immunol.* **140**, 4378–4387.
3. Weber, B., Harms, F., Selb, B., and Doerr, H. (1992) Improvement of rotavirus isolation in the cell culture by immune peroxidase staining. *J. Virol. Meth.* **38**, 187–194.
4. Ditzel, H., Erb, K., Nielsen, B., Borup-Christensen, P., and Jensenius, J. (1990) A method for blocking antigen-independent binding of human IgM to frozen tissue sections when screening human hybridoma antibodies. *J. Immunol. Meth.* **133**, 245–251.
5. Li, C., Lazcano-Villareal, O., Pierre, R., and Yam, L. (1987) Immunocytochemical identification of cells in serous effusions. *Am. J. Clin. Pathol.* **88**, 696–706.
6. Van Bogar, L. (1985) Present status of estrogen-receptor immunohistochemistry. *Acta Histochem.* **76**, 29–35.

12

Processing of Tissue Specimens*

Gary L. Bratthauer

1. Introduction

One of the areas in which the use of immunocytochemistry has had the greatest impact is in the examination of tissue in the medical pathology laboratory. Immunocytochemistry, actually immunohistochemistry, in the pathology laboratory enhances the study of diseased tissue. It is important in studying disease to obtain and process the tissue as quickly as possible. The reason for this is so that the cellular constituents can be preserved as completely as possible. As is the case with any solid piece of tissue, the smaller the piece, the easier it is to preserve the cellular constituents. The varied types of tissue also determine, to an extent, preparation protocols, because some types require a more specialized form of preservation than do others. Some specimens like bone require many days for proper fixation, whereas other looser connective tissues are preserved in a matter of hours by simple immersion fixation. It should be mentioned, though, that depositing large organs like whole brains in buckets of fixative may provide fine cellular detail but will probably result in the loss of some labile brain proteins desired for study *(1)*. The individual who actually obtains the specimen is an important participant in tissue processing. This individual needs to quickly remove the sample, and before much autolysis occurs, immediately place the correctly sized (small) sample into the desired fixative. It is then incumbent on the laboratory to process this specimen into paraffin as soon as possible after the requisite fixation time has ended. Aberrant immunoreactivity often is the result of improper fixation time *(2)*. This

*The opinions or assertions contained herein are the private views of the author and are not to be construed as official or as reflecting the views of the Department of the Army or the Department of Defense.

From: *Methods in Molecular Biology, Vol. 115: Immunocytochemical Methods and Protocols*
Edited by: L. C. Javois © Humana Press Inc., Totowa, NJ

chapter deals with the intricacies involved in preparing a tissue sample for analysis with antibodies.

2. Materials

2.1. Production of Tissue Slides

1. Gloves.
2. Small wide-mouth vials.
3. Process/embedding cassettes.
4. Tissue processor.
5. Paraffin.
6. Clean glass slides.
7. Microtome.
8. Flotation water bath.
9. 50-mL Centrifuge tubes.
10. Dissecting needle and brush.
11. Staining racks and dishes.
12. 60°C oven.
13. Stirring block with stir bars.
14. Microwave oven.

2.2. Processing of Tissue

1. Neutral buffered formalin: add 100 mL of 37–40% formalin to 900 mL deionized water, then add 4 g monobasic sodium phosphate and 6.5 g dibasic sodium phosphate.
2. Bouin's fluid: Combine 1500 mL picric acid (saturated in deionized water at 21 g/L), 500 mL formalin, and 100 mL glacial acetic acid.
3. Ethanol/acid fixative: Add 50 mL glacial acetic acid to 950 mL ethanol.
4. B5 fixative: Add 12 g mercuric chloride to 200 mL of deionized water; add 2.5 g sodium acetate; add 20 mL formalin.
5. Elmer's Glue-All (Borden, Columbus, OH): 15% v/v in deionized water dispensed into 50-mL centrifuge tubes.
6. Phosphate buffered saline (PBS): 0.01 M sodium phosphate, 0.89% sodium chloride, pH 7.40, ±0.05.
7. Xylene.
8. Ethanol.
9. Digestion solution: 0.05% protease VIII (*see* **Note 1**) in 0.1 M sodium phosphate buffer, pH 7.8, kept at 37°C.
10. Endogenous enzyme blocking solution: 3% hydrogen peroxide (H_2O_2) in methanol (10% solution of 30% H_2O_2).
11. Nonspecific binding blocking solution: 10% normal serum of the secondary antibody generating species in PBS.
12. Sodium phosphate buffer: 0.1 M sodium phosphate, pH 7.80, ±0.05.
13. Methanol.
14. 30 H_2O_2.
15. Antigen recovery solution: 0.01 M citric acid in dH_2O, pH 6.00 (*see* **Note 2**).

3. Methods
3.1. Common Fixation

1. Add approx five times the volume of the appropriate fixative (*see* Chapter 8) to the sample in a small wide-mouth vial and incubate for 4–24 h depending on the optimum for the fixative selected (formalin, 6–12 h; Bouin's, 4 h; ethanol/acid, 24 h) (*see* **Note 3**).
2. Rinse in three changes of 70% ethanol and place in embedding cassettes in 70% ethanol.

3.2. Preparation of Tissue Blocks

1. Place the processing cassette with the tissue in 70% alcohol into the tissue processor for dehydration and infiltration with paraffin (*see* **Note 4**).
2. Orient the specimen in the embedding cassette, embed it in paraffin and cool it to a solid block.

3.3. Preparation of Slides

1. Fasten the tissue block in the block holder of the microtome and adjust the knife.
2. Slice 6-µ sections repeatedly so that a ribbon forms, and gently transfer the paraffin ribbon to the surface of a flotation water bath containing water only.
3. With a dissecting needle and brush, tease the ribbon apart into separate floating sections.
4. Holding a slide by the label end, dip the slide into the centrifuge tube of Elmer's glue (*see* **Note 5**).
5. Wipe the glue off of the back of the slide and enter the water bath under the desired section. Maneuver the slide under the section and lift it up, retrieving the section on the slide surface.
6. Keep the slides flat and allow them to dry overnight at room temperature.

3.4. Preparation of Sections for Immunostaining

1. Place the slides into a slide rack and into the oven and bake for 1 h at 60°C.
2. Remove and air-dry for 30 s only.
3. Place the slide rack into xylene and incubate for 2 min to remove the paraffin. Repeat through four changes of fresh xylene (*see* **Note 6**).
4. After the final xylene incubation, place the slides in ethanol for 2 min. Repeat through four changes of fresh ethanol.
5. Move slides into deionized water and wash for 2 min. Repeat through three changes of deionized water.
6. Place the slides requiring proteolytic digestion in digestion solution at 37°C for 3 min. Slides not requiring digestion may remain in deionized water (*see* **Note 7**).
7. Place the slides requiring microwave irradiation in plastic slide holders in the citric acid solution.
8. Microwave for 20 min at near boiling temperatures (*see* **Note 7**).
9. Remove from oven and allow to cool in the citric acid buffer for 45 min (*see* **Note 7**).

10. Following enzyme digestion or microwave irradiation, rinse the slides in deionized water for 2 min. Repeat through three changes of water.
11. Combine with slides not being digested, and place in ethanol for 2 min. Repeat through three ethanol changes.
12. Place the slides in endogenous enzyme blocking solution and incubate with stirring for 30 min (*see* **Note 8**).
13. Remove the slides and wash in deionized water for 2 min. Repeat through three changes of water.
14. Place the slides in nonspecific binding blocking solution and incubate at 4°C overnight (*see* **Note 9**).

4. Notes

1. Because these sections are cut, the intracellular constituents are generally available for detection without the need for special pretreatments. However, some antigens are masked by the use of certain fixatives and cannot be recognized by the antibody *(3)*. Unique to fixed paraffin-embedded samples is the flexibility to treat the tissue a bit more aggressively in trying to enhance immunoreactivity. This is usually accomplished with some proteolytic enzyme digestion that perturbs the tissue enough to unmask these antigens. Newer microwave-enhanced methods also are effective and with certain antigen/antibody combinations are necessary; however there are still antigen/antibody combinations that react optimally following enzyme digestion. Obviously, this should be done cautiously so as not to destroy the section. Protease type VIII has been used routinely and is a quick-acting enzyme. Careful control of concentration, temperature, and time will provide good reproducible results. A standard treatment is possible, although each tissue and each antigen-antibody combination will have an optimal digestion protocol that can be experimentally determined. Samples fixed in milder fixatives such as ethanol may not need to be digested with proteolytic enzymes. Also, depending on the antigen and antibody used, some specimens fixed in formalin need not be digested either, although digestion is required more often than not.
2. This solution is also commercially available (Citra, Biogenex Corp. San Ramon, CA).
3. As always, when handling fresh tissue, it is imperative to wear gloves and protective clothing. Proper fixation is the most important aspect of obtaining successful results. Strive for small samples properly fixed for the optimum time. A small sample allows for quick and thorough penetration, and using at least 5 vol of fixative should help to ensure proper fixation. Penetration and time are the biggest variables in fixing samples. It is important to optimize the time for best results. Process the tissue as soon as possible after adequate fixation. The type of fixative used does not matter as much as the quality of the fixation, but different antigens may be fixative-dependent *(4)* (*see* Chapter 8). In working with fixatives, one should be reminded of the hazardous nature of the materials, and protective clothing such as gloves should be worn. The neutral buffered formalin fixative is preferred for pathology specimens because of the superior morpho-

logic preservation, although the others are also effective. These can be treated the same as formalin, but the immunostaining results may differ depending on the fixative used. For instance, samples fixed in the mercurial fixatives often show mercurial pigments under the microscope. These pigments are often confusing and distracting and can be removed by using a solution of alcoholic iodine or Lugol's iodine *(5)*:

a. Prepare solution of 1% iodine in 80% ethanol or Lugol's iodine of 1% iodine, 2% potassium iodide.
b. Incubate slides for 10–15 min.
c. Wash with water and incubate a few minutes in 5% sodium thiosulfate in distilled water before a final water wash.

This treatment may result in less reactivity because of the chaotropic nature of the solution.

4. There are numerous instruments that automatically process specimens for embedding in paraffin. The processor dehydrates to 100% alcohol and then infiltrates the tissue with xylene. Xylene-saturated tissue is then infiltrated with melted paraffin at 60–70°C. The tissue processor accomplishes this quickly with heat and pressure. The processor should not heat above 65°C to avoid destroying heat-labile antigens. Once the block is formed, it must be sectioned with a good sharp knife, properly oriented in the microtome. Good histologic slide preparation will provide a good immunohistochemical test sample.

5. The sections should be cut as thin as possible on glass slides that are coated with a substance that promotes adherence in order to prevent the tissue from coming off later in the process. Immunocytochemical procedures require the sections to be in buffers and solutions for hours or days, and they can float off of the slides during that time if not affixed properly. Also, slides need to be coated for these samples because of the enzyme digestion or microwave treatment that is normally required. Elmer's Glue-All is an older yet reliable method of section adherence that does not cause nonspecific binding and presents less of a problem with section loss during enzymatic digestion than some of the other coating solutions. Elmer's Glue-All used in high concentrations can cause artifacts, and when viewed under the microscope, can have a refractile, cracked appearance. It is necessary to use no more than 15% v/v of the glue. It is usually recommended that the section be allowed to dry as the glue dries (attach a wet section to a wet, glued slide). Predried glued slides can be used with success; however, occasionally sections drying on predried slides may wash off in subsequent steps.

In addition, poly-L-lysine is a popular and simple means of adhering a section to the surface of a slide, but it can cause nonspecific staining with some chromogens:

a. It is very important that the slides be precleaned. Clean glass slides should be placed in a staining rack.
b. Immerse the slides for 30 min in a 1:10 dilution of 0.1% poly-L-lysine solution in distilled water in a large staining dish.
c. Remove the slides and oven dry for 1 h.

When using poly-L-lysine or Silane-coated slides in enzyme digestion protocols, the method of their preparation determines the success of section adherence. The slides must be clean, and must be heated prior to deparaffinization. Experimentation can determine the slide coating that provides clean backgrounds with the technique of choice.

Finally, charged slides such as Plus slides (Fisher Scientific, Pittsburgh, PA) are a good alternative, which provide adequate section adherence with limited background in immunohistochemical analyses. Some positively charged slides may inhibit antibody binding to all areas of the section, however.

When drying these sections, it is necessary to allow them to dry at room temperature. The use of a warming plate or oven while the sections are still wet can result in the loss of some antigens proportional to the drying temperature. Once the slides are dry though, they must be heated in an oven prior to the immunoassay. There is no adverse effect associated with heating dried sections.

There are instances when glue coatings are not acceptable. Certain automated instruments for immunostaining require capillary action in the process (*see* Chapter 52). These instruments are dependent on slides with painted surfaces that are assayed in tandem, creating a gap of determined width for reagents to traverse. In order for the flow of reagents to be uniform and consistent, the slides must have a smooth surface that withstands proteolytic enzyme digestion as can best be provided by poly-L-lysine or Silane. In preparing slides for automation, it is mandatory that the slide surface not be touched, since skin oils inhibit capillary action, and that the section be located near the bottom of the slide. This is to enable the reagents, by capillary action, to cover the entire section.

6. Deparaffinization is a very important part of the procedure as well. The slides must be deparaffinized in xylene after they are oven-heated. The heat helps to anneal the section a bit before it is subjected to the rigors of the procedure, but it also dissolves the paraffin slightly, making it easier for it to go into solution in the xylene. When the slides are removed from the oven, the paraffin will be clear and wet. As soon as the paraffin starts to solidify, the slides should be placed in the xylene.

7. The use of Protease VIII is not exclusive. Trypsin may also be used, is milder, and will take longer to show the same effect (30 min to 3 h). This can be beneficial when dealing with antigens, which are less stable. Other enhancing enzymes include other protease types like proteinase K, pronase (a very rigorous enzyme), and ficin *(6)*. Saponin, Triton X-100, and Tween-20 are detergents that can have much the same effect as some of the enzymes in enhancing antibody-antigen interactions *(7)*. Also, some have used acids such as formic acid, to denature the antigens and make them more accessible to antibody *(8)*. All of the pretreatments are designed to either perturb the section surface enough to allow recognition of antigen or to make the section more permeable to immunoglobulin by reducing surface charges. If an assay is not working properly with a given enzyme or detergent, another may yield better results. Also, the time of digestion may be varied to enhance the effect.

Another pretreatment that is being more widely used with formalin-fixed samples is heat-induced antigen recovery *(9)*. This technique makes use of microwave digestion coupled with a slightly acidic solution to denature available proteins and enhance antigen detectability. The microwave can be calibrated by using the same containers, filled in the same manner with a constant number of slides for every use *(10)*. If not enough sections are to be processed, blank slides should be used instead. The oven is heated on high power with the containers in place, covers slightly askew to permit steam escape (microwaves with turntables work the best; however, if one is not available, the exact same location of the container within the oven should be used every time). The time to full boiling is noted and subtracted from 20 min. For the remaining time, the microwave is operated at a lower power setting. For most applications, 30% power for the remaining time is adequate. The desired effect is for the solution to begin boiling for a few seconds after the microwaving starts. When only the fan is running, the solution should stop boiling. In this way, the solution is kept at a near boiling temperature, and there is no loss due to evaporation. After the 20 min, the sections are removed from the oven, the lid to the containers replaced, and the slides are allowed to cool in the buffer for at least 45 min. This cooling period appears to be important in the detection of the desired antigen. Alternatively, this process can be performed at high power for three times 5 min, each time adding back some buffer lost through evaporation. This process, though, often yields inconsistent results.

Some newer monoclones detect antigens only after microwave heating, while others are greatly enhanced after using this method. Among them are tests for the estrogen, progesterone, and androgen receptors in the nucleus, some cycling proteins such as PCNA, MIB I, or the cyclins, and some cytoplasmic or surface proteins as well, such as inhibin or CD15. There are also antigens that are less likely to be detected when using the heat-induced methods. Also, artifacts and unwanted crossreactivity can be induced using these more sensitive techniques.

It is difficult to use proteolytic enzyme digestion with this treatment, but if needed, the sections should be digested following the microwave heating because boiling enzymatically digested tissue results in tissue loss. A milder form of this method is to boil the citric acid buffer only and incubate the sections in the preheated solution, or steam, for 15 min, or sections can be heated to 80°C in this buffer and kept overnight with similar results *(11)*.

8. The enzyme or the chromogen detection system determines whether any endogenous material must first be destroyed. Wear gloves and exercise caution when handling the 30% H_2O_2, because it is caustic and can cause burns.

9. Cells exist in an electrically charged environment. To prevent antibodies from binding as a result of excess charges on the tissue surface, a proteinaceous solution is used to bind to these sites in advance of the antibody incubations. A 10% normal animal serum from the same species as that providing the secondary or detecting system antibody in PBS is used. The use of other species' serum may cause the antibodies to adhere nonspecifically or crossreact. The

charged protein molecules of the serum will bind to the charged areas on the section, preventing the antibody reagents from binding to these areas non-specifically. An overnight incubation at 4°C is best for the complete removal of available charged sites for nonspecific binding, but, if desired, a 2-h room temperature incubation could suffice.

References

1. Guntern, R., Vallet, P. Bouras, C., and Constantinidis, J. (1989) An improved immunohistostaining procedure for peptides in human brain. *Experientia* **45**, 159–161.
2. Battifora, H. (1991) Assessment of antigen damage in immunohistochemistry. *Am. J. Clin. Pathol.* **96**, 669–671.
3. Login, G., Schnitt, S., and Dvorak, A. (1987) Rapid microwave fixation of human tissues for light microscopic immunoperoxidase identification of diagnostically useful antigens. *Lab. Invest.* **57**, 585–591.
4. Baumgartner, W., Dettinger, H., Schmeer, N., and Hoffmeister, E. (1988) Evaluation of different fixatives and treatments for immunohistochemical demonstration of *Coxiella burnetii* in paraffin-embedded tissues. *J. Clin. Micro.* **26**, 2044–2047.
5. Ambrogi, L. P., ed. (1957) *Manual of histologic and special staining techniques.* Armed Forces Institute of Pathology, Washington, DC, p. 33.
6. Taschini, P. and MacDonald, D. (1987) Protease digestion step in immunohistochemical procedures: ficin as a substitute for trypsin. *Lab. Med.* **18**, 532–536.
7. Stirling, J. (1990) Immuno- and affinity probes for electron microscopy: a review of labeling and preparation techniques. *J. Histochem. Cytochem.* **38**, 145–157.
8. Kitamoto, T., Ogomori, K., Tateishi, J., and Prusiner, S. (1987) Formic acid pretreatment enhances immunostaining of cerebral and systemic amyloids. *Lab. Invest.* **57**, 230–236.
9. Shi, S., Key, M., and Kalra, K. (1991) Antigen retrieval in formalin-fixed, paraffin-embedded tissues: an enhancement method for immunohistochemical staining based on microwave oven heating of tissue sections. *J. Histochem. Cytochem.* **9**, 741–748.
10. Tacha, D. E. and Chen, T. (1994) Modified antigen retrieval procedure: calibration technique for microwave ovens. *J. Histotechnol.* **17**, 365.
11. Man, Y-g. and Tavassoli, F. A. (1996) A simple epitope retrieval method without the use of microwave oven or enzyme digestion. *Appl. Immunohistochem.* **4**, 139–141.

13

Antigen Retrieval for Immunohistochemical Reactions in Routinely Processed Paraffin Sections

Laszlo Krenacs, Tibor Krenacs, and Mark Raffeld

1. Introduction

Immunohistochemistry is an essential adjunct of modern diagnostic pathology. In the majority of cases, pathological tissue samples are fixed in formaldehyde and embedded in paraffin wax for examination of microscopic morphology. A major limitation of routinely processed tissues for immunohistochemistry is that many potentially interesting antigens are altered during tissue fixation and processing. As an alternative, sections of snap-frozen tissues can be used to detect most of these antigens. Nevertheless, paraffin sections offer well-preserved tissue architecture and cytomorphology superior to that obtained in frozen sections and thus allow more accurate antigen localization. Furthermore, paraffin-embedded tissues represent an invaluable source of human tissues, easily accessible for retrospective studies back decades. Thus, for diagnostic and research purposes, paraffin sections are preferred for immunohistochemical analysis by most laboratories.

Alterations of antigens in paraffin-embedded tissues are related to a variety of changes in the three-dimensional structure (conformation) of proteins resulting from crosslinking by formaldehyde and, to a lesser extent, from heating and dehydration during paraffin embedding. As a consequence, the antigenic determinants (epitopes) are destroyed, denatured, or masked, which may diminish or abrogate their detection. A number of antibodies have been found to react with fixation resistant epitopes, and the application of proteolytic *(1,2)* or chemical *(3,4)* pretreatments has also broadened the antibody repertoire for routinely processed tissues.

Despite some success with the above pretreatments, the development of wet heat-induced epitope retrieval (HIER) procedures has been the critical

From: *Methods in Molecular Biology, Vol. 115: Immunocytochemical Methods and Protocols*
Edited by: L. C. Javois © Humana Press Inc., Totowa, NJ

breakthrough in paraffin-section immunohistochemistry (5–7). HIER methods substantially increase the sensitivity of reactions of antibodies directed to paraffin-resistant antigens. Moreover, some antibodies that have never before reacted in paraffin sections show specific staining following HIER pretreatment. Additionally, HIER minimizes the problem of overfixation, since it abrogates the immunostaining differences found between the 24-h fixed material and tissues that have been kept in formalin for days or even weeks. HIER technology also has enabled immunohistochemists to routinely stain a wide spectrum of antigens in epoxy resin-embedded sections for bone marrow diagnosis (8).

There are several variations of HIER. Many laboratories have attempted to improve the original method by altering the buffer solutions as well as the source and mode of heating. Currently, the most popular HIER technologies use stainless steel or plastic pressure cookers, microwave ovens, or autoclaves as the heat source and low-molarity buffers with acidic or alkaline pH (6,7,9–12).

The exact mechanism by which HIER works is unknown. It is thought to reverse the masking effects of formaldehyde fixation and routine tissue processing. Hydrolytic-proteolytic cleavage of formaldehyde-related crosslinks, unfolding of inner epitopes, as well as the extraction of calcium ions from coordination complexes with proteins are among the hypothesized mechanisms (13–15).

Most recently, a new generation of highly specific antibodies directed against peptide sequences of lymphocyte subset antigens (for example, CD4, CD8, CD79a, etc.), oncoproteins (i.e., cyclin D1), and cancer prognostic factors (i.e., estrogen and progesterone receptors) have been introduced that require wet-heat-mediated antigen retrieval. Currently, a large number of monoclonal and polyclonal antibodies are available to detect diverse antigens in paraffin-embedded tissues, and their number is still increasing rapidly. As a consequence, immunophenotyping of paraffin-embedded tissues, coupled with an appropriate HIER technique, is now capable of providing information equal or superior to that achieved in frozen sections. Therefore, HIER has a central role in modern immunohistochemistry. In this chapter we briefly summarize the current principles of HIER and provide protocols for its optimal use in archival tissues.

The wide methodological repertoire available provides great latitude for laboratories using HIER, but also underscores the need for standardization for better intra- and interlaboratory reproducibility.

1.1. Factors that Influence HIER

1.1.1. Composition of Retrieval Buffer

Experimental data suggest that the pH and molarity of the retrieval buffers are among the most important factors for efficacy (5,6,13–16). Citrate buffer (sodium citrate-citric acid) at pH 6.0 *(5,6)* is a very popular retrieval medium and has been used at molarities between 0.01 and 0.1 *M (5,6)*. Detergents (e.g., 0.1% Tween-20) added to the standard citrate buffer may improve the performance *(12)*, but more prominent tissue deterioration may also be experienced. TRIS-HCl buffers at various concentrations (0.1–0.5 *M*) and at alkaline pH (8.0–10.0) have also been found to be effective *(5,9,12,15,17)*. Most recently, 0.1 *M* EDTA-NaOH at pH 8.0 has been introduced *(13,14)* and has been found to be very effective for general applications *(17)*. Of the commercially available antigen retrieval buffers, Target Retrieval Solution (TRS) (Dako Corporation, Carpinteria, CA) performs surprisingly well *(18,19)*. TRS can be used to retrieve certain epitopes that are not otherwise detectable in formalin-fixed, paraffin-wax sections (i.e., CD5/Leu1, CD35/To5), and it can substitute for enzymatic digestion in the detection of other antigens (e.g., CD21/1F8, CD35/Ber-MAC-DRC, and BerEP4) *(18)*. Other solutions have also been used in specific HIER protocols *(5,6)*, but they are not as widely used as those described above.

1.1.2. Heating Devices

A wide variety of heating devices have been introduced for use in HIER. Among these, the microwave oven (MWO), commercial pressure cooker (PC), and autoclave (AC) have proven to be the most employable.

The use of a commercial MWO for HIER is very popular, as it is easily accessible and an inexpensive source of heat. The most important factors to be considered in choosing a MWO are:

1. Presence of a digital timer for precise time adjustment.
2. Presence of a turntable for uniform heating of the retrieval solution.
3. 700–1000 W of power.

Metallic tools should never be used in a MWO; therefore, heat-resistant plastic Coplin jars or containers and heat-resistant plastic slide holders with a capacity of at least 15 sections are required for this application. MWO-mediated HIER can also be performed in a plastic pressure cooker designed for household kitchen application (*see* **Subheading 3.2.**), which appears to eliminate some of the drawbacks of the standard MWO protocols.

The standard pressure cooker is the simplest and least expensive way for achieving reproducible HIER. The heating device is usually an inexpensive commercial electric hot plate with at least 1 kW of power. Careful use of a PC provides very uniform heating and allows one to treat larger batches of slides *(7,11,17)*. The PC method avoids the need for careful monitoring during the retrieval that is necessary in the standard MWO method to prevent accidental drying of the sections as a result of evaporation during microwaving.

Wet (hydrated) autoclave treatment represents the most uniform heating method and is claimed to be preferable over MWO irradiation by some authors *(10,15)*. The main disadvantages of the autoclave method are that one must have access to an autoclave, and it can be time-consuming.

1.1.3. Temperature and Duration of Treatment

Most methods employ temperatures near or beyond 100°C. Heating above the atmospheric boiling temperature is possible in traditional or in microwave pressure cookers as well as in autoclaves (*see* **Subheading 1.1.2.**). In a commercial PC, the operating pressure is about 103 kPa/15 psi, which results in 120°C temperature *(19)*. The same temperature is employed in wet autoclave HIER protocols *(10,15,19)*.

A higher temperature usually performs better, although pronounced tissue deterioration may also be experienced. If tissue deterioration occurs, reduction of the duration of heating and/or subsequent cooling period, or a decrease in the concentration of the retrieval buffer, may improve the cytomorphological preservation. In some cases, inferior nuclear morphology is found, which can be partially corrected by extending the time of counterstaining. Some antigens may be destroyed by the higher temperature (i.e., glycophorin C and surface immunoglobulin light chains). Therefore, careful titration is needed when a new antibody is tested in these applications.

Firm attachment between the tissue and the glass slide is crucial to prevent detachment of the section during HIER. Mounting the sections on 3-amino-propyltriethoxysilane (APES)-coated glass slides provides suitable gluing technique following heat activation. Alternatively, charged slides available from several of the large supply companies (e.g., Fisher Scientific) may be used.

The efficacy of the antigen retrieval procedure also depends on the treatment duration and the cooling conditions (*see* **Subheadings 3.1.–3.5.**). The power of the heating device and the buffer volume are among the major factors that determine the duration of a particular HIER protocol.

The most rapid HIER procedures utilize the standard MWO or the household pressure cooker, whereas the more time-consuming procedures are the MWO/pressure-cooker combination or the autoclave. In the standard MWO procedures, treatment is performed in two or three 5-min heating cycles. Each cycle is interrupted to replenish evaporated buffer in the Coplin jars containing the slides. The household pressure cooker procedure may take up to 40 min, while the combined use of the MWO and pressure cooker can exceed 60 min (*see* **Subheading 3.2.**). However, the application of a "hot start" modification to the later method (*see* **Subheading 3.2.**) can shorten the time considerably. Wet-autoclave HIER is probably the most time-consuming method, since the total time that may be required from switching on the autoclave to taking out the sections is up to 2 h *(15)*.

2. Materials

1. Deparaffinized, rehydrated, methanolic peroxidase-blocked (optional) tissue sections (*see* Chapter 12).
2. Standard domestic microwave oven rated at 700–1000 W, with turntable for constant temperature and electronic digital timer control.
3. *Plastic* Coplin jars or other suitable microwaveable slide containers.
4. HIER buffer: 0.01 *M* citrate, pH 6.0, containing 0.1% (v/v) Tween-20.
5. Distilled water.
6. Glass slides. (Also Superfrost slides, Fisher Scientific, Pittsburgh, PA.)
7. Post-HIER immunostaining buffer: 0.05 *M* Tris-buffered saline (TBS, pH 7.6) containing 1–2% fetal calf serum (FCS) or normal goat serum.
8. Microwaveable pressure cooker (MWPC) (Nordic Ware, Minneapolis, MN).
9. Standard pressure cooker (e.g., Prestige Model 6193, Prestige Group UK plc., Lancashire, UK).
10. Electric hot plate rated at 1000 W power.
11. Standard laboratory autoclave.

3. Methods (*see* Note 1)

3.1. Standard MWO Antigen Retrieval Method—Basic Method (5,6,15,17,19)

1. Transfer the deparaffinized, rehydrated, and methanolic-peroxide blocked (optional) slides (*see* **Notes 2–6**) into plastic Coplin jars or containers filled with HIER buffer (*see* **Note 7**). (*See* Chapter 12 for preparation of tissue sections.)
2. Fill remaining positions in the Coplin jars or plastic slide racks with blank slides (*see* **Note 8**).
3. Place the Coplin jars or containers in the center of the microwave's turntable, cover containers with loose-fitting lids or screw caps, and heat at maximum power (700–1000 W) (*see* **Note 9**). The time of irradiation depends on the power setting of the microwave, the type of container used, and the volume of buffer. The solution should boil for 3–5 min. A large capacity microwaveable plastic container containing as much as 600 mL of citrate buffer may require a 30-min heating cycle *(11)*.
4. After the heating cycle, add distilled water to the container to replenish any evaporated buffer (*see* **Note 10**). Repeat the heat cycle.
5. Following two to three heating cycles, allow the slides to cool for approx 20 min (*see* **Note 11**).
6. Proceed immediately with the immunostaining protocol (*see* Chapters 24–27, 29, 43–45) (*see* **Notes 12–16**).

3.2. Microwave Pressure Cooker Method (12)

1. Place the dewaxed, rehydrated sections (*see* **Notes 2–6**) in an microwaveable pressure cooker (MWPC) filled with 1500 mL HIER buffer (*see* **Note 7**).
2. Place the MWPC inside the MWO. Heat at maximum power (900 W) for 40 min (*see* **Note 9**). The conditions should also be kept stable in this method

(*see* **Note 8**). Under these conditions, it will take about 20 min to reach maximum temperature. The slides should then remain at maximum temperature for no longer than 20 min.

3. Release the pressure carefully and immediately cool the sections in post-HIER immunostaining buffer, and keep in the same buffer until they are used (15–20 min) (*see* **Note 11**). Alternatively, the slides can be allowed to cool to room temperature in the open pressure cooker (approx 20–30 min).
4. Proceed immediately with the immunostaining protocol (*see* **Notes 12–16**).

3.3. "Hot Start" Variation for Microwave Pressure Cooker (12)

This variation is performed as described above in **Subheading 3.2.**, except that the sections are placed into buffer preheated to 95°C, and irradiation is performed for only 8 min. Under these conditions, maximum heat is attained in 5–6 min.

3.4. Standard Pressure Cooker (PC) Method (11)

1. Preheat 3 L of buffer (*see* **Note 7**) to boiling in a stainless steel 5.5-L-capacity pressure cooker (Prestige) without sealing the lid, using an electric hot plate as the heat source.
2. Place the deparaffinized, rehydrated slides in metal slide racks and immerse in the hot buffer (*see* **Notes 2–6** and **8**). Seal the PC and bring to full pressure. Full pressure is attained when both the "rise-n-time" indicator and the safety plug are in the upright position. Treat the sections at full pressure for 2–3 min.
3. Release the pressure and cool the PC under running tap water for approx 10 min (*see* **Note 11**).
4. Transfer slides to post-HIER immunostaining buffer, and continue with the immunohistochemical staining procedure (*see* **Notes 10** and **12–16**).

3.5. Autoclaving (10,19)

1. Place the deparaffinized, rehydrated slides in metal or heat-resistant plastic slide racks and immerse in an autoclaveable incubation container filled with 250 mL HIER retrieval buffer (*see* **Notes 2–7**). Cover the container with a lid to avoid evaporation (*see* **Note 10**).
2. Autoclave at 120°C (2 atm) for 20 min.
3. As soon as the autoclave can be opened, the slides should be rinsed in post-HIER immunostaining buffer and used immediately in the immunostaining procedure (*see* **Notes 11–16**).

4. Notes

1. The procedures described represent guidelines; optimal methods should be determined in individual laboratories. This is because of the variability of tissue fixation and processing and the variety and stability of antigen targets.
2. Since HIER protocols are very harsh on tissue, it is essential to apply an adhesive to the glass slide to allow a strong attachment of the tissue. Paraffin sections of

5 μm should be mounted onto 3-amino-propyltriethoxysilane-coated (APES; Sigma, St. Louis, MO) or charged (Plus Superfrost slides; Fisher Scientific, Pittsburgh, PA) glass slides and attached by overnight drying at 58°C.

3. If epoxy resin-embedded tissue is used, cut 2-μm-thick sections with a glass knife, mount on APES-coated slides, and dry as described in **Note 2**. Deplasticize the sections by immersing them in sodium eth(meth)oxide for 15 min *(8)*. Wash the sections twice with equal parts of methanol (or IMS) and xylene, twice with methanol, for 3 min each, and rehydrate. Afterward, the same HIER and immunohistochemical protocols are employed as in paraffin sections.

4. APES has some nonspecific protein binding capacity and may bind immunoglobulins used as primary and secondary antibodies. This may result in nonspecific background staining and reduce the intensity and contrast of the immunoreaction. Blocking of nonspecific protein binding should be performed after HIER. Cooling the slides in protein (e.g., fetal calf serum or bovine serum albumin) containing buffer (e.g., tris-buffered saline) after HIER may be beneficial in certain applications.

5. APES may interfere with silver salt solutions in intensification steps of immunogold silver-staining techniques; therefore, for those applications, poly-L-lysine or gelatin-coated slides are preferable.

6. Optional endogenous peroxidase blocking with methanolic peroxide can be performed either before or after HIER.

7. DAKO TRS (Dako Corp., Carpinteria, CA) is an excellent alternative retrieval buffer. For some antigens, 0.05 *M* Tris buffer (pH 10.0) is useful. This buffer is highly proteolytic; therefore, if tissue deterioration occurs, reduce the concentration of buffer and/or the heating time, and/or employ a "hot start method" as described in **Subheading 3.3.**

8. Standardization of the conditions for a specific HIER procedure is prudent, since several factors have great impact on the results. The wattage of the MWO, the power setting, the number of the slides, the volume of the retrieval medium, and the duration of the treatment are all related. Therefore, well-controlled conditions should be applied in order to obtain reproducible results.

9. In the MWO-based protocols, it is recommended to always use maximum power and titrate the procedure by adjusting the duration of retrieval and the buffer conditions.

10. Sections should **never** be allowed to dry during or after the HIER procedure, as this may result in artifactual staining.

11. The cooling-down time following HIER may affect the detection of some antigens and should be considered when testing new antibodies. Longer cooling times may improve or diminish staining quality.

12. Some structural damage to the sections is inevitable because of HIER. This is most prominent in tissues rich in connective tissues (e.g., skin, breast) and may influence the morphological assessment. Tissue damage is more extensive with the higher temperatures achieved using the pressure cooker or autoclave methods, with longer retrieval times, and when detergents have been added to the retrieval buffers.

13. All detection systems can be employed following HIER and noticeable improvement of most antigen detection will be observed. However, we recommend using sensitive detection systems, such as streptavidin-biotin-peroxidase methods when antigen density is low.

14. Diffuse false-positive cytoplasmic staining may be found in HIER-treated sections, particularly with the avidin-biotin detection system. This false-positive staining can usually be eliminated by blocking endogenous biotin using commercially available blocking reagents (e.g., Dako Biotin Blocking System (X0590); Dako). False-positive staining may also be a consequence of overtreatment by HIER; therefore, reduction of the antigen retrieval intensity (e.g., reducing buffer concentration or shortening heating time) may also be helpful.

15. Weak false-positive nuclear staining may occur with some monoclonal or polyclonal antibodies; therefore, careful assessment of the reaction pattern with particular attention to the staining of the normal cells is recommended. False-positive nuclear staining may be accentuated if the slides are allowed to dry during the staining procedure.

16. The combination of proteolytic enzyme digestion and heat-based treatments has been reported; however, this increases the susceptibility of the tissues for disintegration. Significant shortening of the digestion time and reduction of the enzyme concentration should be undertaken when testing such protocols. Furthermore, reproducible proteolytic enzyme treatments require optimal conditions for individual tissues, which make these protocols difficult to perform.

References

1. Huang, S. N., Minassian, H., and Moore, J. D. (1976) Application of immuno-fluorescent staining on paraffin sections improved by trypsin digestion. *Lab Invest.* **35,** 383–390.
2. Battifora, H. and Kopinski, M. (1986) The influence of protease digestion and duration of fixation on the immunostaining of keratins. A comparison of formalin and ethanol fixation. *J. Histochem. Cytochem.* **34,** 1095–1100.
3. Krenacs, T., Stiller, D., Krenacs, L., Bahn, H., Molnr, E., and Dux, L. (1990) Sarcoplasmic reticulum (SR) Ca^{2+}-ATPase as a marker of muscle-cell differentiation: immunohistochemical investigations of rhabdomyosarcomas and enhancement of immunostaining after sodium methoxide treatment. *Acta Histochem.* **88,** 159–166.
4. Krenacs, L., Tiszlavicz, L., Krenacs, T., and Boumsell, L. (1993) Immunohistochemical detection of CD1a antigen in formalin-fixed and paraffin-embedded tissue sections with monoclonal antibody O10. *J. Pathol.* **171,** 99–104.
5. Shi, S-R., Key, M. E., and Kalra, K. L. (1991) Antigen retrieval in formalin-fixed, paraffin embedded tissues: an enhancement method for immunohistochemical staining based on microwave oven heating of tissue sections. *J. Histochem. Cytochem.* **39,** 741–748.
6. Cattoretti, G., Pileri, S., Parravicini, C., Becker, M. H. G., Poggi, S., Bifulco, C., Key, G., DíAmato, L., Sabattini, E., Feudale, E., Reynolds, F., Gerdes, J., and

Rilke, F. (1993) Antigen unmasking on formalin-fixed, paraffin-embedded tissue sections. *J. Pathol.* **171,** 83–98.

7. Norton, A. J., Jordan, S., and Yeomans, P. (1994) Brief, high-temperature heat denaturation (pressure cooking): a simple and effective method of antigen retrieval. *J. Pathol.* **173,** 371–379.

8. Krenacs, T., Krenacs, L., and Bagdi, E. (1996) Diagnostic immunohistochemistry in Araldite-embedded bone marrow biopsies. *J. Cell. Pathol.* **1,** 83–88.

9. Beckstead, J. H. (1994) Improved antigen retrieval in formalin-fixed, paraffin-embedded tissues. *Appl. Immunohistochem.* **2,** 274–281.

10. Bankfalvi, A., Navabi, H., Bier, B., Bocker, W., Jasani, B., and Schmid, K. W. (1994) Wet autoclave pre-treatment for antigen retrieval in diagnostic immunohistochemistry. *J. Pathol.* **174,** 223–228.

11. Miller, K., Auld, J., Jessup, E., Rhodes, A., and Ashton-Key, M. (1995) Antigen unmasking in formalin-fixed routinely processed paraffin wax-embedded sections by pressure cooking: a comparison with microwave oven heating and traditional methods. *Adv. Anat. Pathol.* **2,** 60–64.

12. Krenacs, L., Harris, C. A., Raffeld, M., and Jaffe, E. S. (1996) Immunohistochemical diagnosis of T-cell lymphomas in paraffin sections. *J. Cell. Pathol.* **1,** 125–136.

13. Morgan, J. M., Navabi, H., Schmid, K. W., and Jasani, B. (1994) Possible role of tissue-bound calcium ions in citrate-mediated high-temperature antigen retrieval. *J. Pathol.* **174,** 301–307.

14. Morgan, J. M., Jasani, B., and Navabi, H. (1997) A mechanism for high temperature antigen retrieval involving calcium complexes produced by formalin fixation. *J. Cell. Pathol.* **2,** 89–92.

15. Taylor, C. R., Shi, S-R., and Cote, R. J. (1996) Antigen retrieval for immunohistochemistry: status and need for greater standardization. *Appl. Immunohistochem.* **4,** 144–166.

16. Shi, S-R., Imam, S. A., Young, L., Cote, R. J., Taylor, C. R., Key, M. E., and Kalra, K. L. (1995) Antigen retrieval immunohistochemistry under the influence of pH using monoclonal antibodies. *J. Histochem. Cytochem.* **43,** 193–201.

17. Pileri, S. A., Roncador, G., Ceccarelli, C., Piccioli, M., Briskomatis, A., Sabattini, E., Ascani, S., Santini, D., Piccaluga, P. P., Leone, O., Damiani, S., Ercolessi, C., Sandri, F., Pieri, F., Leoncini, L., and Falini, B. (1997) Antigen retrieval techniques in immunohistochemistry: Comparison of different methods. *J. Pathol.* **183,** 116–123.

18. Leong, A. S. -Y., Milios, J., and Leong, F. J. (1996) Epitope retrieval with microwaves. A comparison of citrate buffer and EDTA with three commercial retrieval solutions. *Appl. Immunohistochem.* **4,** 201–207.

19. A guide to demasking of antigen on formalin-fixed paraffin-embedded tissue. 2nd ed. Dako (1997).

III

LIGHT MICROSCOPIC DETECTION SYSTEMS

14

Overview of Fluorochromes

J. Michael Mullins

1. Introduction

Since its inception in the 1940s, the technique of immunofluorescence microscopy has provided a sensitive, high-resolution method for determining the presence of, and distribution of, an antigen within a specimen. Fluorescent molecules, termed fluorochromes, can be conjugated directly to antibodies by covalent linkage, or coupled indirectly to antibodies via conjugation to proteins A and G or through an avidin–biotin bridge (*see* Chapters 6, 7, and 25). A fluorochrome coupled to an antibody or other probe may be termed a fluorophore; here, the term fluorochrome is used throughout. The basic features of immunofluorescence are straightforward, but a working knowledge of the commonly used fluorochromes is of value in obtaining maximum performance from immunofluorescence microscopy and flow cytometry. The discussion below focuses on fluorochromes commonly used for immunolabeling, and does not encompass fluorochromes that provide molecule- or organelle-specific labeling without the use of antibodies.

2. Fluorescence

Fluorescence, a form of photoluminescence, occurs in response to absorption of light by a fluorochrome, producing an excited state in which an electron from the highest occupied orbital is elevated to a higher energy state in an unoccupied orbital. Some of the absorbed energy is dissipated through rotational or vibrational changes, and so does not contribute to photon emission. Return of the electron to its ground state, however, results in emission of a photon whose wavelength is determined by the wavelength of the absorbed photon. Since some of the energy absorbed by the fluorochrome is dissipated in nonfluorescent ways, the emitted photon will be of longer wavelength and lower energy than the one that was absorbed. Fluorescence is distinguished by the

From: *Methods in Molecular Biology, Vol. 115: Immunocytochemical Methods and Protocols*
Edited by: L. C. Javois © Humana Press Inc., Totowa, NJ

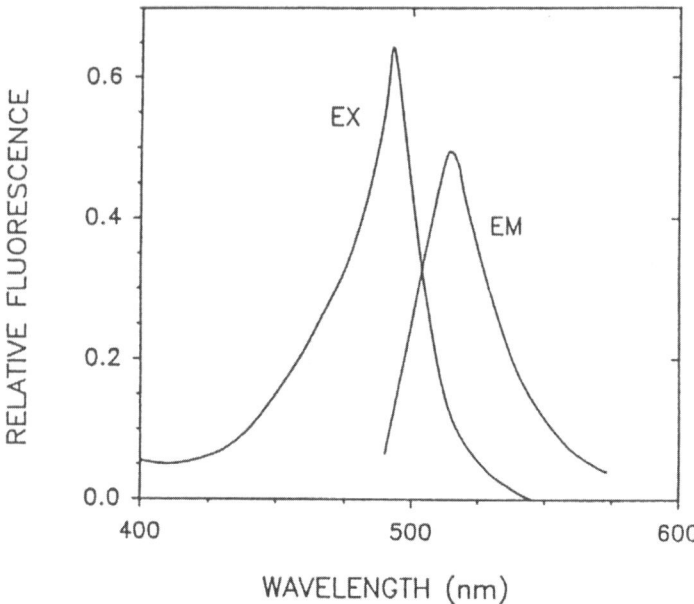

Fig. 1. Excitation and emission spectra of the fluorescein derivative DTAF. Modified from **ref. *1***. EX, excitation spectrum; EM, emission spectrum.

immediate dissipation of the absorbed energy, as opposed to phosphorescence, in which the excited state persists for some interval before photon emission.

Representative absorption (excitation) and emission spectra of a fluorochrome are provided in **Fig. 1**. Some degree of overlap between the two spectra is typical, and often the excitation and emission spectra are mirror images of each other. The separation between the wavelengths at which excitation and emission maxima occur is referred to as the Stokes shift.

3. Brightness and Detection of the Fluorescence Signal

Specific detection of a fluorochrome depends primarily on the brightness, or intensity, of its fluorescence, and on the ability of the detection system to distinguish the emitted wavelengths from background. Mathematically, fluorescence intensity is proportional to the product of the extinction coefficient (ε; units: $cm^{-1} \, M^{-1}$), a measure of the capture of excitation light by the fluorochrome, and the quantum yield (Q; maximum value of 1.0), a ratio of photons fluoresced to photons absorbed *(2)*. In actual practice, however, a number of factors may diminish the fluorescence intensity of a fluorochrome or limit its detection. Quantum yield, for instance, decreases noticeably for many fluorochromes on conjugation to an antibody, producing a concomitant decrease in brightness.

The Stokes shift, or separation between the excitation and emission maxima, is an important factor influencing both brightness and detection of a fluorochrome. Interference with the dim fluorescence image owing to scattering of the intense, excitation light is minimized with a longer Stokes shift. Additionally, the decreased overlap of excitation and emission spectra that attends a longer Stokes shift allows wider band pass optical filters to be employed. The use of wider segments of the excitation and emission spectra provides greater light intensity and, therefore, a brighter image. By contrast, for a fluorochrome with a short Stokes shift and considerable overlap of the excitation and emission spectra, clean detection of the weak fluorescence signal requires the use of narrow band pass optical filters to limit excitation and emission wavelengths to nonoverlapping portions of the spectra. Use of such narrow segments of the spectra diminishes the brightness of the fluorescence signal.

Brightness can be augmented by increasing the numbers of fluorochromes conjugated to each antibody molecule, but there are modest limits to such increases. Interactions between closely positioned fluorochromes on an antibody may reduce fluorescence intensity through self-quenching. In the case of fluorescein, for example, maximum brightness is attained with 2–4 fluorochromes/ antibody *(2)*. An additional consequence of an increased fluorochrome-to-antibody ratio is a greater nonspecific binding of the antibody, which produces background fluorescence that degrades the contrast of the fluorescence image.

Naturally occurring fluorescent compounds in cells and tissues provide another source of background fluorescence, such autofluorescence in mammalian cells arises principally from excitation of the flavin compounds FAD and FMN (excitation at 450, emission at 515 nm) and NADH (excitation at 340, emission at 460 nm) *(3,4)*. Porphyrin compounds, including hemoglobin and chlorophyll, also fluoresce, absorbing in the UV to blue range and emitting in the orange to red range. Additional tissue fluorescence may arise from interactions of additive fixatives, such as glutaraldehyde, with cell constituents to produce fluorescent compounds. Interference from autofluorescence background can be minimized by an appropriate selection of fluorochromes and optical filters to avoid overlap with the spectra of autofluorescing compounds (e.g., **ref. 5**). Fluorescence arising from glutaraldehyde fixation can be minimized by treating fixed cells with 0.1% sodium borohydride *(6)*.

In practical application, the intensity of the fluorescence image depends also on the configuration of the fluorescence microscope or other detection system. The light source, the selection of filters to limit excitation and emission wavelengths, and the choice of objectives, have major influences on the brightness of the microscope image *(7)*. Similarly, for confocal microscopy or flow cytometry, excitation light is restricted to the few wavelengths produced by a given laser. In such cases, fluorescence intensity will often be limited to that

Table 1
Spectral Properties of Some Fluorochromes Employed
for Immunofluorescence
(Arranged by ascending order of excitation wavelengths)

Fluorochrome	Excitation maximum, nm	Emission maximum, nm
SITS	336	438
DAMC	354	441
AMCA	355	450
Cascade Blue	396	410
Oregon Green 488	490	514
Alexa 488	491	515
FITC	494	519
BODIPY FL	503	512
B-PE	545	575
TRITC	547	572
Cy3.18	554	568
R-PE	565	578
RB-200-SC	568	583
Cy3.5	581	596
XRITC	582	601
Alexa 594	585	610
Texas Red	589	615
BODIPY TR	592	618
Cy5	649	670

Abbreviations: AMCA, 7-amino-4-methylcoumarin; B-PE, B phycoerythrin; Cy, cyanine; DAMC, diethylaminocoumarin; FITC, fluorescein isothiocyanate; RB-200-SC, lissamine rhodamine sulfonylchloride; R-PE, R phycoerythrin; SITS, 4-acetamido-4'-isothiocyanato-stilbene-2,2'-disulfonic acid; TRITC, tetramethyl rhodamine isothiocyanate; XRITC, rhodamine X isothiocyanate. Information obtained from **refs. 2**, **9**, and **10**.

which can be obtained by exciting fluorochromes at wavelengths other than their excitation maxima *(8,9)*.

Photobleaching of fluorochromes on exposure to high-intensity excitation light progressively reduces image brightness. This situation presumably arises from the fact that a fluorochrome in an excited state tends to be more chemically reactive, and so may be altered by oxidative or other processes. Susceptibility to photobleaching varies among fluorochromes, and so may be a factor in the choice of a fluorochrome for a given application. Reagents that retard photobleaching are discussed in Chapter 22.

4. Characteristics of Commonly Employed Fluorochromes

Of the many known fluorochromes, only a few are routinely employed for immunofluorescence. Spectral maxima for such fluorochromes are provided in **Table 1**. Conjugation of the fluorochromes to antibodies is typically achieved through isothiocyanate, sulfonyl chloride, succinimidyl ester, or other reactive groups that provide bonding to protein amines. Fluorescein isothiocyanate, for example, is coupled to antibodies by its reactive isothiocyanate group. Since most investigators purchase fluorochromes already conjugated to antibodies, the chemistry of these and other reactive groups will not be further addressed (*see* Chapter 6 for details). Additional fluorochrome information is summarized below.

4.1. Green-Emitting Fluorochromes—Fluorescein

The pioneering immunofluorescence studies of Albert Coons and colleagues in the 1940s and 1950s (reviewed in **ref. *11***) established the effectiveness of fluorescein for immunofluorescence microscopy. The green emission of fluorescein isothiocyanate (FITC) was shown to provide a strong signal, well separated from blue cellular autofluorescence. Continued wide use of fluorescein attests to its utility.

Conjugation of fluorescein to an antibody reduces its quantum yield by half (from 0.85 to 0.5–0.3) *(2)*. Nonetheless, conjugated fluorescein provides good fluorescence intensity, and its small size and hydrophilic nature are advantageous. Drawbacks include a short Stokes shift and a susceptibility to photobleaching *(8)*. Fluorescein is also pH-sensitive, fluorescing maximally in the range of pH 8.0–9.0. This is not a problem for immunofluorescence of fixed cells, since buffers of optimal pH can be used, but does limit fluorescence intensity in procedures by which fluorescein-labeled molecules are introduced into living cells.

Fluorochromes have been introduced that offer excitation and emission spectra similar to those of fluorescein, but that overcome some of fluorescein's limitations. BODIPY FL has a short Stokes shift, but offers higher fluorescence intensity, and is claimed to be more photostabile and less pH-sensitive than fluorescein. Oregon Green 488 and the newly introduced Alexa 488 fluorochromes have spectra nearly identical to those of fluorescein, but are considerably more photostabile, and produce less quenching of fluorescence with higher numbers of fluorochromes per antibody than does fluorescein.

4.2. Orange- and Red-Emitting Fluorochromes—Rhodamines and Cyanines

The red-emitting rhodamine derivatives are constructed around the same basic xanthene framework as is fluorescein *(2)*. Tetramethylrhodamine isothiocyanate (TRITC) has been widely employed for immunofluorescence. Additional derivatives of rhodamine available for conjugation to antibodies include lissamine rhodamine sulfonyl chloride (RB-200-SC), rhodamine B isothiocyanate (RBITC), rhodamine X isothiocyanate (XRITC), and Texas

Red. The spectra of XRITC and Texas Red are shifted to longer wavelengths compared to those of other rhodamines, which makes them particularly useful for dual labeling procedures in combination with fluorescein (*see* **Subheading 5.**). Of the two, Texas Red, which is more hydrophilic and less likely to precipitate proteins upon conjugation *(12)*, is more commonly employed.

Rhodamine conjugates are less sensitive to pH and are less prone to photobleaching than are those of fluorescein. Their fluorescence intensity is generally lower than that of fluorescein conjugates under comparable conditions of excitation, but the intense 546-nm excitatory light provided by the mercury lamp of a fluorescence microscope may make rhodamine appear brighter *(7)*.

Recently, cyanine fluorochromes covering a wide spectral range have become available for immunofluorescence *(13,14)*. The red-emitting fluorochrome Cyanine 3.18, which was shown to give a significantly brighter image than TRITC, lissamine rhodamine, Texas Red, or fluorescein under specific conditions of microscopy *(7)*, provides a useful alternative to the rhodamines. Other useful substitutes for the rhodamines include the BODIPY TR and TMR, and Alexa 568 and 594 fluorochromes. The latter are newly introduced and appear to offer superior photostability.

4.3. Phycobilliproteins

Phycobilliproteins, components of algal photosynthetic systems, are naturally occurring fluorescent molecules *(15,16)*. Each phycobilliprotein has several (as many as 30) tetrapyrrole groups covalently bonded to it. In living algae, the tetrapyrrole groups contribute to photosynthesis by absorbing light, and then transferring the absorbed energy to chlorophyll through fluorescence. The presence of several tetrapyrrole groups per protein, coupled with high extinction coefficients and quantum yields, gives the phycobilliproteins very high fluorescence intensities. Fluorescence of some phycobilliproteins is said to equal that resulting from comparable excitation of 30 fluorescein or 100 rhodamine molecules *(2)*.

Three major groups of phycobilliproteins, termed phycoerythrins, phycocyanins, and allophycocyanins, are available for conjugation to antibodies. Their fluorescence is insensitive to pH, and is characterized by emission maxima at relatively long wavelengths and by short Stokes shifts (**Table 1**). Excitation spectra of the phycobilliproteins span a broad range of wavelengths. This feature, coupled with their high fluorescence intensity, has proven useful for flow cytometry and confocal microscopy, techniques in which excitation wavelengths are limited to those of the laser light source, and so may be far from the maximum for a given fluorochrome.

Most use of phycobilliproteins has involved B and R phycoerythrins, which yield the highest fluorescence intensities. They are also the largest phycobilliproteins, with respective mol wt of 1960 and 2410 kDa *(2)*. Such large size poses problems of stearic interference with binding of antibody to antigen and the possibility of background fluorescence because of inadequate washing of large, unbound antibody conjugates from fixed cells and tissues. Nonetheless, phycoerythrins have been successfully used for immunofluorescence microscopy. Phycoerythrin-conjugated antibodies have been used, for example, to identify T-cell subsets in lymph nodes *(17)*. Corsetti et al. *(5)* used R phycoerythrin to localize apolipoprotein B in cultured rat hepatocytes. Phycoerythrin was chosen for this work since, unlike fluorescein, its spectral properties allowed minimal interference from the strong, green autofluorescence of hepatocytes.

5. Multiple Fluorescent Labeling of Antigens

The sensitivity and high resolution of immunofluorescence microscopy can be exploited to advantage through techniques that allow more than one antigen to be localized in the same cell or tissue section. This is possible when two or more primary antibodies differ in some basic characteristic, such as species of origin or of isotype, allowing secondary antibodies of corresponding specificities to deliver different fluorochromes to each primary antibody. The distribution of each antigen is thus revealed by a different color of fluorescence.

Effective multiple labeling requires that excitation and emission spectra of each fluorochrome have minimal overlap with those of the other fluorochromes. If this is not so, then discrimination of the distribution of one fluorochrome from that of another may be impossible. Double labeling, using a fluorochrome combination of fluorescein and a rhodamine derivative, is the most common multiple-labeling technique. Although excitation and emission spectra of fluorescein and rhodamines overlap to some extent, proper selection of optical filters provides readily distinguishable red and green signals. Texas Red has the advantage of less spectral overlap with fluorescein *(12)*.

Localization of more than two antigens in the same cell is possible if a set of fluorochromes with minimal spectral overlap can be assembled. Successful triple labeling has been achieved by the addition of blue-emitting fluorochromes to the green–red combination of fluorescein and rhodamine. Initially, SITS was employed as the blue-emitting fluorochrome *(18)*. Subsequently, improved results were obtained by replacing SITS with the coumarin derivatives diethylaminocoumarin, DAMC *(19–21)*, or 7-amino-4-methylcoumarin-3-acetic acid, AMCA *(22)*. Cascade Blue, which has a number of useful properties, including less spectral overlap with fluorescein than the coumarins, should prove valuable for triple labeling *(23)*.

Typically, the distribution of each fluorochrome is viewed or photographed separately by selection of the appropriate optical filter set. Alternatively, filter sets are now available which allow simultaneous excitation and observation of two to four different fluorochromes. With such filters, the brightness of the fluorescence for a given fluorochrome is less than that obtained with a filter set designed for its spectra only. In combination with color photography, however, the simultaneous localization of two or more fluorescence signals can provide striking and informative photomicrographs. Additionally, simultaneous detection obviates problems that arise from small changes in the alignment of different filter cubes, and the attendant shifts in the position of the fluorescence images that may make critical comparisons of separate photographs difficult.

The extent to which multiple fluorescence signals can be distinguished is demonstrated by resolution of five separate fluorochromes within individual, living 3T3 cells *(24)*. None of the fluorochromes employed in this work was conjugated to an antibody, but the potential for extensive multiple labeling with a suitable set of fluorochromes and an appropriately equipped microscope system is clear.

References

1. Blakeslee, D. and Baines, M. G. (1976) Immunofluorescence using dichlorotriazinylaminofluorescein (DTAF). I. Preparation and fractionation of labelled IgG. *J. Immunol. Methods* **13,** 305-320.
2. Hemmila, I. A. (1991) *Applications of Fluorescence in Immunoassays.* Wiley, New York.
3. Aubin, J. E. (1979) Autofluorescence of viable cultured mammalian cells. *J. Histochem. Cytochem.* **27,** *36–43.*
4. Benson, R. C., Meyer, R. A., Zaruba, M. E., and McKhann, G. M. (1979) Cellular autofluorescence—Is it due to flavins? *J. Histochem. Cytochem.* **27,** 44–48.
5. Corsetti, J. P., Way, B. A., Sparks, C. E., and Sparks, J. D. (1992) Immunolocalization, quantitation and cellular heterogeneity of apolipoprotein B in rat hepatocytes. *Hepatology* **15,** 1117–1124.
6. Bacallao, R., Morgane, B., Stelzer, E. H. K., and DeMey, J. (1989) Guiding principles of specimen preservation for confocal fluorescence microscopy, in *Handbook of Biological Confocal Microscopy* (Pawley, J. B., ed.), Plenum, New York, pp. 197–205.
7. Wessendorf, M. W. and Brelje, T. C. (1992) Which fluorophore is brightest— A comparison of the staining obtained using fluorescein, tetramethylrhodamine, lissamine rhodamine, Texas Red, and cyanine 3.18. *Histochem.* **98,** 81–85.
8. Haugland, R. P. (1990) Fluorescein substitutes for microscopy and imaging, in *Optical Microscopy for Biology* (Herman, B. and Jacobson, K., eds.), Wiley-Liss, New York, pp. 143–157.
9. Tsien, R. Y. and Waggoner, A. (1989) Fluorophores for confocal microscopy: photophysics and photochemistry, in *Handbook of Biological Confocal Microscopy* (Pawley, J. B., ed.), Plenum, New York, pp. 169–178.

10. Haugland, R. P. (1996) Handbook of Fluorescent Probes and Research Chemicals, 6th ed. (Spence, M. T. Z., ed.), Molecular Probes, Eugene, OR.

11. Kasten, F. H. (1989) The origins of modern fluorescence microscopy and fluorescent probes, in *Cell Structure and Function by Microspectrofluorometry* (Kohen, E. and Hirschberg, J. G. eds.), Academic, San Diego, pp. 3–50.

12. Titus, J. A., Haugland, R., Sharrow, S. O., and Segal, D. M. (1982) Texas Red, a hydrophilic, red-emitting fluorophore for use with fluorescein in dual parameter flow microfluormetric and fluorescence microscopic studies. *J. Immunol. Methods* **50,** 193–204.

13. Mujumdar, R. B., Ernst, L. A., Mujumdar, S. R., and Waggoner, A. S. (1989) Cyanine dye labeling reagents containing isothiocyanate groups. *Cytometry* **10,** 11–19.

14. Southwick, P. L., Ernst, L. A., Tauriello, E. W., Parker, S. R., Murumdar, R. B., Mujumdar, S. R., Clever, H. A., and Waggoner, A. S. (1990) Cyanine dye labeling reagents - carboxymethylindocyanine succinimidyl esters. *Cytometry* **11,** 418–430.

15. Kornick, M. N. (1986) The use of phycobilliproteins as fluorescent labels in immunoassay. *J. Immunol. Methods* **92,** 1–13.

16. Oi, V. T., Glazer, A. N., and Stryer, L. (1982) Fluorescent phycobilliprotein conjugates for analyses of cells and molecules. *J. Cell Biol.* **2,** 981–986.

17. Pizzolo, G. and Chilosi, M. (1984) Double immunostaining of lymph node sections by monoclonal antibodies using phycoerythrin labeling and haptenated reagents. *Am. J. Clin. Pathol.* **82,** 44–47.

18. Rothbarth, Ph.H, Tanke, H. J., Mul, N. A. J., Ploem, J. S., Vliegenthart, J. F. G., and Ballieux, R. E. (1978) Immunofluorescence studies with 4–acetamido-4'-isothiocyanatostilbene2,2'-disulphonic acid (SITS). *J. Immunol. Methods* **19,** 101–109.

19. Meister, B. and Hökfelt, T. (1988) Peptide- and transmitter-containing neurons in the mediobasal hypothalamus and their relation to GABAergic systems: possible roles in control of prolactin and growth hormone secretion. *Synapse* **2,** 585–605.

20. Small, J. V., Zobeley, S., Rinnerthaler, G., and Faulstich, H. (1988) Coumarinphalloidin: a new actin probe permitting triple immunofluorescence microscopy of the cytoskeleton. *J. Cell Sci.* **89,** 21–24.

21. Staines, W. A., Meister, B., Melander, T., Nagy, J. I., and Hökfelt, T. (1988) Three-color immunofluorescence histochemistry allowing triple labeling within a single section. *J. Histochem. Cytochem.* **36,** 145–151.

22. Wessendorf, M. W., Appel, N. M., Molitor, T. W., and Elde, R. P. (1990) A method for immunofluorescent demonstration of three coexisting neurotransmitters in rat brain and spinal cord, using the fluorophores fluorescein, lissamine rhodamine, and 7–amino-4-methylcoumarin-3–acetic acid. *J. Histochem. Cytochem.* **38,** 1859–1877.

23. Whitaker, J. E., Haugland, R. P., Moore, P. C., Hewitt, P. C., Reese, M., and Haugland, R. P. (1991) Cascade blue derivatives: Water soluble, reactive, blue emission dyes evaluated as fluorescent labels and tracers. *Analyt. Biochem.* **198,** 119–130.

24. DeBiasio, R., Bright, G. R., Ernst, L. A., Waggoner, A. S., and Taylor, D. L. (1987) Five-parameter fluorescence imaging: Wound healing of living Swiss 3T3 cells. *J. Cell Biol.* **105,** 1613–1622.

15

Direct Immunofluorescent Labeling of Cells

Lorette C. Javois

1. Introduction

In their original immunofluorescent labeling technique, Coons and colleagues employed a one-step, direct labeling of antigen using a fluorescein-conjugated antibody *(1,2)*. Although this was not the first published report of a chemically modified antibody, it was the first time the antibody was used as a tool to examine an antigen *in situ*. Shortly thereafter, this method was adapted, giving rise to the indirect labeling technique *(3)*. Rather than conjugating a marker molecule to the primary antibody, in this technique, a secondary antibody raised against the γ globulin of the primary species is conjugated with the marker molecule (**Fig. 1**). Although the direct labeling technique has the advantage of being quick, requiring a single incubation with the labeled reagent and one subsequent wash step, there are disadvantages. Each different primary antibody must be fluorescently labeled, and the resulting fluorescence is weak since only one labeled primary antibody binds to each antigen. The indirect labeling technique adds one more incubation and wash step, but it has the advantage of amplifying the fluorescent signal because several fluorescently labeled secondary antibodies can bind to each primary antibody. In addition, labeled secondary reagents specific for various species' immunoglobulin classes are readily available commercially, stable in the lyophilized or frozen state, and relatively inexpensive.

Despite these factors, the direct labeling technique is sometimes required by experimental circumstances. For example, if one wishes to use a mouse monoclonal antibody to label an antigen in mouse tissue significant background will result using a fluorescent antimouse globulin secondary reagent since it will react with endogenous mouse globulin. Under these circumstances, the direct

From: *Methods in Molecular Biology, Vol. 115: Immunocytochemical Methods and Protocols*
Edited by: L. C. Javois © Humana Press Inc., Totowa, NJ

Direct Labeling   **Indirect Labeling**

Fluorophore-
Labeled Antibody

Tissue Antigen

Fluorophore-
Labeled Antibody

Primary Antibody

Tissue Antigen

Fig. 1. Diagram illustrating the molecular interactions of the direct and indirect fluorescent labeling techniques. With direct labeling, the fluorophore-labeled antibody (人*) reacts with the tissue antigen in a one-step staining procedure. Indirect labeling involves a two-step staining procedure. First, a primary, unlabeled antibody (人) reacts with the tissue antigen; then, a secondary, fluorophore-labeled antiglobulin (人*) reacts with the primary antibody. The signal is amplified with indirect labeling since more than one fluorophore-labeled secondary antibody can bind to each primary antibody.

Table 1
Sequence of Reagent Application for a Double-Labeling Experiment

Apply mouse monoclonal anti-X antibody (IgG)
Wash with PBS
Apply fluorescein-labeled goat antimouse IgG
Wash with PBS
Apply unlabeled, normal mouse IgG
Wash with PBS
Apply rhodamine-labeled mouse monoclonal anti-Y antibody
Wash with PBS

labeling technique is the only viable option. Another application for the direct labeling technique is double-labeling experiments involving two fluorophores. Some species combinations are exclusive of each other, but for the most part, double-labeling experiments using two indirect sequences will crossreact at some point, resulting in the same labeling pattern for both fluorophores. This complication can be avoided by using a direct labeling step to apply one of the fluorophores. For example, **Table 1** lists the order of application of reagents for double labeling two antigens, X and Y. Antigen X is visualized using an indirect labeling technique with a primary mouse anti-X monoclonal reagent followed by a secondary fluorescein-labeled goat antimouse IgG. Application of unlabeled normal mouse IgG blocks any unbound antimouse sites on the secondary reagent. The second antigen, Y, is visualized by directly applying a rhodamine-labeled mouse anti-Y monoclonal antibody. Specific, unambiguous double labeling results.

In cases where the direct labeling technique results in too weak a signal, avidin–biotin technology (*see* Chapter 25) can be employed to increase signal strength. Rather than using a fluorophore-labeled primary antibody, the primary antibody can be labeled with biotin (*see* Chapter 7). A secondary fluorophore–avidin reagent can then be used to localize the primary reagent. Functionally, this labeling sequence is one step longer than the direct labeling technique.

If one wishes to avoid the use of fluorophore-labeled reagents altogether, antiserum can be digested with pepsin producing Fab' fragments, which can be conjugated to the enzyme, horseradish peroxidase *(4)*. This enzyme conjugate can then be used in a direct labeling technique in which the labeled antigen is exposed to a peroxidase substrate solution in one additional step (*see* Chapter 23).

Finally, the direct labeling technique has been applied to the imaging of tumors through the use of radiolabeled tumor-specific antibodies *(5–7)*. Iodine, indium, technetium, and rhenium have all been conjugated to antibodies, and the resulting reagents have been demonstrated to be stable in vivo. Although some nonspecific uptake by normal tissues (thyroid, stomach, gut, and liver) is observed, improvements are being made in the ways in which radiolabels are being attached to antibodies, reducing nonspecific background levels. The ultimate goal is a simple, direct method of cancer radioimmunotherapy.

The direct labeling technique may be applied to either unfixed cells or fixed tissue, provided that the antigen is stable to fixation. Often this must be determined empirically (*see* Chapter 8). The following protocol presents the simplest approach: direct labeling of a surface antigen on unfixed cells in suspension.

2. Materials

1. Phosphate-buffered saline (PBS): For a 1-L 10X stock solution of PBS, dissolve 2.56 g $NaH_2PO_4 \cdot H_2O$ and 11.94 g Na_2HPO_4 in 500 mL distilled H_2O (dH_2O) by stirring and then adjust the pH to 7.2. Add 87.66 g NaCl, and bring to a final volume of 1 L. Store the stock solution in the refrigerator and warm to dissolve the salts before dilution. Dilute 10-fold with dH_2O for use.
2. Cells in the appropriate medium: for this example, isolated mouse spleen cells washed and resuspended in PBS at approx 1×10^7 cells/mL.
3. Fluorescein isothiocyanate (FITC)-conjugated goat antimouse immunoglobulin diluted 1:20–1:100 in PBS (*see* **Note 1**).
4. 15-mL Conical centrifuge tubes.
5. Low-speed centrifuge.
6. Pipetors: 10–100 µL with tips.
7. Ice bath.
8. Microslides.
9. 12-mm Round coverslips.

3. Method

1. Add 1 mL of the washed spleen cell suspension to a 15 mL conical centrifuge tube and pellet the cells by centrifuging for 5 min at 200*g*.
2. Decant the supernatant, and add 50 µL of FITC-conjugated goat antimouse immunoglobulin.
3. Gently resuspend the cells in this small volume by flicking the bottom of the tube with a finger.
4. Incubate on ice for 30 min.
5. Wash the cells three times by adding 10 mL PBS, centrifuging for 5 min at 200*g*, and decanting the supernatant. After the last centrifugation, resuspend the cells in two drops of PBS (*see* **Note 2**).
6. Mount the cells on a slide with a coverslip, and examine on the fluorescent microscope with the appropriate filters for fluorescein excitation and emission (*see* **Note 3**).

4. Notes

1. Antibody concentration should be determined by titration of the stock solution and testing on a known positive specimen. Usually, working concentrations are in the range of 10–20 µg/mL. However, depending on the source this concentration could vary significantly. For detailed instructions on titrating antibodies, *see* Chapter 24.
2. It is important to keep the cells concentrated at this stage to facilitate examination of an adequate number in the next step.
3. Antibody molecules will not penetrate the cell membrane of living cells, so any cell showing fluorescence throughout its interior has died. B-cells expressing surface immunoglobulin will bind the antibody and show surface fluorescence only. Approx 50% of the spleen cell suspension will display this pattern. If the cells are not kept cold or maintained with 0.1% sodium azide, capping will occur; antibody–antigen complexes will be clustered into one region of the cell membrane resulting in a crescent-shaped fluorescent labeling pattern.

References

1. Coons, A. H., Creech, H. J., and Jones, R. N. (1941) Immunological properties of an antibody containing a fluorescent group. *Proc. Soc. Exp. Biol. Med.* **47,** 200–202.
2. Coons, A. H. and Kaplan, M. H. (1950) Localization of antigen in tissue cells. *J. Exp. Med.* **91,** 1–13.
3. Coons, A. H., Leduc, E. H., and Connolly, J. M. (1955) Studies on antibody production. 1. A method for the histochemical demonstration of specific antibody and its application to a study of the hyperimmune rabbit. *J. Exp. Med.* **102,** 49–59.
4. Ishikawa, E., Imagawa, M., Hashida, S., Yoshitake, S., Hamaguchi, Y., and Ulno, T. (1983) Enzyme-labeling of antibodies and their fragments for enzyme immunoassay and immunohistochemical staining. *J. Immunoassay* **4,** 209.

5. Hnatowich, D. J. (1990) Recent developments in the radiolabeling of antibodies with iodine, indium, and technetium. *Sem. Nuclear. Med.* **20,** 80–91.
6. Goldenberg, D. M. (1991) Imaging and therapy of gastrointestinal cancers with radiolabeled antibodies. *Am. J. Gastroenterology* **86,** 1392–1403.
7. Griffins, G. L., Goldenberg, D. M., Knapp, F. F., Jr., Callaghan, A. P., Chang, C. H., and Hansen, H. J. (1991) Direct radiolabeling of monoclonal antibodies with generator-produced rhenium-188 for radioimmunotherapy: labeling and animal biodistribution studies. *Cancer Res.* **51,** 4594–4602.

16

Fluorescence Labeling of Surface Antigens of Attached or Suspended Tissue-Culture Cells

Mark C. Willingham

1. Introduction

The surfaces of living cells contain a wide variety of molecular species that are characteristic of the type and physiological role of the cell. These surface molecules mediate cell-recognition events, receptor–ligand interactions and hormonal signaling events, surface homeostasis and transport regulation, attachment to surrounding structures, and a host of other physiologically vital functions. The types of surface molecules present include integral, as well as peripheral elements, and include proteins, carbohydrate moieties, and lipids. The number of each molecular species that serves a physiological function can vary from only a few hundred molecules per cell on the surface to many millions of molecules per cell. The unique nature of many of these molecules was recognized early in the development of immunologic techniques for cell identification, and this serves as the basis for many highly useful diagnostic tools in clinical medicine. The identification of surface antigenic molecules using antibodies can be performed by several methods, including immunofluorescence microscopy, enzyme and discrete marker microscopy, enzyme and radioactive labeling of mass cultures, and flow cytometry using fluorescence dyes. In this chapter, methods will be described that concentrate on the immunofluorescence microscopy approach to surface antigen identification in cultured cells *(1,2)*.

The method described here assumes certain things about the nature of the antigen and the antibody reagents to be used. First, the epitope reactive with the antibody must be accessible on the exterior of the cell. Some transmembrane molecules have multiple epitopes that are present on the intracellular domains of the molecule, and these would not be applicable using the method described. Second, the antibody must react with this epitope when the molecule

From: *Methods in Molecular Biology, Vol. 115: Immunocytochemical Methods and Protocols*
Edited by: L. C. Javois © Humana Press Inc., Totowa, NJ

is in its native state, that is, when it is either alive or fixed in a way that preserves a native conformation. Some antibodies to surface antigens react only when the molecule is denatured or unfolded, such as the conditions of proteins after SDS gel electrophoresis and Western blotting. This type of epitope may not be detected by the protocol presented here, since the molecule will not be unfolded in this way by these methods. Third, the number of molecules present on the cell surface must be high enough to be detected using microscopy. This generally means that the molecules must be present at over 1000 sites/cell surface, at least in some cells in the culture, since below this level, fluorescence microscopy is not sensitive enough to allow detection of the surface reaction. Such low-level epitopes can be more easily detected using mass detection techniques, in which large numbers of cells are combined into a single detectable signal, or by flow cytometry, which can detect very low levels of surface reactivity.

Even though these points define the limitations of this method, it should be pointed out that direct visualization using immunofluorescence can detect surface antigens present in single cells that represent a minor population in a heterogeneous culture, and that the distribution of that antigen on that surface can be interpreted in the context of cell shape and surface morphologic domains, such as attachment points to substratum or surface patching and capping. Fluorescence detection, because of its point-light-source nature, also allows the visualization of very small objects, such as individual viral particles, that are below the resolution of light microscopy refractile methods. For these and other reasons, immunofluorescence microscopic detection of surface antigens is a powerful and frequently useful tool.

2. Materials

1. 35-mm Plastic tissue-culture dishes (Falcon, Costar, and so forth, Baxter, McGaw Park, IL). Subculture cells onto these dishes at least 24 h before.
2. Dulbecco's phosphate-buffered saline (PBS) (with calcium and magnesium) (GIBCO/BRL, Grand Island, NY).
3. Crystalline bovine serum album (BSA) (Pentex, Miles Labs., Kankakee, IL). Prepare solution of BSA at 2 mg/mL in PBS (BSA-PBS) (*see* **Notes 1** and **2**).
4. Primary antibody: polyclonal or monoclonal of a specific species, e.g., mouse, diluted to 10 μg/mL in BSA-PBS (can be stored at –70°C for extended periods).
5. Secondary antibody: affinity-purified fluorescent antiglobulin conjugate reactive with the species globulin of the first step, e.g., affinity-purified rhodamine-conjugated goat antimouse IgG(H + L chains) (Jackson ImmunoResearch, West Grove, PA), diluted to 25 μg/mL in BSA-PBS. Can be stored at –70°C and reused (*see* **Notes 3** and **4**).
6. 3.7% Formaldehyde freshly diluted into PBS: 1:10 stock solution (37%) "Formalin" (Fischer, Pittsburgh, PA) into PBS.

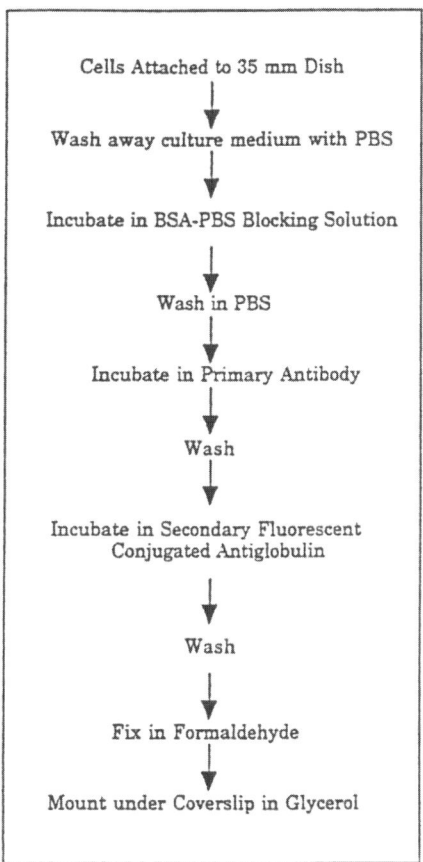

Fig. 1. Flowchart of the surface labeling procedure. Abbreviations as in text.

7. Glycerol mounting medium (Difco) or 90% glycerol, 10% PBS.
8. Epifluorescence microscope equipped with appropriate filters (e.g., Zeiss Axioplan epifluorescence microscope with rhodamine filters).

3. Methods

1. Wash cultured cells attached to 35-mm plastic tissue-culture dishes in Dulbecco's PBS, then incubate in a blocking buffer consisting of BSA-PBS for 5 min, and cool to 4°C (*see* protocol flow chart in **Fig. 1**). Cooling prevents subsequent endocytosis of any added antibody reagents, as well as minimizing lateral mobility of bound antibody in the plane of the plasma membrane (*see* **Notes 5** and **6**).
2. Add the primary antibody to the living cells at 4°C in BSA-PBS, and incubate for 30 min (1-mL vol for a 35-mm dish). **Do not pipet directly onto the cells**, but add antibody solutions at the edge of the dish, and add wash solutions from a wide-mouth bottle or beaker to minimize the potential of removing cells by too vigorous a fluid

stream. **Do not allow the cells to dry at any step.** Rock the dish back and forth to maintain coverage over the cells if the antibody solution volume is too small (*see* **Note 7**).

3. Harvest the primary antibody solution from the dish using a Pasteur pipet and save. Wash the dish again in BSA-PBS at 4°C five times.
4. Add a secondary fluorescent conjugate antiglobulin in BSA-PBS at 4°C for 30 min.
5. Remove the indirect conjugate and save. Wash the dish with BSA-PBS at 4°C five times, followed by washing briefly in PBS alone at room temperature.
6. Fix the cells attached to the dish in 3.7% formaldehyde–PBS at room temperature for 15 min (*see* **Note 8**).
7. Wash the dish in PBS at room temperature five times, cover the cells with mounting medium, and overlay with a no. 1 coverslip (*see* diagram in Chapter 17). Then view on an upright epifluorescence microscope (using rhodamine filters). **Figure 2** demonstrates an example of the type of image seen for surface labeling of attached cells (*see* **Notes 9–11**).

4. Notes

1. The use of competitor proteins is important in these procedures to minimize non-specific labeling owing to sticking of protein. Bovine serum albumin is a convenient, purified, inexpensive protein available for this purpose, but other proteins, such as normal globulin of the same species as the second step, or normal calf serum or plasma, are also useful for this purpose.
2. The use of any detergent-like molecule is inappropriate for surface labeling, since it may remove portions of the surface membrane and dissolve the attachments for surface markers, as well as permeabilizing the cell to antibodies.
3. The use of an affinity-purified second-step reagent is important, since it dramatically reduces background. Although some other companies offer this reagent, Jackson ImmunoResearch, Inc. has a large series of specialty second-step reagents conjugated with several different labels that are all affinity-purified.
4. Rhodamine is a preferred fluorochrome over fluorescein because of its slower bleach rate and its emission in a spectrum that shows less cellular autofluorescence. Also, this spectrum produces less autofluorescence in plastic substrata (*see* Chapter 14). Rhodamine requires a mercury vapor light source, since other sources, such as xenon, do not have sufficient emission in the green spectrum.
5. Attached cultured cells are easiest to work with, since the wash steps can be performed directly on the dish without additional manipulation. For suspended cells, the cells can be incubated in suspension, and the wash steps performed using centrifugation in an microfuge (Eppendorf, Hamburg, Germany) to separate cells at each incubation and washing step. This is much more time-consuming and may lead to a significant loss of cell number during the procedure.
6. Another approach with suspended cells is to convert them into attached cells by either primarily culturing them on a highly adhesive surface, such as BioGlue (Sigma, cat. #A2707), when they are subcultured the day before, or to wash them free of protein-containing medium and attach them using poly-L-lysine to a dish

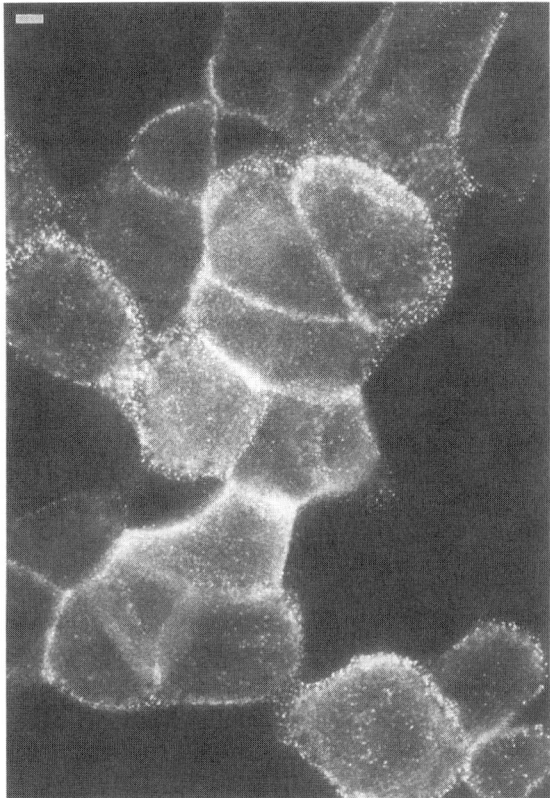

Fig. 2. Fluorescence image of surface localization in cultured cells. OVCAR-3 human ovarian cancer cells were incubated with monoclonal antibody K1, followed by antimouse IgG conjugated to rhodamine as described in the protocol *(3)*. The intense diffuse speckled surface fluorescence image is shown on multiple cells in this area of the culture, with reinforced bright images at the vertical cell margins generating a cobblestone appearance (bar = 4 µm).

surface. This is performed by incubating tissue-culture plastic dishes with 1 mg/mL poly-L-lysine (Sigma, St. Louis, MO) of high mol wt (> 50,000) in PBS for 5 min at room temperature. The dish is then washed free of poly-L-lysine using PBS and left wet. The cell suspension in protein-free medium is then added, and the cells are either allowed to settle at 1g or spun onto the coated surface in a large centrifuge on a swinging bucket adapted to hold dishes. This is analogous to the cytoprep approach (without drying) and works quite well. Unlike the routine "Cytoprep" (Shandon, Inc., Pittsburgh, PA) method, however, it is important to avoid drying of the cells for good preservation of surface membrane integrity. The cells are then processed through the same protocol as if they were originally an attached cell type.

Small amounts of surface labeling are sometimes more easily seen if the cells are rounded, in which the surface labeling appears as a "ring" around the cell periphery. For suspended cells, this labeling can be performed completely in suspension using a centrifuge to separate wash steps, or cells can be attached as above, but later disattached using mechanical means by vigorous pipeting. The observation is then performed while still in suspension in PBS under a coverslip. This method makes it more difficult to obtain photographs of the cells because of cell drift during exposure.

7. The most sensitive approach to surface antigen labeling is to incubate cells at 4°C while alive with antibodies. This prevents internal labeling background and also minimizes artifacts seen with some fixation. However, there are situations in which living cells are not available or the antigen of interest fails to react without some form of fixation. In these settings, formaldehyde fixation or other forms of fixation can be performed prior to incubation with antibodies. However, it should be noted that fixed cells are much more fragile and sticky compared to living cells, and some fixatives, such as ethanol or acetone, render the interior of the cell permeable, as well. This internal labeling results in background unrelated to the cell surface, as well as the chance of removing the elements of the cell membrane that stabilize the surface antigen of interest (*see* Chapter 8). As a result, prefixation of cells is best performed when cells are already attached to an immobile surface, such as in attached cultured cells or with suspended cells preattached to plastic or glass. Another point to note is that formaldehyde and glutaraldehyde fixation do not inherently permeabilize the plasma membrane, but they do induce a surface "blistering" artifact that produces random holes in the cell surface. Such artifactual holes can sequester antibody in a moth-eaten pattern that is nonspecific and not reflective of actual surface or intracellular distribution of the antigen.

8. The final formaldehyde fixation step links all of the antibody steps in place, preventing their dissociation. Immediate viewing of cells may not require this step, but after fixation, mounted cells can be kept at 4°C for several weeks with little loss of signal. Drying at any stage will severely affect the label, either causing it to be displaced or drastically altering morphology of the surface.

9. Weak fluorescence signals can be preserved for photography by using antioxidants in the mounting medium, such as *p*-phenylenediamine *(4)* or *N*-propyl gallate *(5)*. This allows long exposure times without loss of signal resulting from bleaching of the fluorochrome. Color film is not optimal for recording fluorescence images, since color film is usually less sensitive. Also, it shows an artifactual yellowing of localization in areas of higher intensity, which may be the result of background intensity and nonspecific signals. Further, color images are darker and harder to project for presentations and are much more expensive to process and print. The best films for recording fluorescence images are high-speed black and white films, such as Tri-X (Kodak) developed in Diafine (rated speed ASA 1600). (*See* Chapter 22.) A more recent alternative is the use of direct digital image capture using a cooled CCD camera (e.g., SPOT camera, Diagnostic Instruments, Sterling Heights, MI).

10. The specificity of fluorescence localization is dependent on the specificity of the primary antibody, a property that must be tested and controlled by other methods, such as immunoprecipitation or immunoblotting. Controls for the labeling procedure described include deletion of the primary antibody step, which controls for the second-step reagent, or inclusion of a similar, but nonreactive antibody as a first step. In the case of the availability of purified primary antigen, competition controls can be used, but they only control for the reactivity of the antibody with one antigen, and do not rule out the possibility of a crossreactive, but unrelated antigen.

11. In the case of cell-surface antigens, carbohydrate epitopes are very frequently present in large amounts that can produce highly reactive antibodies that will not necessarily appear on analysis for protein antigens. Thus, glycolipid and other carbohydrate antigens must be kept in mind in analysis of the specificity of a primary antibody. This is especially true of polyclonal antibodies raised to complex mixtures of antigens, such as whole cells. Also, in the case of polyclonal antibodies, affinity purification using purified antigen is a very useful step in preparing reagents for immunofluorescence, since it may eliminate many unwanted antibodies to highly immunogenic contaminants.

References

1. Willingham, M. C. and Pastan, I. (1985) Immunofluorescence methods, in *An Atlas of Immunofluorescence in Cultured Cells,* Academic, Orlando, FL, pp. 1–13.
2. Willingham, M. C. (1990) Immunocytochemical methods: useful and informative tools for screening hybridomas and evaluating antigen expression. *FOCUS* **12,** 62–67.
3. Chang, K., Pastan, I., and Willingham, M. C. (1992) Isolation and characterization of a monoclonal antibody, K1, reactive with ovarian cancers and normal mesothelium. *Int. J. Cancer* **50,** 373–381.
4. Platt, J. L. and Michael, A. F. (1983) Retardation of fading and enhancement of intensity of immunofluorescence by *p*-phenylenediamine. *J. Histochem. Cytochem.* **31,** 840–842.
5. Giloh, H. and Sedat, M. (1982) Fluorescence microscopy: reduced photobleaching of rhodamine and fluorescein protein conjugates by *N*-propyl gallate. *Science* **217,** 1252–1255.

17

Fluorescence Labeling of Intracellular Antigens of Attached or Suspended Tissue-Culture Cells

Mark C. Willingham

1. Introduction

Cells in culture offer a unique opportunity to visualize intracellular organelles. Because of the ability of some cultured cells to attach and spread on a substratum, the cell's cytoplasm can be spread over a large surface area, resulting in a thin, broad cytoplasmic layer. In this layer, optical methods can resolve details in much the same way as can be accomplished by sectioning of embedded cells, but without the need for embedding and sectioning. The patterns of different organelles are, in this setting, very easy to interpret using immunofluorescence, and the subcellular distribution of an antigen can frequently be more clearly and easily detected by this method than with any other. An important feature of this approach is the ease of performing the experiment, in that the organelle localization of an unknown antigen in a cultured cell can be revealed using a specific antibody in less than an hour by immunofluorescence, a result that might take weeks using cell fractionation methods *(1,2)*. On the other hand, there are several important methodological considerations that affect the accuracy and interpretability of the results.

One major problem area in immunofluorescence is the choice of the primary fixative. For intracellular antigens, unlike surface antigens, the access of a large protein, such as an antibody, requires permeabilizing the cell membrane, and the preservation of cell architecture and antigen distribution after membrane permeabilization requires precise structural preservation. This fixation step, usually a chemical treatment, will frequently determine the appearance and authenticity of organelle structure and antigen immobilization. Some fixatives, such as organic solvents, function by rendering molecules, such as pro-

From: *Methods in Molecular Biology, Vol. 115: Immunocytochemical Methods and Protocols*
Edited by: L. C. Javois © Humana Press Inc., Totowa, NJ

teins, insoluble. Others, such as aldehydes, result in specific crosslinking of some, but not all, molecular species in cytoplasmic, organelle, and nuclear matrices. The choice of fixative is often a compromise among structural preservation, accessibility of antibodies to antigen locations, and the preservation of antigen chemical structure in a form that can still react with antibody (*see* Chapter 8).

Antibodies recognize antigens, usually proteins or carbohydrates, generally in the state in which they were exposed to the immunized animal. Thus, proteins injected into an animal in their native conformation will produce antibodies that often react with this same native conformation. On the other hand, proteins that are originally denatured and unfolded, such as after SDS treatment, may result in antibodies that preferentially recognize these denatured forms. Frequently, commercially available antibodies react with proteins in one, but not both, of these situations. Antibodies produced to synthetic peptides will frequently recognize these small peptide region epitopes, often in unfolded proteins, but may fail to recognize them in the complex tertiary form of a native folded protein. This is usually indicated by reactivity of antibodies with antigen in Western immunoblots from denatured SDS gels, but not by immunoprecipitation of native protein from detergent cell extracts. In general, epitopes that are most readily utilized for immunofluorescence are those that represent the native, folded conformation of proteins, and not those that recognize only small peptide determinants in an unfolded, denatured preparation. There are, of course, exceptions to every rule, and some antibodies recognize epitopes in all situations. Suffice it to say that the type of epitope recognized and the quality of the antibody preparation are the single most important aspects of these procedures that determine the ability to interpret the distribution of an antigen in an intracellular site.

The patterns of antigen distribution in different organelles are quite characteristic in cultured cells, and examples of such patterns have been published in many journal articles and books (e.g., **ref. 2**). From immunofluorescence images, antigens restricted to intracellular membranous organelles, cytoskeletal elements, cytosolic compartments, and nuclear components are readily identified. Such information is vital in the understanding of functional and biochemical potential roles of antigens. The procedure described here is an indirect immunofluorescence procedure that is both practical and simple, and results in the highest degree of structural preservation, especially for membranes, while still maintaining accessibility of antigens in most intracellular sites.

2. Materials

1. 35-mm Plastic tissue-culture dishes (Falcon, Costar, and so forth, Baxter, McGaw Park, IL). Subculture cells into these dishes at least 24 h before.

2. Dulbecco's phosphate-buffered saline (PBS) (without calcium and magnesium) (GIBCO/BRL, Grand Island, NY).
3. Fixative solution. (Formalin stock solution ["Formalin," Fischer, Pittsburgh, PA].) Freshly dilute formalin (37%) stock solution 1:10 in PBS (final concentration = 3.7% formaldehyde in PBS).
4. Prepare competitor protein buffer detergent solution: 4 mg/mL normal goat globulin or other competitor protein, (such as fetal calf serum, bovine serum albumin, bovine plasma, and so forth), and 0.1% saponin (Sigma, St. Louis, MO) in phosphate-buffered saline (NGG-sap-PBS).
5. Primary antibody (polyclonal or monoclonal of a specific species, e.g., mouse). Prepare solution (e.g., 10 µg/mL mouse monoclonal) in NGG-sap-PBS, minimum vol of 1 mL for each dish of cells to be examined. Diluted primary antibody in NGG-sap-PBS can be stored frozen, harvested back from the dish, and refrozen and reused several times.
6. Secondary antibody (affinity-purified fluorescent antiglobulin conjugate reactive with the species globulin of the first step, e.g., affinity-purified rhodamine-conjugated goat antimouse IgG (H + L chains) (Jackson ImmunoResearch, West Grove, PA) diluted in NGG-sap-PBS (25 µg/mL). This can be stored in this form frozen, harvested back from the dish, and reused the same as the first antibody solution. Minimum volume = 1 mL for each dish to be incubated.
7. Glycerol mounting medium (90% glycerol, 10% PBS), or commercial glycerol mounting medium (Difco, Detroit, MI).
8. Epifluorescence microscope equipped with appropriate filters (e.g., Zeiss Axioplan epifluorescence microscope with rhodamine filters).

3. Method

The sequence of steps is summarized in the flow chart in **Fig. 1**.

1. Cultured cells attached to 35-mm plastic tissue-culture dishes are washed in PBS, then fixed in 3.7% formaldehyde in PBS for 10 min at room temperature. The dishes are then washed in PBS five times (*see* **Notes 1–3**).
2. Incubate dishes with NGG-sap-PBS solution for 10 min at room temperature. The normal goat globulin serves as a blocking protein to minimize nonspecific binding, and the 0.1% saponin renders the fixed cell membranes reversibly permeable to proteins (*see* **Notes 4** and **5**).
3. The primary antibody in NGG-sap-PBS (e.g., 10 µg/mL mouse monoclonal antibody) is then added to the fixed cells in NGG-sap-PBS and incubated for 30 min at room temperature. **Do not pipet directly onto the cells,** but add antibody solutions at the edge of the dish, and add wash solutions from a wide-mouth bottle or beaker to minimize cell disattachment. The minimum volume to cover a 35-mm dish surface completely is 1 mL.
4. Harvest and save the primary antibody solution, and immediately wash the dish with PBS five times. **Do not let the cells dry at any step.** Especially during washings, handle each dish individually, since leaving a washed dish without medium for even a few seconds can allow drying in the center of the dish.

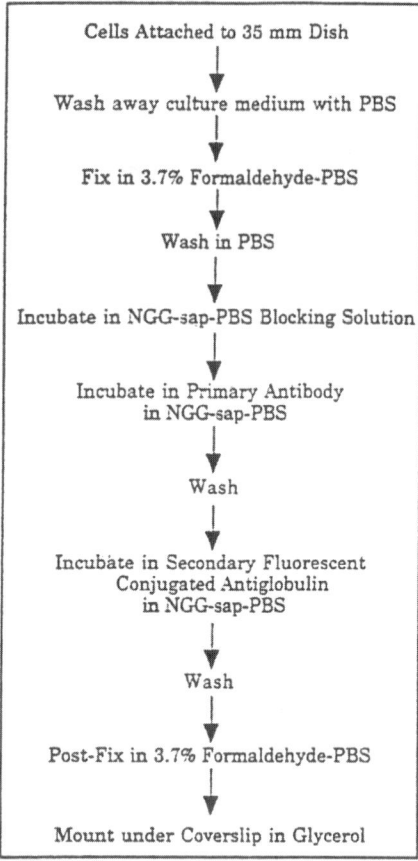

Fig. 1. Flowchart of an intracellular antigen labeling protocol. Abbreviations as in text.

5. Add NGG-sap-PBS competitor protein solution to the dish again briefly, and then pour off.
6. Add the secondary antibody conjugated to rhodamine in NGG-sap-PBS for 30 min at room temperature. Saponin does not render membranes irreversibly permeable, so it must be present in all antibody incubations.
7. Harvest the secondary antibody and save. Wash the cells in PBS five times.
8. Fix the cells again using 3.7% formaldehyde freshly made as performed in the initial fixation. The purpose of the second fixation is to crosslink the antibodies in place and prevent subsequent diffusion of label. If not postfixed in this way, the localization may not be stable for more than a few hours.
9. Wash the cells free of fixative using PBS.
10. Mount the cells under glycerol mounting medium under a no. 1 coverslip as shown in **Fig. 2** (*see* **Notes 6–9**).

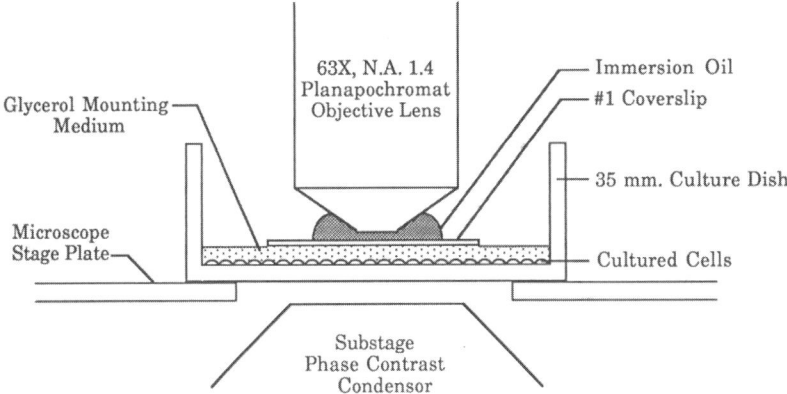

Fig. 2. Diagrammatic summary of the method of mounting and viewing cells in culture dishes on an upright microscope.

4. Notes

1. Physical methods of processing immunofluorescence samples other than plastic tissue-culture dishes are commonly used. The plastic tissue-culture dish is convenient in that cells can be easily grown and processed for immunofluorescence in the same container. Dishes have the slight disadvantage that they possess some autofluorescence, especially in the shorter wavelength ranges. Special chamber slides are available for creating separate wells on glass slides (Baxter), and other devices for creating large numbers of small wells for use in immunofluorescence screening procedures have been devised *(3)*. Coverslips or glass slides can also be placed in the bottom of individual tissue-culture dishes for cell growth when a slide format is desired. All of these approaches work well, but involve tradeoffs of convenience, size, or optical quality. For routine use, direct labeling in 35-mm plastic dishes seems to be the most practical approach. Multiwell tissue-culture dishes often have dimensions that limit their ability to be used on a microscope, either because of too small an opening for large objectives or too large a base to fit easily on a microscope stage.

2. It is possible to label cells directly in suspension for intracellular antigens, but the detail of rounded cells for intracellular sites may be obscured by the cell shape, except for gross distributions, such as nuclear or cell-surface patterns. To discern the detail of intracellular distribution in suspended cells, it is best to attach them and, if possible, cause them to spread or flatten onto a substratum so that intracellular detail is visible. Such an approach is similar to that used for surface labeling, in which an adhesive substrate, such as poly-L-lysine (Sigma) or BioGlue (Sigma, cat. #A2707), is fashioned to attach cells to the surface of a dish. The cells can then be handled the same as adherent cultured cells. For cells that do not spontaneously flatten, cytocentrifugation is useful, although the amount of intracellular detail is still somewhat limited. However, mitotic cells that are rounded in many

cultured cell lines are an example of the possibility to visualize detailed intracellular structures even when cells are rounded. Tubulin localization in the mitotic spindle, for example, is very easy to visualize and has a characteristic appearance, even though the cell is completely round. Therefore, it is clear that the amount of detail obtainable in rounded cells will vary with the concentration and distribution of the antigen. The approaches to fixation and processing of such cells once attached to a substratum are essentially the same as for flattened cultured cells.

3. Formaldehyde works well as a primary fixative in this setting because it preserves cell morphology well and the time of exposure to the fixative is short. Formaldehyde fixation at room temperature is very effective, but at 4°C, it is a very poor fixative. Longer fixation times (30–60 min) may be helpful for some antigens that are difficult to preserve.

Fixation in glutaraldehyde produces better morphology, but induces a great deal of autofluorescence, and limits cytoplasmic and nuclear permeability. A protocol utilizing glutaraldehyde followed by borohydride treatment has been previously described that is applicable also for electron microscopy of cultured cells *(4)*. Some areas of the cell are relatively impermeable with this approach, but this is an excellent choice of fixative protocol for microtubule morphology.

Another fixative approach is the use of organic solvents, such as ethanol, methanol, and acetone. These precipitating fixatives also produce membrane permeability and generally yield a poorer quality of preservation, although with a high degree of permeability. Since pure acetone will dissolve styrene plastic dishes, this fixative is usable only with specially resistant plastic dishes (Permanox) or with glass substrates, such as glass coverslips. Acetone mixed with water (80% acetone, 20% water) will fix cells in styrene plastic dishes without dissolving the plastic. Methanol at –20°C is also a commonly used fixative for this purpose. Ethanol is generally a rather poor fixative for cultured cells.

Other fixatives, such as water-soluble carbodiimides or di-imidates, have been used as fixatives, either alone or in combination with aldehydes, but except in special cases, they have no major advantage over aldehyde fixation *(5)* (*see* Chapter 8).

4. Since intracellular antigens are located inside the plasma membrane barrier, some treatment must be used in intact cells to permeabilize cell membranes. Organic solvent fixation produces permeability directly as a consequence of the fixation process. Formaldehyde and glutaraldehyde do not. Extended fixation in either of these aldehydes leads to fixation blister artifacts in the plasma membrane, which, if they pop, lead to artifactual holes in the cell surface and a moth-eaten appearance in subsequent antibody labeling. A common approach to producing membrane permeability is the use of detergents after the primary fixative step. Triton X-100 (0.1%), NP40, or Tween-20 have been used to permeabilize fixed membranes. These detergents remove the cell's phospholipid barriers and render all membrane-limited compartments, including mitochondria and the nucleus, permeable. Saponins are cholesterol-like sugar-containing detergents that intercalate into membranes that contain cholesterol. Such molecules surround antibody molecules and allow the transfer of antibody across fixed, cholesterol-containing

membranes, as long as saponin is continuously present. Saponin-treated membranes are not permeable, however, to antibody molecules never exposed to saponin in solution. Therefore, for the protocol described above, saponin is present in all antibody incubation steps. The presence of saponin does not, however, render membranes permeable if they have no cholesterol. These include mitochondrial membranes and the inner nuclear envelope. Therefore, the protocol above is useful for cytoplasmic antigens in the cytosol or in membranous organelles that contain cholesterol, including lysosomes, the endoplasmic reticulum, and the Golgi, but will not be useful for antigens inside the nucleus or in mitochondria. For access to these sites, a treatment with a detergent, such as Triton X-100, is necessary. Saponin permeabilization, however, is the mildest form of membrane permeabilization and, as seen by electron microscopy, leaves an intact membrane structure at the cell surface into which many antigens are anchored. The effects of extraction of some membrane protein antigens by detergents other than saponin has been previously demonstrated *(6)*.

5. The inclusion of normal goat globulin in this protocol is important in reducing nonspecific binding of globulins to fixed cell sites that are rendered sticky by chemical treatments. Thus, the high concentration of competitor protein that is not detected by the labeling method yields very low background levels in fixed cells. The globulin of the same species as the second step reagent would be the least likely protein to be detected by this antiglobulin reagent. Other proteins at high concentrations (1–4 mg/mL) will also serve this purpose, and may be more available and cheaper. Care must be taken to minimize the content of proteases or other elements that might affect the preservation of the cell or the retention of the antigen of interest. A side benefit of including high concentrations of carrier (competitor) protein is that the solutions may be harvested back from the dish and reused many times with little loss in specific antibody concentration. Also, the high concentrations of carrier protein allow freezing and thawing of these solutions without denaturation of the small amount of specific antibody present. Freezing is a much safer method of storage and does not require the presence of antibacterial agents to prevent overgrowth of organisms.

6. The pattern of organelles and their distribution in cultured cells of various types have been extensively studied, and books are available that demonstrate these patterns (e.g., **ref.** *1*). These patterns include membrane-associated antigens, organelle-selective antigens, cytoskeletal antigens, and diffusely distributed antigens. The information gained from these images can immediately categorize a particular antigen as being associated with a specific structure, such the microtubule pattern shown in **Fig. 3**. By observing many cells at a time in a culture dish, one can also search for localization that indicates antigens expressed in only a small percentage of the cells in a culture or in different cellular conditions, such as during various stages of mitosis. It is this interpretive power that makes immunofluorescence such a useful tool. Because of its point-light-source nature, fluorescence can detect objects smaller than the limit of refractile resolution of light, such as individual viral particles. Difficulty in preservation of such objects and

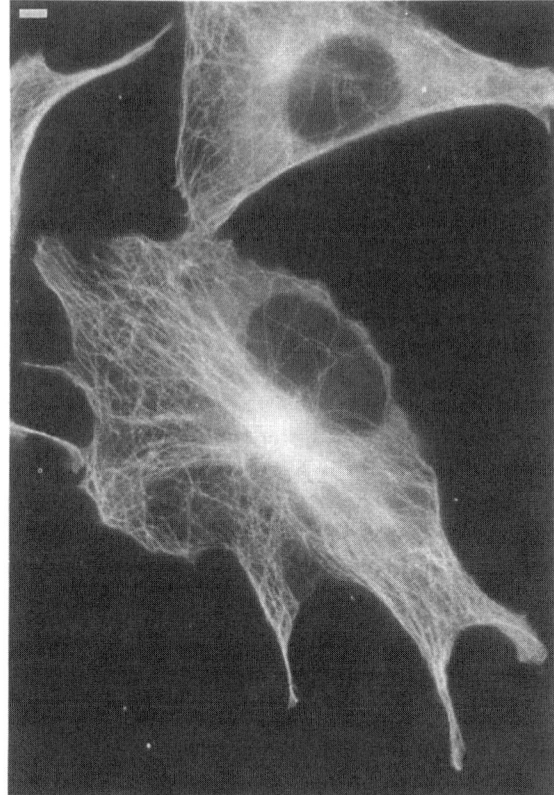

Fig. 3. Immunofluorescence localization of tubulin in microtubules in Swiss 3T3 cultured fibroblasts. Swiss 3T3 cells were fixed in glutaraldehyde followed by treatment with sodium borohydride *(4)*. Tubulin was detected using a rat monoclonal antitubulin antibody (YL1/2), followed by goat antirat IgG conjugated to rhodamine, all steps in the continuous presence of 0.1% saponin. Note the individual microtubules visible under the dark nuclear region (bar = 4 μm).

   their infrequent location in a culture, for example, might make this detection
   impossible or impractical using electron microscopy.
7. The classical methods of demonstrating specificity of antibody reactions with
   antigens, such as immunodiffusion plates, are obsolete. Immunofluorescence is a
   very sensitive method that requires controls for specificity of equal sensitivity.
   The most important controls are not just those that demonstrate the presence of an
   antibody that reacts with the antigen of interest, but more importantly, the absence
   of antibodies that react with other things.
      Precise controls for the specificity of antibody reactions reside with other tech-
   niques, such as immunoprecipitation or immunoblotting. Controls for the labeling

methodology include deletion of the first-step antibody (blank control), use of a nonreactive antibody in the first step (normal globulin control), or attempts to absorb the antibody if purified antigen is available (preabsorption control). Blank and normal globulin controls characterize the labeling system. Preabsorption controls can be problematic, require large amounts of purified antigen, and still depend on the purity of the antigen preparation. In the early days of these methods when affinity purification was not common and polyclonal sera were used, these preabsorption controls were the only specificity controls possible. With newer, more sensitive biochemical methods, such as Western blots, these preabsorption controls are sometimes not as useful. Controls in which a cell type completely lacks an antigen or in which the antigen's gene can be introduced into a negative cell type, serve as much better, although not perfect demonstrations of specificity.

Immunofluorescence frequently reveals crossreactions to other major molecular species that were not appreciated in biochemical isolation experiments because of lack of solubility, extractability, or stability of these crossreacting antigens. This is especially true of antibodies made against synthetic peptides, which may crossreact in immunofluorescence in a way not detected by other methods. This is frequently explainable by the nature of preservation of cell structures for immunofluorescence, in that fixation usually preserves molecules in their native conformations, molecular shapes that would not be preserved by some biochemical extraction methods. The observations relating to the effect of detergent treatments on preservation of membrane-associated antigens (*6*) also emphasize that some forms of fixation may not preserve an antigen in place sufficiently well for it to be detected by immunofluorescence.

8. Fixation methods play an important role in creating false-negative results in some cases. Formaldehyde fixation may crosslink cellular elements sufficiently to mask sites of antigen concentration. This is especially true in prolonged fixation of intact tissues, rather than the short incubation times for use with cultured cells. A similar problem exists in cultured cells with the use of glutaraldehyde as a primary fixative, in which many cellular sites are covered up by the tight crosslinking present in a glutaraldehyde-fixed cell. A uniquely accessible area of such cells is the region around microtubules that is accessible even after 2% glutaraldehyde fixation (*4*). The interior of the nucleus, on the other hand, is very easily rendered inaccessible to antibodies by tight fixation. These alterations are usually not related to unique chemical structural changes in the individual antigen, although that situation may theoretically exist in some amine-containing epitopes. Following glutaraldehyde primary fixation, for example, actin is accessible in surface ruffles, but not accessible in the microfilament bundles present in the same cell (*4*). This emphasizes a cardinal rule of immunocytochemistry in general: **Lack of detection of an antigen does not necessarily imply that the antigen is not present in that site.** That is, false negatives are frequent and can be caused by many things, especially by fixation.

9. Photography of fluorescent images is most easily accomplished using black and white, high-speed film, such as Kodak Tri-X. When developed in Diafine, the

relative ASA rating of this film is ~1600. Historically and because of personal preference, many investigators in the past have used color films to record fluorescence images. There are several disadvantages to the use of color film:

a. The film sensitivity is usually lower, requiring longer exposure times;
b. The contrast and brightness of recorded images are usually less, and make presentation of projected 2×2 slides difficult because of their darkness; and
c. The processing of color materials for publication purposes is much more difficult, requires more elaborate darkroom facilities, and is much more expensive for journal presentation (~$1000 for one color plate).

In addition, color film has an inherent artifact in areas of overexposure, in which the color of the film becomes yellow, no matter what the color of the exposing light. This produces apparent areas of enhanced concentration, which may or may not reflect the true intensity of label, especially in areas in which a low-level background image is added to the specific signal. This can lead to misinterpretations about selective areas of concentration of label.

Color recording is important in some settings, especially in those in which either double exposures using labels of different color emission are desired or when filters that allow more than one fluorochrome signal through to the camera are used. Using black and white film, these double-label images can be recorded on two separate negatives using filters that separate both emissions discretely, and the images are then compared as individual panels. Sometimes, however, the inherent clarity of points of reference in an image are enhanced by incorporating both label images in one color image (*see* Chapter 22). The recent availability of highly sensitive cooled CCD cameras that allow direct digital image capture may soon make obsolete the use of film for this purpose.

References

1. Willingham, M. C. (1990) Immunocytochemical methods: useful and informative tools for screening hybridomas and evaluating antigen expression. *FOCUS* **12,** 62–67.
2. Willingham, M. C. and Pastan, I. (1985) *An Atlas of Immunofluorescence in Cultured Cells*, Academic, Orlando, FL.
3. Willingham, M. C. and Pastan, I. (1990) A reversible multi-well chamber for incubation of cultured cells with small volumes: application to screening of hybridoma fusions using immunofluorescence microscopy. *Biotechniques* **8,** 320–324.
4. Willingham, M. C. (1983) An alternative fixation-processing method for pre-embedding ultrastructural immunocytochemistry of cytoplasmic antigens: the GBS procedure. *J. Histochem. Cytochem.* **31,** 791–798.
5. Willingham, M. C. (1980) Electron microscopic immunocytochemical localization of intracellular antigens in cultured cells: the EGS and ferritin bridge procedures. *Histochem. J.* **12,** 419–434.
6. Goldenthal, K. L., Hedman, K., Chen, J. W., August, J. T., and Willingham, M. C. (1985) Postfixation detergent treatment for immunofluorescence suppresses localization of some integral membrane proteins. *J. Histochem. Cytochem.* **33,** 813–820.

18

Fluorescent Labeling of Surface or Intracellular Antigens in Whole-Mounts

Lorette C. Javois

1. Introduction
1.1. Historical Perspectives

During the last century, before the advent of microtomes capable of producing thin, even sections, whole-mounts and thick sections were used to study the peripheral nervous system. Tissues were either viewed directly, if sufficiently thin, or further separated into layers by acid maceration *(1)*. Initially, histochemical staining, particularly methylene blue *(2)*, was applied to whole-mounts to characterize neuronal cell types, distributions, and innervation of various organs. In more recent times precise innervations were traced through the use of histochemical techniques specific for neuronal components. Histofluorescence techniques were introduced by Falk *(3)* for the localization of nonadrenaline-containing nerves, and the discovery of neuropeptides led to the application of immunohistochemical techniques in the late 1970s and early 1980s *(4,5)*. Most recently, the application of monoclonal antibody technology has opened a new era in research and diagnosis such that *in situ* normal and pathological tissue constituents may be detected in whole-mounts ranging from tissues to entire organisms.

1.2. General Methodology

The general immunocytochemical principles as applied to whole-mounts do not differ substantially from those applied to cells (*see* Chapters 16 and 17). However, several additional factors must be taken into account. Foremost is the problem of tissue penetration by reagents. Unlike sectioned material where intracellular components are directly exposed to the reagents, in whole-mount

From: *Methods in Molecular Biology, Vol. 115: Immunocytochemical Methods and Protocols*
Edited by: L. C. Javois © Humana Press Inc., Totowa, NJ

preparations, cells have intact membranes that act as barriers. Membranes may be permeabilized in a number of ways, such as by freezing and thawing, treatment with detergents (Triton X-100), or treatment with ethanols (dehydration/rehydration to phosphate-buffered saline [PBS]). Although some tissues will be sufficiently thin to view directly, in other cases, it may be necessary to separate the tissues into layers using acid maceration, enzymatic digestion, or manual dissection. This may be done prior to fixation or after fixation *(6)*. The whole-mount technique has been successfully applied in a wide range of systems. For example, dissected tissues *(7,8)*, isolated organs *(9)*, or entire organisms *(10–14)* have been utilized.

Finally, careful mounting technique is essential. Variations in thickness, introduction of bubbles, or folding will severely impair both observation and photography. Additionally, the possibility of nonspecific background staining is increased because of the thickness of the specimen. In conjunction with the appropriate controls and in the context of an understanding of morphology obtained from routine histological sections, the application of immunofluorescence to whole-mounts can be a powerful means for obtaining detailed information within the three-dimensional context of a tissue or organism.

1.3. Advantages and Disadvantages to the Whole-Mount Approach

1.3.1. Advantages

In addition to the superior resolution afforded by immunofluorescence, one of the greatest advantages of using the whole-mount immunofluorescent approach is the ability to examine relatively large areas of tissue for the pattern and distribution of a small population of cells. Labeled cells relative to the entire tissue or organism are readily apparent, and if these cells happen to have very fine processes, as in the nervous system, the processes are also readily apparent (**Fig. 1A** and **B**). In tissue sections, these same cells would be widely scattered, and many negative sections would have to be examined to locate the few with positive cell bodies. A tedious three-dimensional reconstruction would be necessary to obtain detailed information regarding the distribution of the cells and individual cell morphology, whereas this information is readily apparent from the whole-mount preparation (**Fig. 1B**). Additionally, processes would appear as tiny immunofluorescent points in cross-section, whereas with the whole-mount technique, not only are networks of processes visible, but fine histological details, such as smooth fibers, beaded fibers, or vesicular swellings, are obvious *(15)* (**Fig. 1C**).

Another example of how whole-mount technology has facilitated the study of three-dimensional arrays of filamentous networks is the visualization of cytoskeletal elements within the context of tissues. It is only through whole-

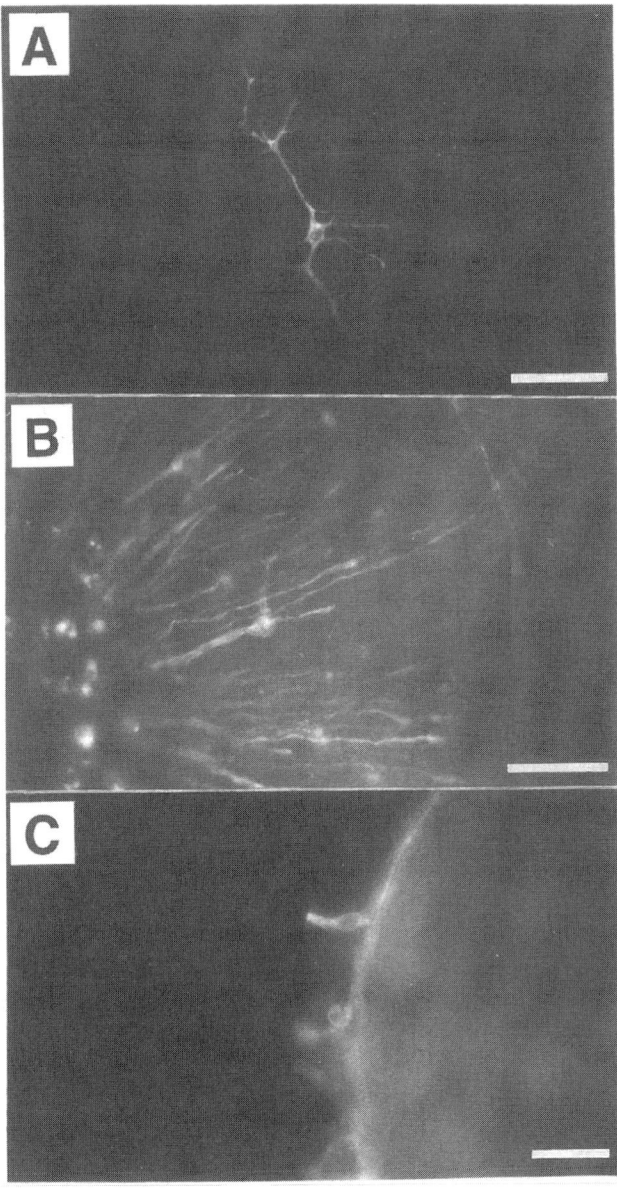

Fig. 1. *Hydra oligactis* whole mount labeled with monoclonal antibody JD1. (**A**) Isolated ganglionic neuron in the body column, scale bar = 50 µm; (**B**) hypostomal nerve net with sensory neurons of the mouth at the left and ganglionic neurons of the perihypostomal ring to the right, scale bar = 50 µm; (**C**) cell bodies of hypostomal sensory neurons extending from the mesoglea (processes) to the surface of the ecto-derm, scale bar = 25 µm.

mount analysis that the three-dimensional relationship of microfilaments, microtubules, and intermediate filaments relative to the changes in the morphology of tissues or organisms has been visualized *(16)*. In all cases, the simplicity of the whole-mount procedure compared with routine histological or cryostat sectioning and serial reconstructions facilitates the examination of large numbers of specimens. Variation between specimens and throughout a developmental time-course is readily apparent.

1.3.2. Disadvantages

It is important to be aware of the potential disadvantages associated with whole-mount preparations to avoid misinterpreting results. The effective penetration of reagents (primary and secondary antibodies) is difficult to predict. A negative result on the first labeling attempt should not be interpreted as an absence of antigen. Several other fixation methods should be tried before ruling out the presence of a particular antigen (*see* Chapter 8).

In addition, nonspecific background autofluorescence of thick tissue resulting from fixation can mask faint labeling of an antigen. Whenever the specimen is thicker than the depth of the optical field, the perceived image will include both in-focus and out-of-focus labeling and autofluorescence. The out-of-focus light will greatly reduce sharpness and resolution of the specific image. This can sometimes be overcome by fine focusing through the thickness of the preparation (**Fig. 1C**) or by doing optical sectioning using a confocal microscope (**Fig. 2**; *see also* Chapter 20). It should be noted that given the three-dimensional arrangement of labeled cells within a tissue, at high magnification, many labeled cells will appear out of focus. In some situations, the high resolution, short working distance lenses typically used for immunofluorescence will not be useful at all. In whole embryos that are 0.5 to 1 mm thick, only superficial layers can be observed. Using a lower-power objective with a longer depth of focus will alleviate some of this effect (**Fig. 3**). The alternative, again, is to do optical sectioning with the confocal microscope (**Fig. 3**; *see also* Chapter 20).

With the exception of tissues prepared as a single sheet, the presence of loose collagen or elastic fibers following the separation of layers may result in the trapping of antibodies and nonspecific labeling. These antibody-positive fibers may be mistaken for neuronal processes. The use of appropriate blocking reagents prior to antibody labeling will reduce nonspecific labeling.

It should be noted that an alternative approach that avoids some of the disadvantages of immunofluorescence is the use of enzyme-conjugated secondary antibodies (*see* Chapter 23). Although this approach sacrifices the resolution of a light-emitting source, low-power objectives compatible with thicker whole-mounts give optimal images *(13)*.

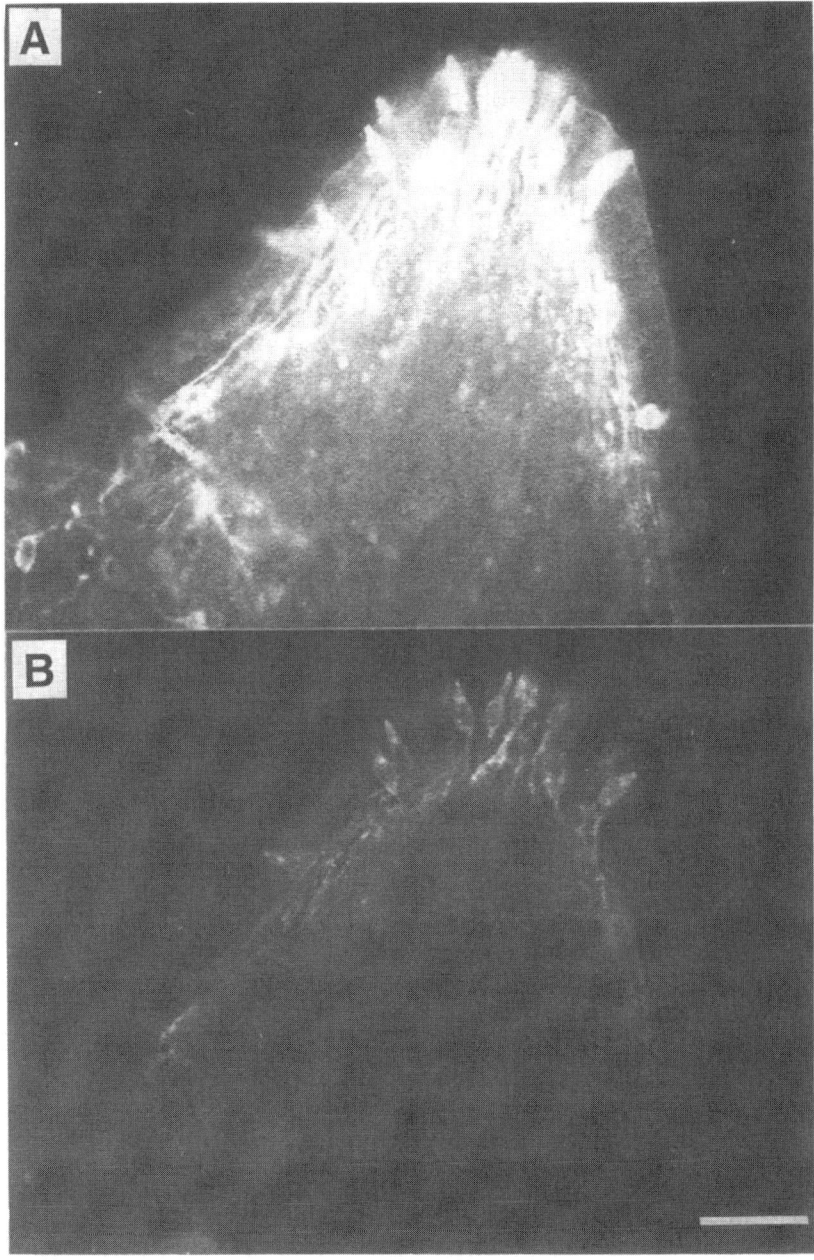

Fig. 2. Hypostome (mouth) of a *Hydra oligactis* whole mount labeled with monoclonal antibody DB5. **(A)** Nonconfocal image of sensory neurons and processes; **(B)** confocal optical section of the same field as (A) illustrating details of neuronal cell bodies and processes. Scale bar = 25 μm.

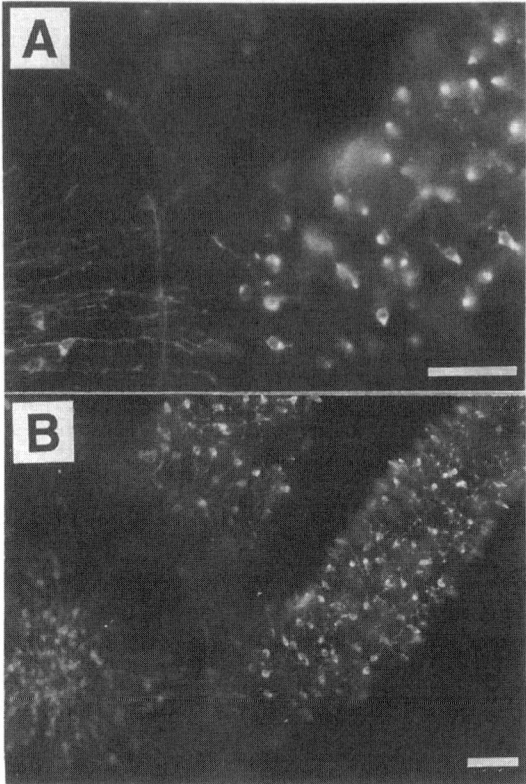

Fig. 3. Head of a *Hydra oligactis* labeled with monoclonal antibody DB5. **(A)** Neurons of hypostome and perihypostomal ring (left) and base of tentacle (right) at high magnification (× 20 objective); note considerable out-of-focus labeling of the tentacle; **(B)** same specimen as in (A) but at lower magnification (× 10 objective); although the hypostome is out of focus, the neuronal net of the tentacles is much clearer. Scale bars = 50 μm.

2. Materials

All chemicals are available from Sigma (St. Louis, MO) unless otherwise noted.

1. Lavdowsky's fixative: 95% ethanol, distilled H_2O, formalin, and acetic acid in 50:40:10:2 proportions v/v/v/v (*see* **Note 1**).
2. 70% Ethanol.
3. PBS: For a 1-L 10X stock solution of PBS, dissolve 2.56 g $NaH_2PO_4·H_2O$ and 11.94 g Na_2HPO_4 in 500 mL distilled H_2O (dH_2O) by stirring, and then adjust the pH to 7.2. Add 87.66 g NaCl, and bring to a final vol of 1 L. Store the stock solution in the refrigerator, and warm to dissolve salts before dilution. Dilute 10–fold with dH_2O for use.

4. 24–Well tissue-culture plate (*see* **Note 2**).
5. Pasteur pipets and pipet bulb.
6. Blocking solution: 10% normal goat serum in PBS with 0.1% sodium azide as a preservative (*see* **Note 3**).
7. Primary antibody, ascites fluid of monoclonal antibody CP8 diluted 1:500 in blocking solution (*10*) for this example (*see* **Note 4**).
8. Secondary antibody, FITC-conjugated goat antimouse IgM (Antibodies, Davis, CA) diluted 1:250 in blocking solution for this example.
9. 22 mm^2 and 24 × 60-mm Glass coverslips.
10. Mounting medium: 70% glycerol, 30% 0.1 M Tris buffer, pH 9.0, and 5% w/v *n*-propyl gallate (*see* **Note 5**).
11. Fluorescence microscope equipped with appropriate excitation filter, dichroic mirror, and barrier filter. For this example, an Olympus BHT compound microscope equipped with a BP490 excitation filter, BH-2DM500 dichroic mirror, and a LP515 barrier filter is used.
12. Variable-speed rotating platform capable of 100 rpm.

3. Method

1. Fix the specimen. For this example, the live coelenterate *Hydra oligactis* is first relaxed using an anesthetic solution of 2% urethane in hydra culture medium for 1 min to prevent contraction into a ball on fixation (*17*). 0.5 mL is used in a well of a 24–well tissue-culture plate.
2. Remove and discard the urethane solution, and quickly add 1 mL of Lavdowsky's fixative. Place the culture plate on a rotating platform, and gently (100 rpm) agitate for 30 min.
3. Remove and discard the fixative. Rinse the animal with 1 mL 70% ethanol. Discard the ethanol and rinse the animal with two changes of PBS (*see* **Note 6**).
4. Remove and discard the PBS. Add blocking solution, and agitate the tissue-culture plate for 30 min (*see* **Note 7**).
5. Remove and save the blocking solution. Add enough primary antibody to cover the animal and to allow it to agitate freely when submerged in the fluid (*see* **Note 8**).
6. Remove and save the primary antibody. Rinse with PBS, and then fill the well with PBS and wash with agitation for 30 min (*see* **Note 9**).
7. Remove and discard the PBS. Add enough secondary antibody to cover the animal, and incubate for 30 min with agitation.
8. Remove and save the secondary antibody. Rinse with PBS, then fill the well with PBS, and wash with agitation for 30 min (*see* **Note 10**).
9. Transfer the animal to the larger coverslip (24 × 60 mm), and pipet away the PBS. Add one drop of mounting medium. Lower the smaller coverslip (22 mm^2) onto the animal. Use forceps first to put one side of the coverslip into the mounting medium and then to lower the coverslip slowly allowing any air bubbles to be pushed to the edge. Avoid dropping the coverslip onto the animal since there will surely be many air bubbles trapped beneath it (*see* **Note 11**).

10. View the specimen on a fluorescence microscope equipped with a filter combination appropriate for the fluorochrome conjugated to the secondary antibody (*see* **Note 12**).

4. Notes

1. *See* Chapter 8 for a detailed discussion of choosing a fixative. In general, ethanol-based fixatives will allow for good penetration and labeling of intracellular antigens. An alternative is 4% paraformaldehyde (w/v in PBS heated to 60°C to solubilize the paraformaldehyde) for 30 min to overnight. After three washes with PBS, paraformaldehyde-fixed specimens can be further permeabilized by treatment with 0.1% protease K in PBS at 37°C for 1–3 min. Quickly remove the protease solution and replace it with 95% ethanol; rinse two more times with 95% ethanol. Once fixed, specimens should not be allowed to dry before mounting since this will increase nonspecific autofluorescence.

2. The size of the well and volume of solutions can be adjusted depending on the size of the specimen and the need to conserve reagents; for example, a 96–well microtiter plate with a total vol of 200 µL/well can be substituted.

3. The serum used in the blocking solution should be normal serum from the species in which the secondary antibody was raised. Store refrigerated (*see* **Note 7**). Alternatives include 3% BSA or 10% nonfat dry milk in PBS.

4. Polyclonal serum (produced in a rabbit) will contain approx 10–15 mg/mL of total IgG of which only about 1% may be directed against the antigen of interest. The effective concentration of specific antibody is therefore about 100 µg/mL. The affinity of high-avidity antibodies correlates with a binding constant in the range of 1 µg/mL. Therefore, whole serum may be diluted about 1:100 in blocking solution. Unpurified ascites fluid can have a monoclonal antibody concentration that ranges from 100 µg/mL to 20 mg/mL with 1 mg/mL being common. Thus, ascites fluid can be diluted 1:1000 and still have a 1 µg/mL specific antibody concentration. In general, it is best to keep specific antibody concentrations in the range of 10–50 µg/mL to assure saturation binding. Store refrigerated (*see* **Note 8**).

5. Glycerol enhances the fluorescence of most fluorochromes. Such compounds as *n*-propyl gallate, 1,4–diazobicyclo-2,2,2–octane, or *p*-phenylenediamine reduce the photo-bleaching that occurs during exposure of the fluorochrome to excitation light. The *n*-propyl gallate will go into solution over several hours with agitation. This may be speeded up with gentle heating.

6. The volume of PBS used in washes is not critical. It may be applied from a squirt bottle, and the well may be filled (about 2 mL). The animal will sink in the PBS solution when all of the ethanol has been removed from the tissue (1–2 min). Before each incubation period, the animal should be checked to make sure it is submerged in the solution. If it floats on the surface of the fluid, the tissue not exposed to the fluid will not be fixed or, in later steps, labeled with reagent.

7. Nonspecific background problems are the result of primary or secondary antibody binding to the tissue through interactions that do not involve the antigen combin-

ing site. Saturating amounts of nonspecific proteins, which will not be recognized by the secondary reagent, will block these nonspecific interactions (*see* **Note 3**).

8. There is no need to rinse off the blocking solution since the primary antibody is diluted in blocking solution. Blocking solution can be saved, stored refrigerated, and reused for upward of a month. Despite the presence of sodium azide as a preservative, the solution should be visually inspected for evidence of a white precipitate, which can be indicative of bacterial contamination. If present, the solution should be discarded.

9. As with the blocking solution, primary and secondary antibodies may be saved and reused for about a month. If labeling intensity decreases because of adsorption of the antibodies from the solution during this period, discard the solutions. Experience will dictate how long the solutions can realistically be used.

10. *See* Chapter 22 for counterstaining options that reduce yellow autofluorescence when preparing color photographs.

11. Depending on the whole-mount specimen, fine forceps or needles can be used to arrange the specimen so that tissue is not folded prior to applying the top coverslip. Two coverslips, rather than a slide and coverslip, are used so that both sides of the whole-mount may be examined.

12. If nonspecific labeling is a problem, reduce the incubation times in primary and secondary antibody, and/or increase the number and duration of washes.

Acknowledgment

I would like to acknowledge Dr. Jack Dunne, who introduced me to monoclonal antibody technology and immunocytochemistry on whole-mounts and with whom this protocol was originally developed.

References

1. Meissner, G. (1857) Uber die Nerven der Darmwand. *Z. Rationelle Med.* **8,** 364–366.
2. Erlich, P. (1886) Uber ide Methylenblau reaktion der lebenden Nervensubstang. *Deutsch Med. Wschr.* **12,** 49–52.
3. Falk, B. (1962) Observations on the possibilities of the cellular localization of monamines by a fluorescence method. *Acta Physiol. Scand.* **56,** 19–25.
4. Costa, M., Patel, Y., Furness, J. B., and Arimura, A. (1977) Evidence that some intrinsic neurons of the intestine contain somatostatin. *Neurosci. Lett.* **6,** 215–222.
5. Costa, M., Buffa, R., Furness, J. B., and Solcia, E. L. (1980) Immunohistochemical localization of polypeptides in peripheral antonomic nerves using whole-mount preparations. *Histochemistry* **65,** 157–165.
6. Terada, M., Iwanaga, T., Takahasi-Iwanaga, H., Adachi, I., Arakawa, M., and Fujita, T. (1992) Calcitonin gene-related peptide (CGRP)-immunoreactive nerves in the tracheal epithelium of rats: an immunohistochemical study by means of whole-mount preparations. *Arch. Histol. Cytol.* **55,** 219–233.
7. Liou, W. (1990) Whole-mount preparations of mouse lens epithelium for the fluorescent cytological study of actin. *J. Microscopy* **157,** 239–245.

8. Tam, P. K. and Boyd, G. P. (1990) Origin, course, and endings of abnormal enteric nerve fibres in Hirschspring's disease defined by whole-mount immunohistochemistry. *J. Ped. Surg.* **25**, 457–461.

9. Konig, N., Wilkie, M. B., and Lauder, J. M. (1988) Tyrosine hydroxylase and serotonin containing cells in embryonic rat rhombencephalon: a whole-mount immunocytochemical study. *J. Neurosci. Res.* **20**, 212–223.

10. Javois, L. C., Wood, R. D., and Bode, H. R. (1986) Patterning of the head in hydra as visualized by a monoclonal antibody. I. Budding and regeneration. *Dev. Biol.* **117**, 607–618.

11. Kuratani, S., Tanaka, S., Ishikawa, Y., and Zukeran, C. (1988) Early development of the hypoglossal nerve in the chick embryo as observed by whole-mount nerve staining method. *Am. J. Anat.* **182**, 155–168.

12. Plickert, G. and Kroiher, M. (1988) Proliferation kinetics and cell lineages can be studied in whole-mounts and macerates by means of BrdU/anti-BrdU technique. *Develop.* **103**, 791–794.

13. Dent, J. A., Polson, A. G., and Klymkowsky, M. W. (1989) A whole-mount immunocytochemical analysis of the expression of the intermediate filament protein vimentin in Xenopus. *Develop.* **105**, 61–74.

14. Gustafsson, M. K. (1991) Skin the tapeworms before you stain their nervous system: A new method for whole-mount immunocytochemistry. *Parasitology Res.* **77**, 509–516.

15. Tamaki, M., Iwanaga, T., Takeda, M., Adachi, I., Sato, S., and Fujita, T. (1992) Calcitonin gene-related peptide (CGRP)-immunoreactive nerve terminals in the whole-mount preparations of the dog urethra. *Arch. Histol. Cytol.* **55**, 1–11.

16. Vielkind, U. and Swierenga, S. H. (1989) A simple fixation procedure for immunofluorescent detection of different cytoskeletal components within the same cell. *Histochem.* **91**, 81–88.

17. Javois, L. C. and Tombe, V. K. (1990) Head activator does not qualitatively alter head morphology in regenerates of *Hydra oligactis*. *Roux's Arch. Dev. Biol.* **199**, 402–408.

19

TUNEL Assay for Apoptotic Cells

Virginia M. Heatwole

1. Introduction

Cell death may occur via two different mechanisms: apoptosis or necrosis. Necrosis is caused by the cell's inability to maintain homeostasis and is characterized by loss of plasma membrane integrity, cell swelling and lysis, random degradation of DNA, and lack of an energy requirement or macromolecular synthesis. Necrotic cell death may be observed in localized areas of a tissue, probably as the result of physical damage. An inflammatory response is usually elicited. In contrast, apoptosis is genetically driven and is characterized by chromatin condensation, nuclear shrinkage and eventual loss of nuclear membrane, membrane blebbing that produces apoptotic bodies containing cellular organelles and chromatin, DNA degradation into distinct nucleosomal units, an energy and protein synthesis requirement, and a lack of an inflammatory response. Whereas it is common to observe localized clusters of cells in a tissue dying via necrosis (for example, as a result of tissue injury), usually only individual or isolated cells are engaged in the apoptotic death pathway (for example, as a genetically programmed step in normal development of a tissue) *(1,2)*.

In the earliest events of apoptosis, before any detectable changes in the nucleus, phosphatidyl serine (PS) is translocated from the inner to the outer leaflet of the cytoplasmic membrane. Translocation of PS to the outer leaflet is detectable by Annexin-V, a protein with a strong binding affinity to PS. In the presence of calcium, Annexin-V-FITC binding to the PS on the outer leaflet of the cytoplasmic membrane can be observed using fluorescence microscopy or quantitated using flow cytometry. The Annexin-V binding assay does not require fixation of the cells and can be performed on living cells. DNA dyes

From: *Methods in Molecular Biology, Vol. 115: Immunocytochemical Methods and Protocols*
Edited by: L. C. Javois © Humana Press Inc., Totowa, NJ

such as propidium iodide are used to eliminate false positives in this assay. When the cytoplasmic membrane integrity is lost, DNA dyes are able to pass to the nucleus and stain DNA, and Annexin-V is able to pass through the membrane and bind to PS on the inner leaflet, where it normally resides (3–5).

The controlled degradation of nuclear DNA into nucleosomal units is a hallmark of apoptosis (although not all cells undergoing apoptosis exhibit this activity). Degradation of the genomic DNA occurs in two steps. First, the chromosomes are cleaved into large fragments (50–300 kb). Then, as DNA degradation proceeds, endogenous Ca^{+2}- and Mg^{+2}-dependent endonucleases cleave the chromatin at the linker DNA sites between the nucleosomes, resulting in DNA fragments of approx 180- to 200-bp multimers (the amount of DNA in one nucleosome). DNA degraded into nucleosomal units appears as a "ladder" of approx 180-bp increments when visualized by agarose gel electrophoresis. Visualization of the DNA "ladder" is a useful method to identify cell death via apoptosis, but only when a large number of cells in a sample are engaged in the apoptotic death pathway. Agarose gel analysis is not a practical technique to detect DNA degradation when only a few cells in a sample or tissue are apoptotic, and thus a different method of identifying these apoptotic cells is required (6).

The TUNEL assay (TdT-mediated dUTP Nick End Labeling) (6) was developed as a method to identify individual cells that are undergoing apoptosis by labeling the ends of the degrading DNA with the polymerase terminal deoxynucleotidyl transferase (TdT). TdT catalyzes the template-independent addition of deoxynucleotide triphosphates to the 3'-OH ends of DNA (**Fig. 1**). When labeled nucleotides (biotin-, fluorescein-, or DIG-labeled) are incorporated by TdT, nuclei with degrading DNA can be easily detected by standard immunohistochemical or immunofluorescent techniques (7). FITC-labeled nucleotides can be detected directly by immunofluorescence, biotin-labeled nucleotides can be detected using avidin conjugated to a reporter (FITC, peroxidase, or alkaline phosphatase) and DIG-labeled nucleotides can be detected by anti-DIG antibody conjugated to a reporter molecule (8,9). In in situ end labeling (ISEL), an assay similar to TUNEL, Klenow is used to label 5' overhangs on the degraded DNA in a template-dependent reaction using labeled nucleotides (10,11).

Identification of TdT-labeled degrading DNA in the nucleus of cells is not sufficient to demonstrate that the cell is engaged in the apoptosis pathway, as chromosomal DNA degradation also occurs in necrotic cells. Further examination of morphological characteristics and location of cells that are labeled by the TUNEL assay is required. Necrotic nuclei are generally clustered and swollen and the labeling that identifies degraded DNA appears diffuse. In contrast, apoptotic cells generally appear isolated (**Fig. 2**), and the nuclei are small and have condensed chromatin. Eventually the nuclear membrane disappears, and

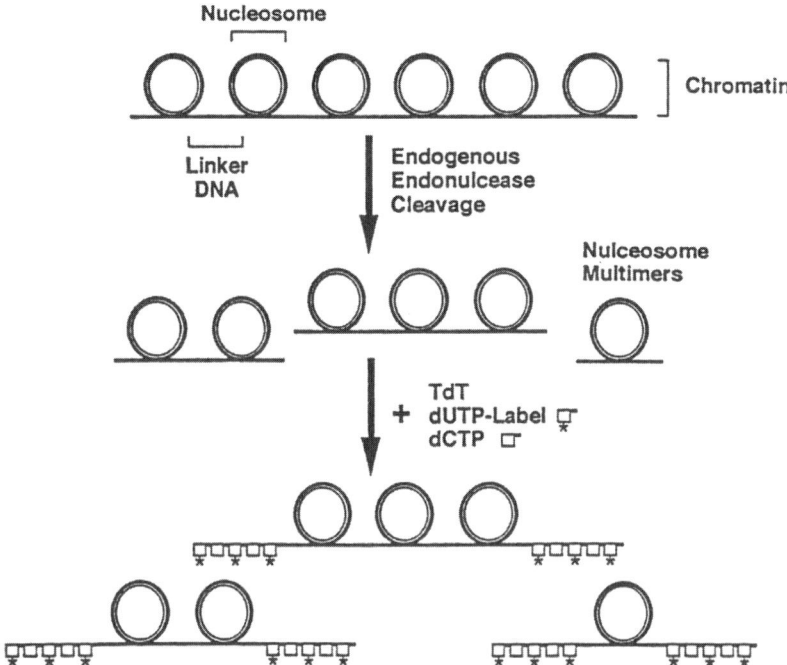

Fig. 1. Labeling of degraded chromatin by the TUNEL assay. During apoptosis endogenous, endonucleases cleave chromatin in the linker region between nucleosomes. The resulting nucleosome multimers are labeled by TdT and a dUTP analog with a detectable label (biotin, DIG, or FITC) shown as *. The additional nucleotide in the reaction (here shown as dCTP) may be any dNTP and serves to extend the labeling reaction by preventing steric hindrance by two adjacent labeled dUTPs. (Abbreviations are as in text.)

membrane blebbing produces apoptotic bodies containing cellular organelles and chromatin. In apoptosis, TdT-labeled degrading DNA can be observed on the condensed chromatin in the small apoptotic nuclei and in the released apoptotic bodies (which may be found in the cytoplasm of nearby phagocytic cells).

2. Materials

1. Coplin jars.
2. Microscope slides: poly-L-lysine-coated or silanized.
3. Equilibration buffer: 200 mM potassium cacodylate, pH 6.6, 25 mM Tris-HCl, pH 6.6, 0.2 mM DTT, 0.20 mg/mL BSA, 2.5 mM cobalt chloride (Promega, Madison, WI) (*see* **Note 2**).
4. Nucleotide mix: 50 µM fluorescein-12-dUTP, 100 µM dATP, 10 mM Tris-HCl pH 7.6 (Promega) (*see* **Note 3**).
5. TdT (Promega) (15 U/µL).

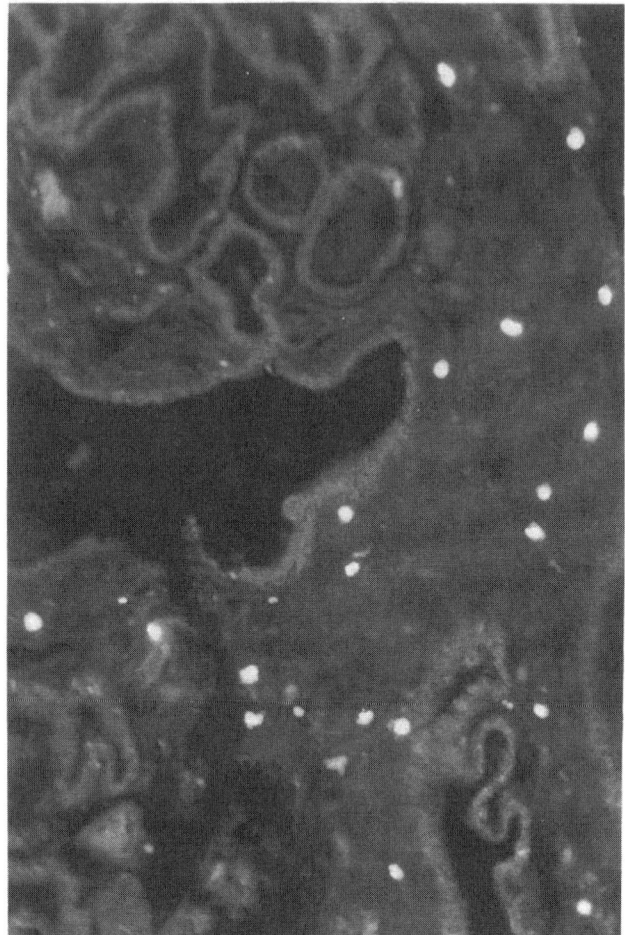

Fig. 2. Apoptotic cells fluorescently labeled using the TUNEL assay. Fixed paraffin-embedded rat mammary glands, 4 d postweaning (obtained from Oncor), were labeled using the TUNEL assay to detect apoptotic cells. Reagents were from Oncor's Apoptag Direct In Situ Apoptosis Detection Kit (Fluorescein), which directly incorporates fluorescein-labeled dUTP in the TdT end-labeling reaction. All reagents were used as described by the manufacturer (*see* **Note 1**). (Abbreviations are as in text.)

6. DNaseI: Sigma no. D5793 (Sigma, St. Louis, MO).
7. DNaseI buffer: 40 mM Tris-HCl, pH 7.8, 10 mM NaCl, 6 mM MgCl$_2$, 10 mM CaCl$_2$.
8. Humid chambers.
9. Fluorescent microscope with standard fluorescein filter set.
10. Xylene (Sigma no. X2377).
11. Ethanol: 100, 95, 85, 70, 50%.

12. 1X PBS: 1.37 mM NaCl, 2.68 mM KCl, 1.47 mM KH$_2$PO$_4$, 8.1 mM Na$_2$HPO$_4$.
13. 2X SSC: 0.3 M NaCl, 0.03 M Na citrate, adjust pH to 7.0 with HCl.
14. Proteinase K (Sigma no. P2308).
15. 4% paraformaldehyde solution: Mix 35 mL PBS, 12.5 mL of 16% methanol-free formaldehyde, pH to 7.4 with 1 N NaOH. Make up fresh before each use.
16. 1% paraformaldehyde solution: Mix 45 mL PBS, 3.125 mL of 16% methanol-free formaldehyde, pH to 7.4 with 1 N NaOH. Make up fresh before each use.
17. Antifade mounting solution: Molecular Probes no. S7461 (Eugene, OR). (Alternatively, 10% glycerol in PBS can be used as a mounting medium, but it has no ability to prevent quenching of fluorescent signal.)
18. 50 mM EDTA, pH 8.0.
19. 0.2% Triton X-100 in PBS.
20. Propidium iodide (5 μg/mL in PBS) (Sigma no. P4170).
21. RNase A (50 μg/mL in PBS) (Sigma no. R6513).
22. Coverslips.

3. Methods

3.1. Preparation and Permeabilization of Paraffin-Embedded Tissue

1. Deparaffinize tissue by immersing twice in xylene for 5 min.
2. Rehydrate tissue by the following washes (in the order given): two washes for 5 min each in 100% ethanol, then one wash for 3 min each successively in 95, 70, and 50% ethanol.
3. Wash the sample in PBS for 5 min.
4. Drain excess PBS from tissue and incubate for approx 15 min in 20 mg/mL Proteinase K (in PBS) solution (*see* **Note 4**).
5. Terminate the protease treatment by washing cells four times for 2 min each in PBS with gentle agitation (*see* **Note 5**).
6. Proceed with TUNEL Assay for Adherent Cells (**Subheading 3.5.2.**).

3.2. Preparing Cell Suspensions or Cells Grown on Coverslips

3.2.1. Cell Suspensions

1. Wash cells in PBS by centrifugation (300g) and resuspend in PBS so that the final concentration of cells is 1×10^7 cells/mL.
2. Pipet 100 μL of the cell suspension onto treated slides, spread using the edge of a clean slide, and allow to dry. Alternatively, cells can be attached to the slide by cytospin preparation. Continue with **step 2** immediately below (**Subheading 3.2.2.**).

3.2.2. Cells Grown on Coverslips

1. For cells grown on a coverslip, wash cells twice for 5 min each in PBS.
2. Fix cells by immersing in 4% formaldehyde solution (in PBS, pH 7.4) for 20 min at 4°C.

3. Wash by immersing in PBS twice for 5 min each.
4. Permeabilize cells by washing in 0.2% Triton-X 100 for 5 min at 4°C.
5. Wash cells twice for 5 min each in PBS. Proceed with TUNEL Assay for Adherent Cells (**Subheading 3.5.2.**).

3.3. Preparation of Cell Suspensions for Analysis by Flow Cytometry

1. Centrifuge cells ($300g$) to pellet and resuspend 5×10^6 cells in 0.5 mL of PBS.
2. Add 5 mL of 1% paraformaldehyde (in PBS, pH 7.4) and incubate on ice for 15–20 min.
3. Centrifuge the cells ($300g$) to pellet. Remove the supernatant and resuspend in 5 mL PBS. Repeat this wash two more times. After the final wash, resuspend the cells in 70% cold ethanol to permeabilize and keep at –20°C (*see* **Note 6**). Alternatively, resuspend cells in 0.2% Triton-X 100 for 5 min at 4°C to permeabilize.
4. Proceed with TUNEL assay for Cell Suspensions for Flow Cytometry (**Subheading 3.5.1.**)

3.4. Preparation of Control Slides

Treatment of fixed cells with DNaseI will provide positive controls for the TUNEL assay. DNaseI will digest chromosomal DNA and provide 3'-OH ends for the TdT enzyme to label in the TUNEL reaction. Labeling of these control cells will appear diffuse throughout the nucleus.

1. Prepare slides as in **Subheading 3.2.** up to **step 5**.
2. Equilibrate cells in DNaseI buffer for 10 min.
3. Remove buffer and add enough fresh buffer containing DNaseI (0.5 µg/mL) to completely cover sample. Cover with a coverslip and incubate 10 min at room temperature.
4. Remove coverslip and wash cells in five changes of PBS for 2 min each with gentle agitation. It is important to remove all the DNaseI from the sample.
5. Proceed with the TUNEL Assay for Adherent Cells (**Subheading 3.5.2.**). Process all positive control slides in separate coplin jars to avoid introducing DNaseI contamination into experimental samples.

3.5. TUNEL Assay

3.5.1. Cell Suspensions for Flow Cytometry

1. Centrifuge 1×10^6 cells, remove supernatant, and resuspend cells in 1 mL PBS. Repeat this step three more times.
2. After last centrifugation, resuspend cells in 100 µL of equilibration buffer. Incubate at room temperature for 10 min.
3. Prepare TdT labeling reaction buffer by mixing 90 µL of equilibration buffer, 10 µL of nucleotide mix, and 2 µL of TdT enzyme (15 U/µL).
4. Centrifuge cells, remove buffer, and resuspend in 50 µL of TdT labeling reaction buffer. Incubate at 37°C for 60 min. Periodically mix cells gently.

5. Terminate the reaction by adding 1 mL of 50 m*M* EDTA.
6. Centrifuge cells, remove buffer, and resuspend in 1 mL PBS. Repeat wash.
7. Centrifuge cells, remove PBS, and resuspend in 1 mL 5 µg/mL propidium iodide and 50 µg/mL RNase A in PBS. Incubate 10–30 min.
8. Analyze cells by flow cytometry.

3.5.2. Adherent Cells

1. Equilibrate cells/tissues in enough equilibration buffer to completely cover sample. Place plastic coverslip to spread buffer completely over sample and to prevent evaporation. Incubate at room temperature for 10 min.
2. Prepare TdT labeling reaction buffer by mixing 90 µL of equilibration buffer, 10 µL of nucleotide mix and 2 µL of TdT enzyme (15 U/µL).
3. Remove coverslip and blot excess buffer from sample, taking care not to directly touch sample. Add enough TdT labeling reaction buffer to cover sample. Cover with plastic coverslip and incubate in humid chamber for at least 60 min at 37°C.
4. Stop labeling reaction by removing coverslip and washing sample in 2X SSC for 10 min at room temperature.
5. Wash samples four times for 5 min each in PBS.
6. Counterstain sample by incubating in 1 µg/mL propidium iodide in PBS for 10 min (*see* **Note 7**).
7. Wash sample four times for 5 min each in PBS.
8. Add an aqueous mounting medium or an antifade solution, mount a coverslip and analyze using fluorescent microscopy with a fluorescein filter.

4. Notes

1. Tissue sections were deparaffinized and permeabilized as described in **Subheading 3.1.** and then equilibrated in Oncor's reaction buffer (Oncor, Inc., Gaithersburg, MD) for 5 min at room temperature in a humid chamber with a plastic cover slip. The reaction buffer was removed and reaction buffer with TdT was applied. The tissue was covered with a plastic cover slip and incubated in a humid chamber for 1 h at 37°C. The labeling reaction was terminated by removing the coverslip and incubating in Oncor's stop/wash buffer for 10 min at room temperature with periodic agitation. Tissue was counterstained with 1 µg/mL propidium iodide in antifade (antifade supplied in the Oncor Apoptag Direct Detection kit), a coverslip was applied, and the sample was viewed on a fluorescence microscope using an FITC filter set.
2. Equilibration buffer contains potassium cacodylate (dimethylarsinic acid). Avoid contact with skin and eyes. Harmful if swallowed. Wear appropriate protective clothing.
3. Other nucleotides, such as dCTP, can be substituted for dATP at equal concentrations. The addition of unlabeled dNTPs to the reaction allows for a longer tail to be added by TdT without the problem of steric hindrance caused by the modification on the nucleotide (biotin, DIG, or FITC). Fluorescein-labeled nucleotides should be protected from light at all times to avoid loss of signal.

4. The time of protease digestion will have to be optimized for specific tissue types and thicknesses. Overdigestion by protease will result in loss of cellular structure and possible release of tissue section from slide. Underdigestion will result in poor TdT labeling. Make protease solutions fresh just before use.

5. Some tissues require a post-protease digestion fixation by treatment with 4% formaldehyde solution for 5 min. This is followed by two washes for 5 min each in PBS. This step may not be necessary for many tissue samples.

6. Cells can be stored at –20°C in 70% ethanol for up to 1 mo.

7. Concentration of counterstain may have to be adjusted depending on the tissue being stained. Overstaining by propidium iodide may result in difficulty in observing the fluorescein label.

Acknowledgment

I would like to thank Dr. Robin Levis for assistance with the fluorescence microscopy and photography.

References

1. Stewart, B. (1994) Mechanism of apoptosis: integration of genetic, biochemical and cellular indicators. *J. Natl. Cancer Inst.* **86,** 1286–1294.
2. Bortner, C., Oldenburg, N., and Cidlowski, J. (1995) The role of DNA fragmentation in apoptosis. *Trends Cell Biol.* **5,** 21–26.
3. Xhang, G. and Gurlu, V. (1997) Detection of early-stage apoptosis in three adherent mammalian cell lines using the ApoAlert Annexin V Apoptosis Kit. *Clonetechniques* **XII,** 24,25.
4. ApoAlert Annexin V Apoptosis Kit (1996) *Clonetechniques* **XI,** 9–11.
5. Martin, S., Reutelingsperger, C., McGahom, A., Radner, J., van Schie, R., Laface, D., and Green, D. (1995) Early redistribution of plasma membrane phosphatidylserine is a general feature of apoptosis regardless of the initiating stimulus: inhibition by overexpression of Bcl-2 and Abl. *J. Exp. Med.* **182,** 1545–1556.
6. Gavrieli, Y., Sherman, Y., and Ben-Sasson, S. A. (1992) Identification of programmed cell death *in situ* via specific labeling of nuclear DNA fragmentation. *J. Cell Biol.* **119,** 493–501.
7. Gorczyca, W., Gong, J., and Darzynkiewicz, Z. (1993) Detection of DNA strand breaks in individual apoptotic cells by the *in situ* terminal deoxynucleotidyl transferase and nick translation assays. *Cancer Res.* **53,** 1945–1951.
8. Zhang, Z. and Galileo, D. (1997) Direct in situ end-labeling for detection of apoptotic cells in tissue sections. *Biotechniques* **22,** 834–836.
9. Mitra, G. (1996) Detection of apoptotic cells. *Promega Notes Mag.* **57,** 10–15.
10. Lovelace, C. I. P., Zhang, J., Vanek, P. G., and Collier, B. (1996) Detecting apoptotic cells *in situ. Biomed. Prod.* **21,** 76,77.
11. Wijsman, J. H., Jonker, R. R., Keijzer, R., van de Cees, C. J. H., Cornelisse, J., and van Dierendonck, J. H. (1993) A new method to detect apoptosis in paraffin sections: *in situ* end-labeling of fragmented DNA. *J. Histochem. Cytochem.* **41,** 7–12.

20

Overview of Fluorescence Analysis with the Confocal Microscope

Liana Harvath

1. Principles of Confocal Microscopy

When fluorescently labeled biological specimens are viewed with a conventional wide-field microscope, a haze of out-of-focus fluorescence is usually created by the overlapping structures within the sample. As we focus through the specimen, our brains have a remarkable ability to discern substantial structural detail. However, the resolution of the images we record on film is degraded by the out-of-focus fluorescence. The confocal microscope can reject out-of-focus information and enhance the contrast of an image because the illumination and the detection are confined to an identical (small) region of the specimen. An overview of the basic principles of a confocal microscope is presented in **Fig. 1** and outlined below.

The conventional wide-field fluorescence microscope (**Fig. 1**) consists of: an excitation light source (V), an excitation filter (E), a dichromatic mirror (DM), an emission barrier filter (B), an objective lens (⌒), and a detector (D). The excitation light is reflected by the dichromatic mirror and focused by the objective lens onto the specimen. The emitted fluorescence is collected by the objective lens, and because of its longer wavelength, transmitted to the detector by the dichromatic mirror and barrier filter. Since the entire field of view is illuminated by the existing light and visualized by the detector, wide-field microscopy is subject to interferences resulting from stray light produced at all points in the illuminated area. The excitation light also passes through all planes above and below the focal plane, producing fluorescence.

The stray light arising from the simultaneous excitation of multiple regions of the specimen may be substantially reduced by restricting the illumination to a point. This is accomplished by introducing a pinhole (P) into the excitation

From: From: *Methods in Molecular Biology, Vol. 115: Immunocytochemical Methods and Protocols*
Edited by: L. C. Javois © Humana Press Inc., Totowa, NJ

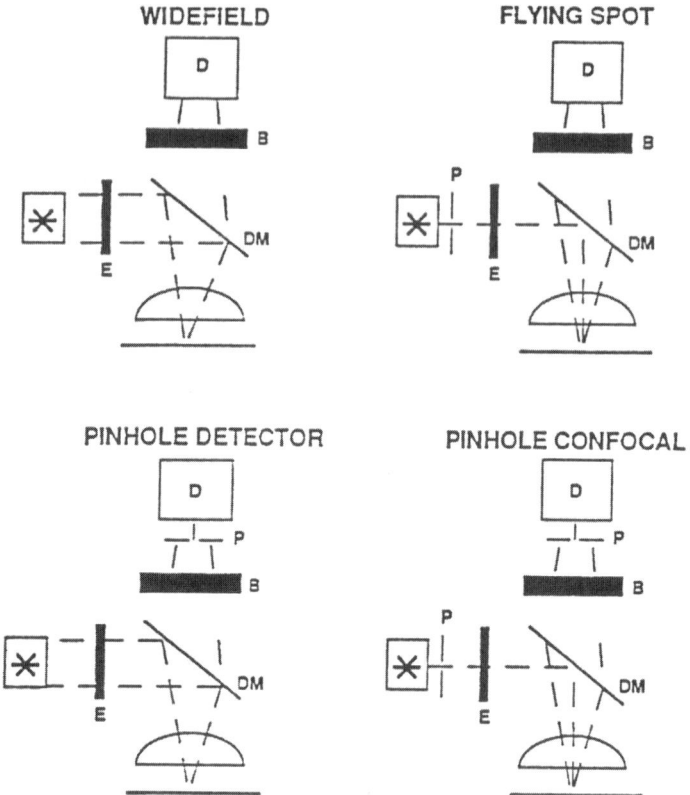

Fig. 1. Comparisons of the wide-field, flying spot, pinhole detector, and pinhole confocal microscopes. Components include: an excitation light source (V), an excitation filter (E), a dichromatic mirror (DM), an emission barrier filter (B), an objective lens (⌒), a detector (D), and a pinhole (P).

path at the appropriate conjugate plane. Moving this pinhole by mechanical means results in "flying spot" (**Fig. 1**) scanning of the specimen *(1)*. At any point in time, only one point is illuminated, and stray light can only arise from out-of-focus planes along the axis of illumination. If the spot is scanned at sufficient speed, the image produced may not appear different from a wide-field image, except for a reduction in stray light arising from the specimen.

The introduction of a pinhole into an image plane of the light path to the detector of a wide-field microscope also leads to a reduction in stray light (**Fig. 1**, pinhole detector). This occurs because light rays arising in regions outside of that defined by the pinhole are unlikely to strike the detector. Recommended practice in microscope photometry *(2)* is a combination of a narrowly delineated area of illumination, defined by a pinhole or a sharply restricted field diaphragm, and a pinhole diaphragm in front of the detector. When the illumination and/or detection is limited to a fixed point, a small field of view is obtained. The combination of point illumination and detection has been extended to allow simultaneous, synchronized movement of both the illumination and detection points (**Fig. 1**, pinhole confocal). When both the illumination and detection sites are identical in space, the microscope optics are confocal.

The primary advantage of confocal microscopy is the improvement in depth discrimination (z-axis resolution). This is accomplished because the illumination pinhole restricts light to a small area of the specimen, thereby reducing light scatter from other areas outside the illumination point, and the detector pinhole permits only a small fraction of the fluorescent light emitted by the sample to be collected by the detector (**Fig. 2**). Light emitted from the specimen that is not in the plane of focus and on the optical axis will not pass through the pinhole in front of the detector and will not be imaged. Confocal microscopes can collect a series of images from various focal planes of the specimen, a process referred to as optical sectioning. Detailed reviews of confocal microscopy instrumentation, theory, and practice are found in several recent texts (*3–5*), review articles (*6–9*), and original publications (*10–15*).

2. Overview of Instrumentation

Most fluorescence confocal microscopes consist of a conventional wide-field fluorescence microscope and a scanning device that contains a laser light source. The confocal microscope has a mechanical requirement for rapid and precise scanning of both the illumination and detection points. Two basic approaches to mechanical scanning have been developed: specimen or stage scanning, and beam scanning. In specimen or stage scanning microscopes, the beam and optics remain stationary while the stage moves the specimen relative to the illumination and detection beams. These instruments are usually not convenient for most biological applications because of their relatively slow scan rate. Beam scanning confocal microscopes are generally used for biological applications. In beam scanning microscopes, the specimen and stage remain stationary while the beam is scanned across the specimen. The beam may be a point or a parallel (slit) source.

Point source beam scanning microscopes provide optimal optical sectioning and lateral resolution, because the source and detector apertures are both restricted to diffraction-limited spots. However, single spot illumination with a laser can reach power densities of hundreds of mW/cm^2 which may saturate and photobleach fluorophores and damage living specimens. A variety of instruments have been developed to speed up the scanning rate of point source beam scanning microscopes, including mirror galvanometer scanning of both the illumination and detection beams, acousto-optic beam displacement, and resonant scanning mirror galvanometers to achieve video rate imaging.

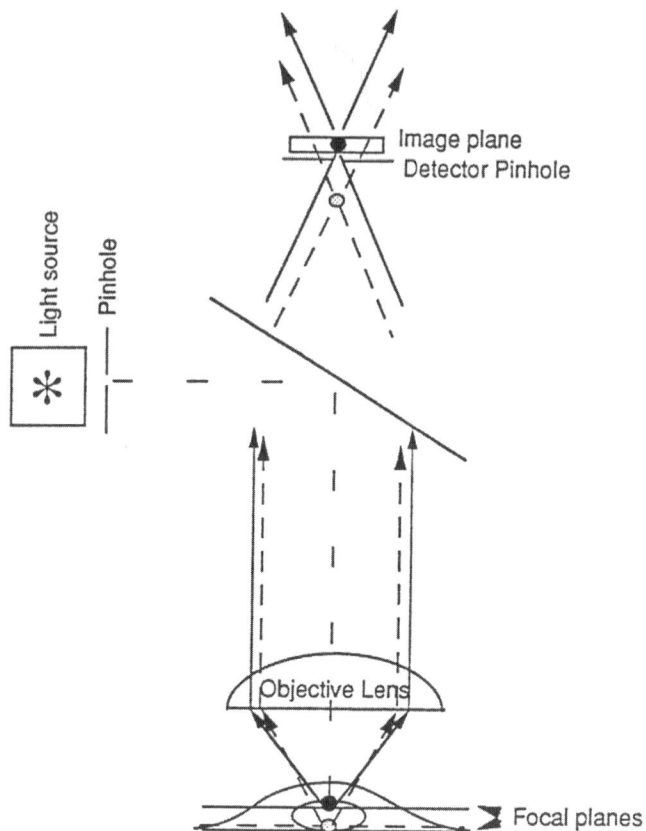

Fig. 2. Depth discrimination (*z*-axis resolution) properties of a confocal microscope. The illumination and detection images in a confocal microscope are diffraction-limited and confined to a small region of the specimen *(1)*. Only light emitted in the plane of focus and on the optical axis will pass the detector pinhole and form an image. Light emitted from other areas of the specimen does not enter the detector pinhole.

Slit scanning microscopes provide another approach to improving scan rate and reducing the illumination intensity. In slit scanning instruments, the illumination and/or detection beams are slit rather than point sources. The major advantages of slit scanning microscopes are the increased scanning speed, and illumination and detection efficiency. However, confocality is sacrificed because the illumination and detection apertures are not diffraction-limited spots. Slit scanning instruments are particularly useful for analysis of living specimens because the illumination intensity is substantially reduced.

When investigators are selecting a confocal microscope for their laboratory, it is very important to gather a series of their "typical" (routinely studied) specimens and compare the image quality from several different commercially available instruments. Investigators should determine whether rapid scanning of living specimens or optimal resolution of fixed specimens will be the most important requirement. Since confocal microscopes are expensive, most manufacturers will arrange hands-on demonstrations of their instruments.

3. Overview of Applications

Confocal microscopy is particularly useful for dissecting complex three-dimensional structural detail. Comparisons of thick-specimen wide-field fluorescence and confocal images dramatically illustrate the improvement in resolution when out-of-focus information is eliminated *(6,16–18)*. For example, the filamentous actin (F-actin) network in a hydra tentacle (**Fig. 3**) is obscured by a blur of out-of-focus fluorescence when viewed in wide-field fluorescence images (**Fig. 3A**); however, the F-actin architectures of the mesoglea level (**Fig. 3B**) and apical nematocyte cnidocil apparatus (**Fig. 3C**) are readily distinguishable in confocal images. Components of a variety of thick specimens, including intact chick embryos *(6)*, intact pancreatic islets of Langerhans *(17)*, and 400-µm thick coronal slices of developing cerebral cortex *(18)* have been fluorescently labeled and optically resolved with confocal microscopes.

The structural relationships of intracellular elements may be determined by optically sectioning a specimen with a confocal microscope. The diagram in **Fig. 4** and the corresponding confocal sections of a neutrophil migrating through a pore in a membrane toward chemoattractant (**Fig. 5**) illustrate the concept of optical sectioning. The directed migration of leukocytes in response to chemoattractants can be evaluated in vitro with chemotaxis chambers (*see* protocol described in Chapter 38, **Subheading 3.3.**, for details). Leukocytes are placed in the upper well of the chamber and separated from the lower well (containing chemoattractant) by a 10-µm-thick polycarbonate membrane that contains 5-µm pores. A chemoattractant gradient is created across the membrane when chemoattractant molecules diffuse through the membrane pores. Chemotactically responsive leukocytes migrate toward an increasing chemoattractant concentration gradient and crawl through the membrane pores to the polycarbonate membrane lower surface. When the migration assay is completed, the chamber is disassembled, and the membrane can be fixed in 3.7% formaldehyde. F-actin can be visualized by staining the sample with rhodamine phalloidin (as described in Chapter 37, **Subheading 3.2.1.**). The membrane is mounted, secured with mounting media and a coverslip, and subsequently analyzed. The images in **Fig. 5** were obtained by optically sectioning the top (Panel A), middle (Panel B), and bottom (Panel C) surfaces of the filter. Each image illustrates the F-actin distribution in a 0.5-µm-thick portion of a single cell migrating through a polycarbonate membrane pore in response to the chemoattractant, *N*-formylmethionyl-leucyl-phenylalanine.

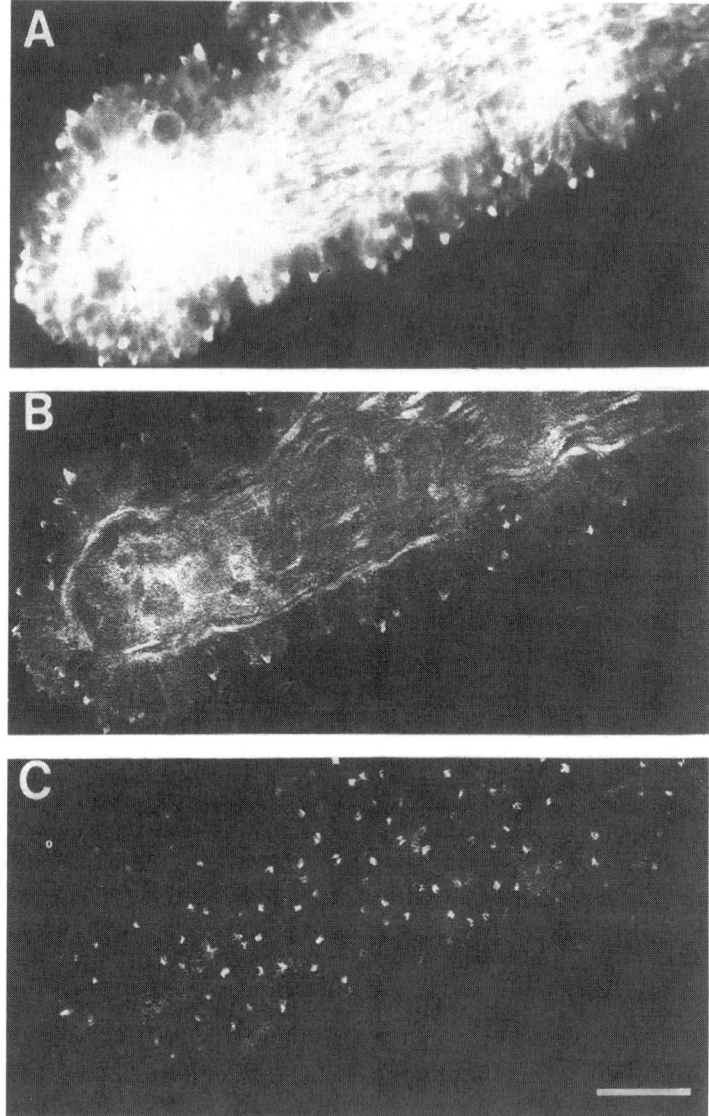

Fig. 3. Comparisons of wide-field (**A**) and confocal fluorescence images (**B**, mesoglea level; **C**, apical) of rhodamine phalloidin-stained F-actin in a whole-mount hydra tentacle. The hydra was fixed and stained as described in Chapter 18. The bar represents 25 μm. All images were collected with a Nikon (New York) Microphot FX microscope (×40 objective lens). Confocal images were collected with the microscope connected to a Bio-Rad (Hercules, CA) MRC600 laser-scanning confocal system.

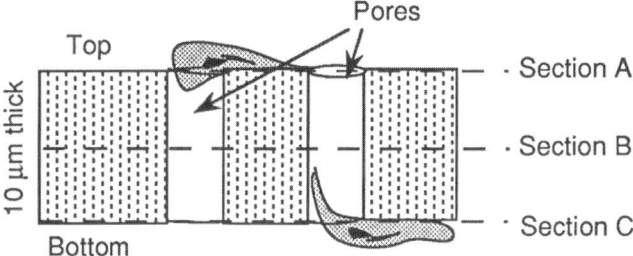

Fig. 4. Diagram of neutrophils migrating through the pores of a polycarbonate membrane. Sections A, B, and C correspond to the areas where optical sections were imaged and are presented in **Fig. 5**A, B, and C, respectively.

Confocal microscopy has been successfully utilized in experimental pathology *(9)*, neuropathology *(7)*, and studies of nuclear structure *(6,12,19)*. DNA of individual chromosomes has been visualized with excellent resolution in fluorescent in situ hybridization specimens *(19)*. Many of the published confocal microscopy studies have been performed with fixed specimens. However, two recent studies have successfully utilized confocal microscopy to monitor living specimens containing fluorescent probes. In one study, the migratory pathways of fluorescently labeled young neurons were mapped in living slices of developing cerebral cortex using time-lapse confocal microscopy *(18)*. In another study, the nuclear localization of oligonucleotides for antisense gene inhibition was monitored in living kidney and fibroblast cell lines *(20)*. Confocal microscopy technology is rapidly expanding to address a diverse array of questions, because living as well as fixed specimens may be analyzed and samples ranging in thickness from whole embryos to cellular organelles are readily studied.

4. Considerations for Sample Preparation and Analysis

When preparing samples for fluorescence confocal analysis, it is important to optimize the signal intensity without destroying important structural and/or functional features of the specimens. The signal intensity can be optimized by selecting objective lenses that have large numerical apertures, and choosing fluorochromes that are photostable and quantum-efficient. Dyes, such as Texas red and rhodamine, are relatively photostable and are frequently selected for labeling living as well as fixed specimens. Antiphotobleach agents, such as *n*-propyl galate *(21)* and 1,4-diazobicyclo-(2,2,2)-octane *(22)*, may be added to fixed-specimen mounting media individually or in combination with *para*-phenylenediamine *(6)* (*see* Chapter 14).

The fixation media and conditions should be optimized for each type of specimen to maintain the normal structural integrity of the specimen, and mini-

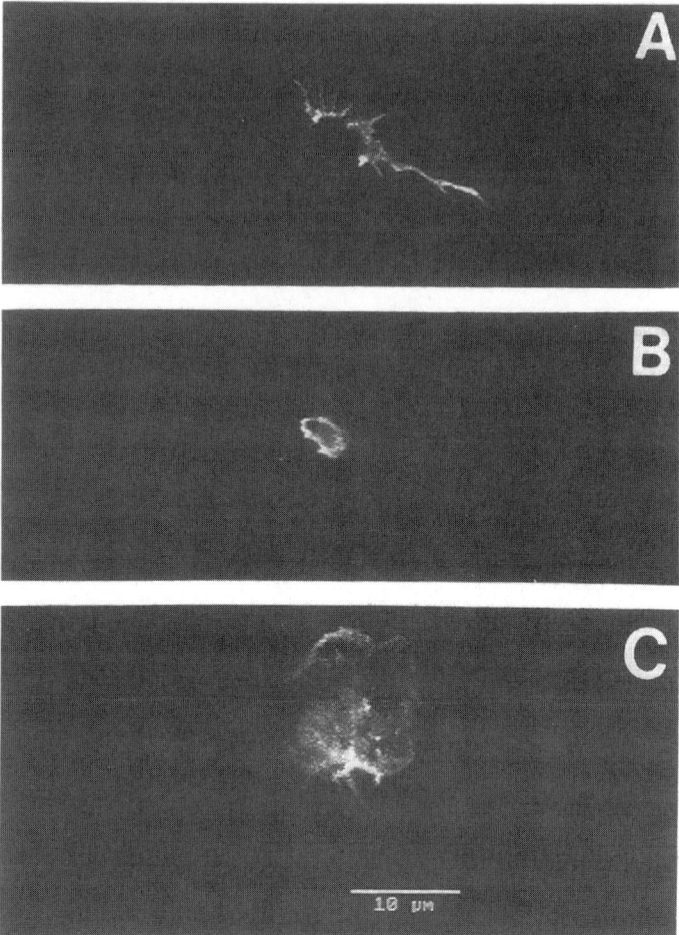

Fig. 5. Optical sectioning of rhodamine phalloidin-stained F-actin in a neutrophil migrating through a 5-μm pore of a polycarbonate membrane. The neutrophil migration is stimulated in response to 10^{-7} M N-formylmethionyl-leucyl-phenylalanine. **(A)**, **(B)**, and **(C)** correspond to 0.5-μm optical sections indicated as sections A, B, and C, respectively, in **Fig. 4**. The bar represents 10 μm. The images were collected with a Nikon Microphot FX microscope (×60 Plan-apochromat lens, numerical aperture, 1.6) connected to a Bio-Rad MRC600 laser-scanning confocal system.

mize specimen shrinkage and autofluorescence. An excellent review of the fixation factors affecting specimen structure has been written by Bacallao et al. *(23)* (*see* Chapter 8). If specimens are mounted with a coverslip, it is important to avoid touching the specimen with the coverslip, because gross distortions in cellular specimen height can occur *(23)*. In addition, coverslip flexing may occur when oil-immersion lenses are used, resulting in image distortion *(6)*. A convenient solution to these problems is to use a coverslip no larger than necessary to cover the specimen, seal the edges of the coverslip with clear nail polish, and use a low-viscosity oil with the oil-immersion objective lenses *(6,23)*. Since confocal microscopes have extremely sensitive z-axis resolution, small vertical displacements of the specimen can cause image deterioration.

The extraction of meaningful data from confocal images of living cells requires the establishment of fluorescent probe loading conditions for each specimen type. Consideration should be given to:

1. The effects of the fluorophore on normal cell physiology and structure;
2. Compartmentalization of the fluorophore as an artifact of the labeling conditions; and
3. The phototoxic effects of laser illumination of the specimen.

It is important to establish the optimal conditions for specimen labeling prior to detailed confocal analysis.

Acknowledgment

Figure 1, the text in **Subheading 1.** regarding **Fig. 1**, and portions of the text in **Subheading 2.** were generously provided by Kenneth Spring, DDS, PhD, National Heart, Lung, and Blood Institute, National Institutes of Health, Bethesda, MD.

References

1. Young, J. Z. and Roberts, F. (1951) A flying spot microscope. *Nature* **167,** 231–234.
2. Piller, H. (1977) *Microscope Photometry.* Springer-Verlag, Berlin.
3. Pawley, J. (ed.) (1989) *The Handbook of Biological Confocal Microscopy.* Integrated Microscopy Resource, Madison, WI. Republished 1990, Plenum, New York.
4. Wilson, T. (1990) *Confocal Microscopy.* Academic, London.
5. Shooton, D. (ed.) (1993) *Electronic Light Microscopy. The Principles and Practice of Video-Enhanced Contrast, Digital Intensified Fluorescence, and Confocal Scanning Light Microscopy.* Wiley-Liss, New York.
6. Shuman, H., Murray, J. M., and DiLullo, C. (1989) Confocal microscopy: an overview. *BioTechniques* **7,** 154–163.
7. Murray, J. M. (1992) Neuropathology in depth: the role of confocal microscopy. *J. Neuropathol. Exp. Neurol.* **51,** 475–487.
8. Stelzer, E. H., Wacker, I., and De Mey, J. R. (1991) Confocal fluorescence microscopy in modern cell biology. *Semin. Cell. Biol.* **2,** 145–152.
9. Smith, G. J., Bagnell, C. R., Bakewell, W. E., Black, K. A., Bouldin, T. W., Earnhardt, T. S., Hook, G. E. R., and Pryzwansky, K. B. (1991) Application of confocal scanning laser microscopy in experimental pathology. *J. Electron Microsc. Tech.* **18,** 38–49.

10. Brakenhoff, G. J., Blom, P., and Barends, P. (1979) Confocal scanning light microscopy with high aperture immersion lenses. *J. Micros.* **117,** 219–232.

11. Wijaendts van Resandt, R. W., Marsman, H. J. B., Kaplan, R., Davoust, J., Stelzer, E. H. K., and Stricker, R. (1985) Optical fluorescence microscopy in three dimensions: microtomoscopy. *J. Micros.* **138,** 29–34.

12. Brakenhoff, G. J., van der Voort, H. T. M., van Spronsen, E. A., Linnemans, W. A. M., and Nanninga, N. (1985) Three-dimensional chromatin distribution in neuroblastoma cell nuclei shown by confocal scanning laser microscopy. *Nature* **317,** 748–749.

13. Carlsson, K., Danielsson, P. E., Lenz, R., Liljeborg, A., Majlof, L., and Aslund, N. (1985) Three-dimensional microscopy using a confocal laser scanning microscope. *Optics Lett.* **10,** 53–55.

14. Suzuko, T. and Horikawa, Y. (1986) Development of a real-time scanning laser microscope for biological use. *Applied Optics* **25,** 4115–4121.

15. Shuman, H. (1987) Contrast in confocal scanning microscopy with a finite detector. *J. Microscopy* **149,** 67–71.

16. White, J. G., Amos, W. B., and Fordham, M. (1987) An evaluation of confocal versus conventional imaging of biological structures by fluorescence light microscopy. *J. Cell. Biol.* **105,** 41–48.

17. Brelje, T. C., Scharp, D. W., and Sorenson, R. L. (1989) Three-dimensional imaging of intact islets of Langerhans with confocal microscopy. *Diabetes* **38,** 808–814.

18. O'Rourke, N. A., Dailey, M. E., Smith, S. J., and McConnell, S. K. (1992) Diverse migratory pathways in the developing cerebral cortex. *Science* **258,** 299–302.

19. Hulspas, R. and Bauman, J. G. J. (1992) The use of fluorescent in situ hybridization for the analysis of nuclear architecture by confocal microscopy. *Cell Biology Int. Rep.* **16,** 739–747.

20. Wagner, R. W., Matteucci, M. D., Lewis, J. G., Gutierrez, A. J., Moulds, C., and Froehler, B. C. (1993) Antisense gene inhibition by oligonucleotides containing C-5 propyne pyrimidines. *Science* **260,** 1510–1513.

21. Giloh, H. and Sedat, J. W. (1982) Fluorescence microscopy: reduced photobleaching of rhodamine and fluorescein protein conjugates by *n*-propyl galate. *Science* **217,** 1252–1255.

22. Johnson, G. D., Davidson, R. S., McNamee, K. C., Russell, G., Goodwin, D., and Holborow, E. J. (1982) Fading of immunofluorescence during microscopy: a study of the phenomenon and its remedy. *J. Immunol. Methods* **55,** 231–242.

23. Bacallao, R., Bomsel, M., Stelzer, E. H. K., and DeMay, J. (1989) Guiding principles of specimen preservation for confocal fluorescence microscopy, in *The Handbook of Biological Confocal Microscopy* (Pawley, J., ed.), Integrated Microscopy Resource, Madison, WI. Republished 1990, Plenum, New York.

Overview of Laser Microbeam Applications as Related to Antibody Targeting

P. Scott Pine

1. Introduction

Laser-based microscopic systems (laser microbeams) are becoming popular tools for investigating various aspects of molecular and cellular biology (*1*). Depending on the wavelength, energy, and beam geometry employed, laser microbeams can be used for fluorescence excitation, microsurgery, cellular ablation, or micromanipulation of cells and organelles. The use of antibodies permits the targeting of specific antigens or cell types for analysis or treatment. Integrating a laser, microscope, and detection system (camera or photomultiplier tube) with a personal computer creates a workstation capable of controlling data acquisition parameters and performing subsequent data analysis. An example of one such workstation is shown in **Fig. 1**.

Several properties of lasers make them an ideal source of photons. For example, laser light is monochromatic, available in wavelengths ranging from the UV to the infrared depending on the medium stimulated to emit light. It has directionality. For all practical purposes, the light emitted by a laser can be considered parallel, making it possible to achieve microscopically focused spot sizes with diameters on the order of one wavelength of light (0.5 µm for lasers emitting in the visible region of the spectrum). Laser light maintains its brightness. The wave fronts of laser light are in phase, a phenomenon known as coherence, and do not destructively interfere with each other. This property allows the beam to retain its intensity over greater distances than the light produced by conventional thermal sources.

The laser beam can be directed along one of several optical paths available in either the upright or inverted microscope configurations. For fluorescence applications, it is often convenient to make use of the standard epi-illumination

From: *Methods in Molecular Biology, Vol. 115: Immunocytochemical Methods and Protocols*
Edited by: L. C. Javois © Humana Press Inc., Totowa, NJ

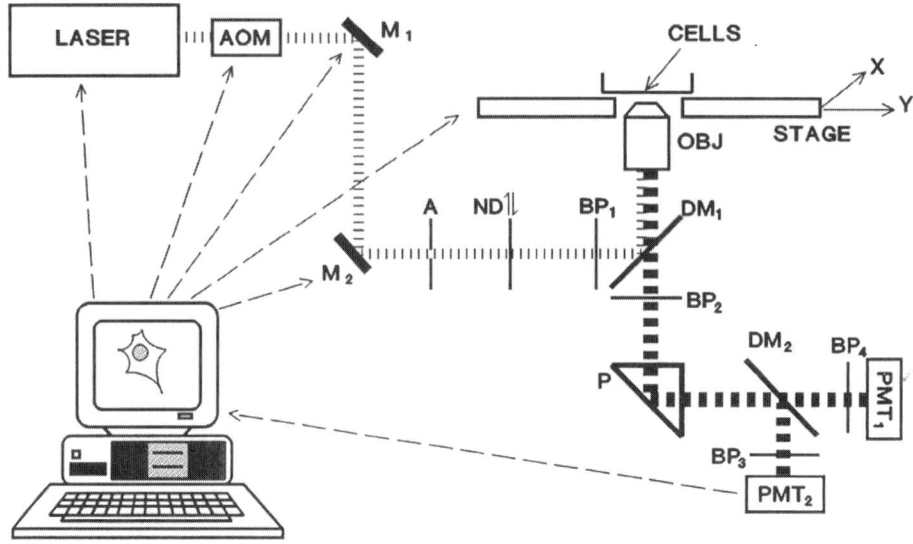

Fig. 1. Schematic diagram of the epi-illumination optical path in the inverted microscope and the computer input and output used to control laser excitation and fluorescence emission detection. The output wavelength is determined by the type of laser. The intensity of the laser output can be controlled by varying the current to the laser, varying the frequency of an acousto-optic modulator (AOM) and selecting for the modulated beam with an aperture (A), and/or placing neutral density filters (ND) in the optical path. Fluorescence excitation and emission detection can be achieved using the combination of filters (excitation bandpass [BP$_1$], dichroic mirror [DM$_1$], and emissions bandpass [BP$_2$]) found in the conventional filter cubes available for fluorescence microscopes. The emissions signal is passed to a photomultiplier housing by a prism (P), that distributes the light to the eyepiece lenses (not shown) and accessory ports. Dual emissions signals can be monitored by two photomultiplier tubes (PMT$_1$ and PMT$_2$) using a combination of bandpass filters (BP$_3$ and BP$_4$) and a dichroic mirror (DM$_2$). The computer can generate two-dimensional images by correlating the PMT signal with the x,y coordinates of the cell relative to the laser beam. This can be achieved by changing the sample position relative to a fixed laser beam using a motorized stage or by changing the position of the spot illuminated by the laser beam using a combination of galvanometric mirrors (M$_1$ and M$_2$).

pathway by simply replacing the arc lamp with a laser. However, some systems use the accessory camera ports, which are usually used for collecting the emitted light. In this case, the laser beam is directed back toward the objective in order to achieve epi-illumination. For such applications as microsurgery or optical trapping, even the transillumination pathway can be used. Precise positioning of the microbeam can be achieved with a combination of mirrors (for the x-y, specimen plane) and lenses (for z-axis, focal-point distance).

The following discussion provides an overview of several applications selected for their use of (or potential use of) antibodies in combination with a laser microbeam to investigate some aspect of molecular or cellular biology.

2. Biophysical Applications

2.1. Introduction

Laser microbeams offer several advantages over other fluorescence excitation techniques. In spectrofluorometry, observations are often made on a population of cells in a cuvette, resulting in a combined signal that lacks information about individual cellular responses. In flow cytometry, many individual cells are measured, but there is no temporal resolution since each cell is observed only once, and there is no spatial resolution since the entire cell is illuminated as it passes through the laser beam (*see* Chapter 30). In conventional fluorescence microscopy, individual cells can be monitored over time, and information about the two-dimensional spatial distribution of fluorescence can be obtained. However, some samples may be more susceptible to photobleaching by the arc lamps used for excitation, and the temporal resolution is limited to video-rate data acquisition (30 frames/s) (*see* Chapter 14).

Fluorescence excitation with a laser microbeam allows for a smaller region to be illuminated. Monitoring fluorescence with a sensitive photomultiplier tube also permits the use of lower intensities of irradiation for shorter periods of time. Therefore, unwanted photobleaching can be significantly reduced. If the spot size is adjusted to illuminate an entire cell, information analogous to spectrofluorometry or flow cytometry can be obtained on an individual cell basis with a high degree of temporal resolution. If the spot size is smaller than the cell, similar information can be obtained from a particular location within the cell.

By varying the location of the laser spot in the *x-y* plane using galvanometric mirrors for positioning (laser scanning) or moving the specimen relative to a fixed laser beam (stage scanning), a two-dimensional array of data points can be generated. This produces an image with a spatial resolution corresponding to the spot size. Laser scanning has the advantage of collecting two-dimensional images at up to video rates, whereas stage scanning permits areas larger than the objective's field of view to be imaged. In addition, laser microbeams provide a point source of illumination with sufficient intensity for use in confocal microscopy, which allows for the construction of three-dimensional images by optical sectioning along the *z*-axis (*see* Chapter 20).

2.2. Epitope-to-Epitope Distance Measurements

Fluorescence resonance energy transfer (FRET) is a technique that has been used to measure distances between pairs of proximal fluorochromes. A suitable pair consists of a donor fluorochrome, which has an emission spectrum that significantly overlaps with the absorption spectrum of an acceptor fluoro-chrome *(2)*. With the availability of monoclonal antibodies to many cell-sur-face determinants, intramolecular distances between nearby epitopes and intermolecular distances between adjacent cell-surface macromolecules can be investigated to analyze molecular interactions influencing important cellular events. Such monoclonal antibodies can be conjugated to fluorescein-isothiocyanate (FITC) as the donor, and either tetramethyl-rhodamine-isothiocyanate (TRITC) or phycoerythrin (PE) as the acceptor.

Two important factors determine the efficiency with which energy is trans-ferred from the donor to the acceptor: the extent of spectral overlap and the distance that separates the donor–acceptor pair (**Fig. 2A**). The spectral overlap for any particular pair (e.g., FITC–TRITC or FITC–PE) is constant. However, the rate of energy transfer is extremely sensitive to changes in distance because it is inversely proportional to the sixth power of the distance separating the two fluorochromes. By using the same donor–acceptor pair, FRET is useful for studying relative changes in either molecular conformation or intermolecular interactions.

The energy-transfer efficiency can be measured in several ways. In the absence of an energy acceptor, a donor fluorochrome in the excited state will return to the ground state with energy being lost in the form of emitted photons (i.e., fluorescence). However, if a suitable acceptor fluorochrome is nearby, the energy will be transferred in a nonradiative manner, and the fluorescence intensity of the donor will be reduced (i.e., quenched) (**Fig. 2B**). In addition, the acceptor fluorochrome will become excited, resulting in the emission of fluorescence with the characteristic spectrum of the acceptor. By measuring the fluorescence intensities of the donor fluorochrome (both quenched and unquenched) and the acceptor fluorochrome (both enhanced and non-enhanced), determinations of energy-transfer efficiencies have been made by spectrofluorometry *(3)*, flow cytometry *(4)*, and fluorescence microscopy *(5)*.

In the microscopic technique, photobleaching FRET *(5)*, the intensity of fluorescence excitation is increased to cause photobleaching of the donor fluorochrome, and the decay kinetics are measured in the absence or pres-ence of an acceptor fluorochrome. If the acceptor is in close proximity to the donor, then the availability of excited-state donors for photobleaching is reduced, thus making the photobleaching process slower.

The photobleaching method has been adapted for use with a laser micro-beam system *(6)*. The advantage of this system is that it allows for the selec-tion of individual cells to be analyzed. The laser beam can be tuned to the wavelength closest to the excitation maximum of the donor (the 488-nm line of an argon ion laser is used for FITC excitation), and the beam can be opti-cally expanded to irradiate an entire cell. A cell density should be used that

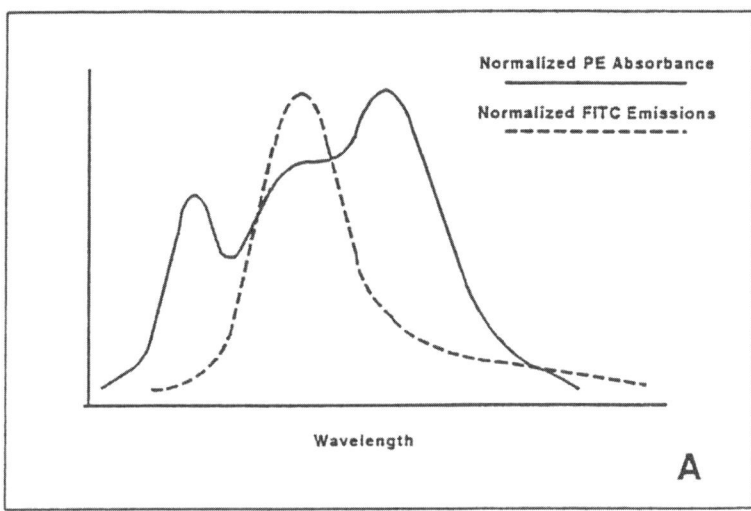

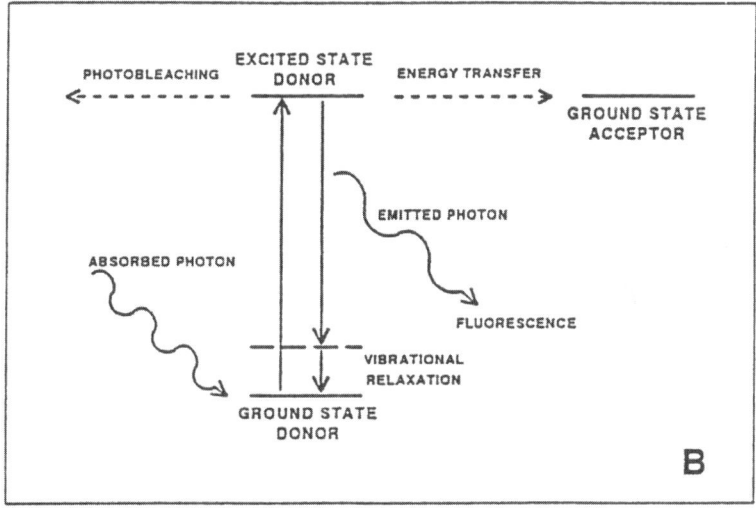

Fig. 2. Parameters affecting the efficiency of energy transfer. (**A**) Overlay of FITC emission spectrum and PE absorbance spectrum normalized to maximum fluorescence intensity and maximum optical density, respectively. FITC fluorescence intensity was measured as a function of emissions wavelength using a fluorimeter with an excitation wavelength of 488 nm. PE optical density was measured as a function of wavelength using a spectrophotometer. (**B**) Schematic representation of energy absorption and the possible pathways for the subsequent energy release (abbreviations as in the text).

permits the irradiation of one cell without any signal contribution from neighboring cells. The intensity of the laser beam can be adjusted to bleach a cell completely in a brief amount of time (<1 s). The photomultiplier tube can resolve thousands of time-points during the photobleaching, which permits an accurate computer fitting of the decay curves.

By comparing the kinetic parameters of a donor-only sample to a donor-plus acceptor sample (50 cells/sample), statistically significant shifts can be determined for donor–acceptor pairs that have an average separation of <17 nm. This technique has been used to measure conformational changes in the CD4 antigen of human peripheral blood T-cells (7), as well as the relationships between various CD3 antigens and nearby accessory molecules (8).

2.3. Measurement of Protein Mobility Within Cell Membranes

Photobleaching may also be used to induce spatial gradients of fluorescence intensity by introducing localized regions of bleached (nonfluorescing) probes. Various parameters can be measured depending on the initial distribution of the probe and the geometry of the laser beam used for bleaching and monitoring. Fluorescence redistribution after photobleaching (FRAP) is a technique most commonly used for measuring the ability of cell-surface proteins and lipids to diffuse laterally within the plane of cellular membranes (9). Variations on this technique have been used to look at gap junctional communication using total cell photobleaching (GAP-FRAP) (10), rotational mobility using polarization photobleaching (pFRAP) (11), and receptor–ligand binding kinetics using total internal reflectance microscopy (TIR-FRAP) (12).

In one FRAP configuration (13), the laser is focused to illuminate a small (1 μm) spot on the surface of a cell labeled with a monoclonal antibody conjugated to a fluorochrome (**Fig. 3**). Initially, the laser power is set at a level low enough to monitor the fluorescent signal emitted from the spot with a minimum amount of photobleaching. Then the laser is pulsed at a higher power to bleach a large proportion of the fluorochromes present within the measured spot. Immediately after the spot is bleached, the power is returned to the lower monitoring level, and the fluorescence within that spot is measured over time. If the protein is free to move within the plane of the membrane, then random diffusion will allow unbleached fluorochrome molecules to diffuse into the spot, and the fluorescent signal will increase. The diffusion coefficient can then be calculated from the rate of recovery, and the proportion of molecules free to diffuse (the mobile fraction) can be determined from the magnitude of the recovery. Some of the molecules may be free to diffuse only a short distance. Changing the spot size may provide some information regarding limitations to the distance a molecule may diffuse. For an additional approach to studying boundary limitations, *see* **Subheading 4.3.**

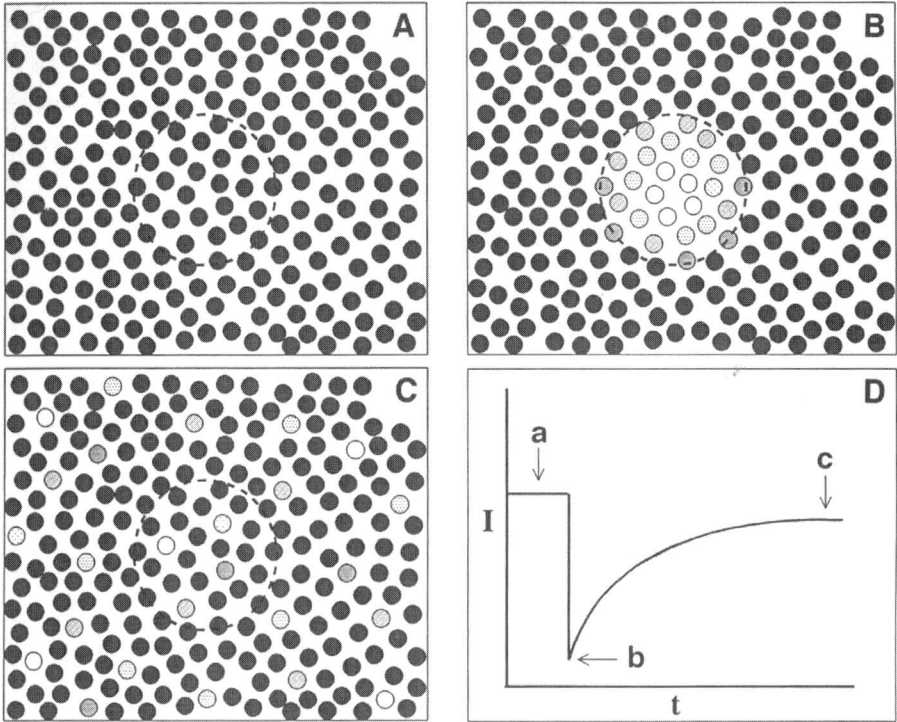

Fig. 3. Schematic diagram of the "spot" photobleaching method of FRAP. (**A**) Darkened circles represent fluorescently labeled molecules evenly distributed over a two-dimensional surface (assumed to be an infinite plane). (**B**) White and light gray circles represent the initial postbleach distribution of photobleached molecules within a 1-μm diameter spot. (**C**) Redistribution of photobleached and unbleached molecules as a consequence of random diffusion over time. (**D**) Curve representing the fluorescence intensity within the 1-μm diameter spot monitored over time; arrows a, b, and c indicate the time-points that correspond to their respective panels. The rate of recovery from point b to point c is used to determine the diffusion constant. The magnitude of the recovery is determined by comparing the fluorescence intensity at point c with the initial intensity at point a, and is used to determine the mobile fraction.

Another variation of the FRAP technique scans the laser in one dimension across the cell producing a "line-scan" profile of fluorescence intensity *(14)*. For a round cell whose surface is evenly labeled with a fluorescently conjugated antibody, a line scan typically produces a profile with two peaks of fluorescence, since the laser beam is illuminating more fluorophores at the edges of the cell (**Fig. 4**). In this approach, one of the two edge peaks is bleached, and then line scans are repeatedly recorded as a function of time. The coefficient of diffusion is calculated from the rate of recovery of the bleached peak, and the mobile fraction is determined from the extent of recovery. In addition, changes in the integrated area under the curve can be used to correct for any minor photobleaching caused by the laser during the postbleach phase of the measurements.

3. Ablative Applications

3.1. Introduction

In addition to being used as sources of fluorescence excitation, lasers can provide a high power source of irradiation for microsurgery and for inducing cellular injury and/or death (*1*). UV wavelengths (200–400 nm) are absorbed by many of the proteins within cells. In addition, some proteins, such as hemoglobin and melanin, absorb strongly in the visible region of the spectrum (400–700 nm). As a result of this absorption, the photon energy is converted to heat, and the proteins become thermally denatured leading to photocoagulation. The extent of photocoagulation can be controlled by varying both the power of the laser and the duration of irradiation. When an extremely short pulse of very high power is used, the temperature rise becomes exceptionally large because there is insufficient time for thermal diffusion to occur. This can result in vaporization at the point of irradiation (i.e., photoablation).

3.2. Specific Damage

Chromophore-assisted laser inactivation (CALI) is a technique that has been used to destroy the function of antibody-targeted proteins selectively (*15*). In this application, the antibody (or some other protein-specific ligand) is conjugated to a chromophore that has an absorption maximum at a wavelength that is not absorbed by any of the endogenous biomolecules. Irradiation of the cell with laser light of this wavelength results in the thermal denaturation of the proteins in close proximity to the chromophore. The effectiveness of this heat transfer decreases substantially with distance so that only the targeted proteins are significantly inactivated (*16*).

In one application, CALI was used to investigate the function of fascilin I, a cellular adhesion molecule (*17*). An antibody to fascilin I was conjugated to malachite green, a chromophore with an absorption maximum at 620 nm, a wavelength that is not strongly absorbed by cellular components. The chromophore-labeled antibodies were introduced into grasshopper eggs by microinjection. The embryos were irradiated with a neodymium:yttrium-aluminum-garnet (Nd:YAG) pumped tunable dye laser to achieve the necessary high-power laser light at a wavelength of 620 nm. During axonal outgrowth, sister

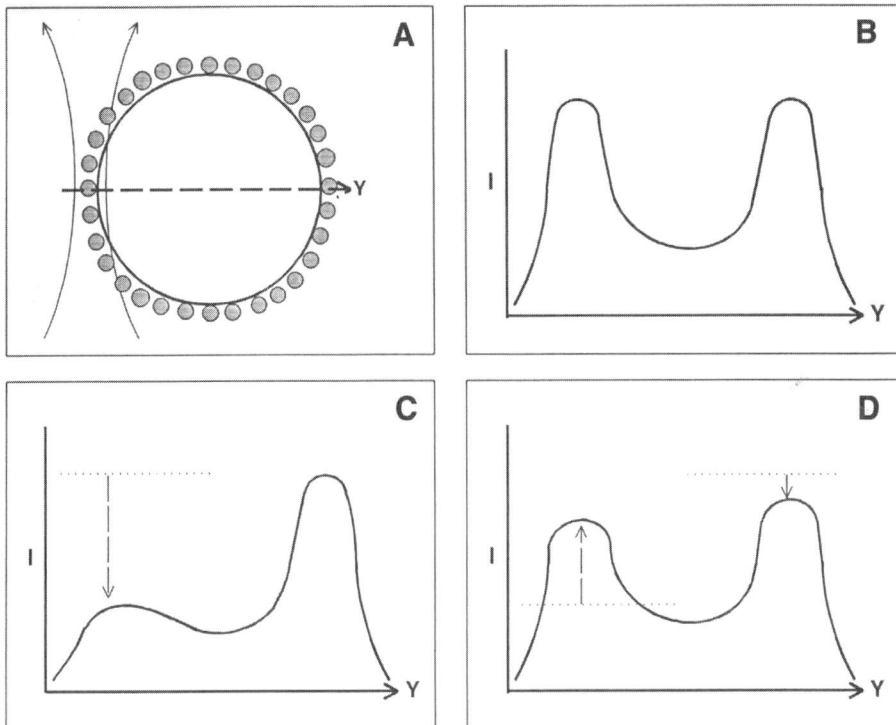

Fig. 4. Schematic diagram of the "line-scan" photobleaching method of FRAP. **(A)** Darkened circles represent fluorescently labeled molecules distributed along the surface of a 1-μm-wide "slice" through a round cell as viewed from the side. The laser beam waist is represented by the two upward curving arrows, which point in the direction of light propagation. The dashed line arrow represents the scanning direction, y, with a fixed x position in the specimen plane. The movement of the cell relative to the laser beam results in a line-scan profile of fluorescence intensity as a function of y. **(B)** The prebleach profile appears bimodal with the two peaks corresponding to the edges of the cell because the laser excites more fluorescent molecules when illuminating the cell tangentially. **(C)** The postbleach profile after one of the edges has been photobleached. Dashed arrow indicates decrease in fluorescence intensity owing to photobleaching. **(D)** The redistribution of fluorescent molecules results in an increase of fluorescence intensity at the bleach edge and a decrease at the opposite edge. Note: The redistribution is not limited to the slice illustrated in panel A, but actually occurs across the entire surface of the cell, i.e., the surface of a sphere. Recovery curves analogous to those of the spot photobleaching method can be made by monitoring the change in fluorescence intensity at the bleach position of the line scan profiles.

axons normally form a single fascicle, but laser irradiation resulted in defasciculation of those neurons that normally express fascilin I on their surface during differentiation. From these experiments, the investigators were able to show that fascilin I performs a function in axon adhesion during limb bud development.

3.3. Specific Killing

Fluorescence-activated cell sorters (FACS) have been used to separate subpopulations of cells for subsequent treatment or analysis (*see* Chapter 30). However, this approach requires that the cells be in suspension. In the case of adherent cells, some cannot be easily suspended, or the treatments used to suspend them may interfere with subsequent analysis. In these situations, a laser microbeam system capable of fluorescence imaging can serve two purposes. At low power, the laser can excite fluorescence to produce an image, and at a higher power, the laser can be used to kill the undesired cells.

For the same laser beam to be used for both fluorescence excitation and killing, the identifying fluorochrome must be excitable by a wavelength of light that can also be absorbed by the cell and converted to thermal energy. However, for cells that do not contain endogenous biomolecules with the requisite absorption spectrum, it is necessary to provide some other method of absorbing the photons from lasers that emit visible light. For this purpose, photoabsorptive dyes have been introduced into cells to make them susceptible to high-intensity irradiation *(18)*, and special substrates have been employed that can absorb light and convert it to thermal energy for transfer to adjacent cells *(19,20)*.

Based on the qualitative and quantitative analyses of a two-dimensional fluorescent image scan using a low laser power, areas within the field can be chosen for rescanning at a higher "killing" laser power. This results in the selective cell death of those cells that meet a specific fluorescence-labeling criteria, i.e., selecting for those cells that are either above or below a certain fluorescence-intensity threshold value (*see* **Fig. 5**).

3.4. "Cookie Cutter"

This technique is based on the use of a proprietary substrate from Meridian Instruments (Okemos, MI) consisting of a special film coating on 35-mm Petri dishes *(19,20)*. This film can be welded to the surface of the dish with high-power visible laser light. This application is particularly useful for rare-event cell selection *(21)*. If a small percentage of a mixed-cell population can be identified fluorescently, then the low-power laser setting and the stage scanning feature of the laser microbeam system can be used to locate those cells. When the cells of interest are found, a high-power laser setting is used to create octagonal welds around the cells (**Fig. 6**). The film is then peeled away from the dish, leaving "cookies"

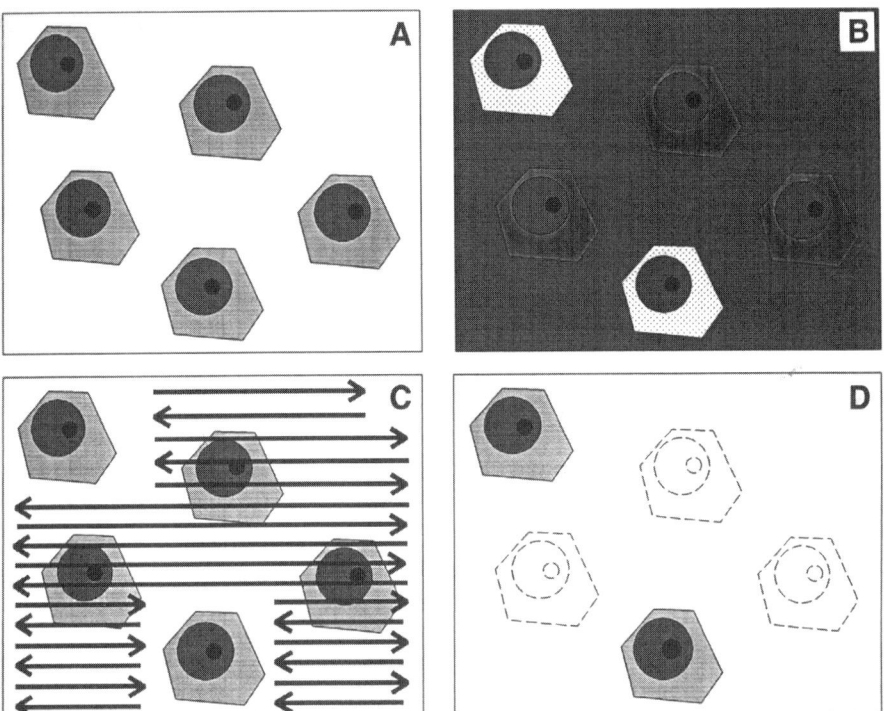

Fig. 5. Schematic diagram of the steps involved in ablative sorting on the basis of fluorescence intensity criteria. In this case, a positive sort is made for cells above a threshold fluorescence-intensity value. (**A**) Field of view representing a light microscope image of undistinguishable cells. (**B**) Two-dimensional image reconstruction based on fluorescence intensity as measured by a stage-scanning laser microscope. (**C**) Using fluorescence intensity criteria, cells are selected for "saving" from subsequent rescanning at the higher "killing" power, which will be activated only in the areas represented by the arrows (arrows point in the direction of stage movement during scanning). (**D**) Dotted outlines represent ablated cells, which usually appear disrupted, blebbed, or shriveled.

containing the specific cells on the dish. If there are any contaminating cells within the cookies, they can be killed using the technique described in **Subheading 3.3.**

4. Optical Trapping Applications

4.1. Introduction

Recently, infrared laser microbeams (wavelengths >700 nm) have been used to produce electromagnetic fields capable of exerting a sufficient force on cells

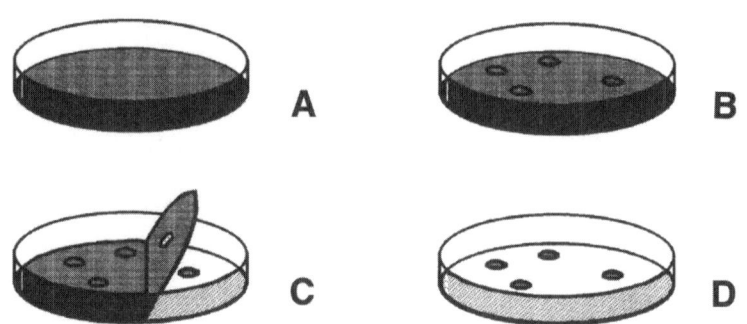

Fig. 6. Schematic diagram of the steps involved in the "cookie cutter" method of cell selection. (**A**) Cells are grown on plastic Petri dishes covered with a darkened nylon film. (**B**) Based on fluorescence intensity image scans using a stage-scanning laser microscope, rare event cells are identified, and octagonal welds are made around those cells to fuse the film to the dish. (**C**) The film is then peeled away from the dish. (**D**) The "cookies" containing the desired cells remain on the dish so that the cells may be analyzed further or subsequently cloned.

or organelles to hold them within the path of the laser beam near the focal point, a process referred to as optical trapping *(22)*. Laser light can exert a force on a particle in the optical path owing to refraction and reflection of the light. This force, known as radiation pressure, results from the exchange of momentum between the photons and the particle. In the case of an inverted microscope with an epi-illuminating optical trap, this force would "lift" the particle as the objective was raised toward the stage, whereas in the case of a transilluminating optical trap, this force would "push" the particle toward the substrate (*see* **Fig. 7**). In both cases, the lateral trapping force (i.e., the force perpendicular to the optical axis) is directed toward the center of the laser beam and "pulls" the particle into the optical trap. This approach provides a sterile, gentle, noninvasive means for the micromanipulation of cells and organelles. Several practical applications of optical trapping (also referred to as optical tweezers) have been demonstrated *(23–29)*, and those making use of antibodies are discussed in **Subheading 4.2.** and **4.3.**

4.2. Micromanipulation of Cells for Sorting Purposes

In one application, an epi-illuminated optical trap was used in conjunction with an inverted fluorescence microscope so that individual cells could be identified with FITC-conjugated antibodies and subsequently trapped for further micromanipulation *(24)*. In the first part of the experiment, as a demonstration of a sorting technique, the trapped cells were transferred into a capillary tube, that was located in the Petri dish along with the cells. The optical trap was used to lift

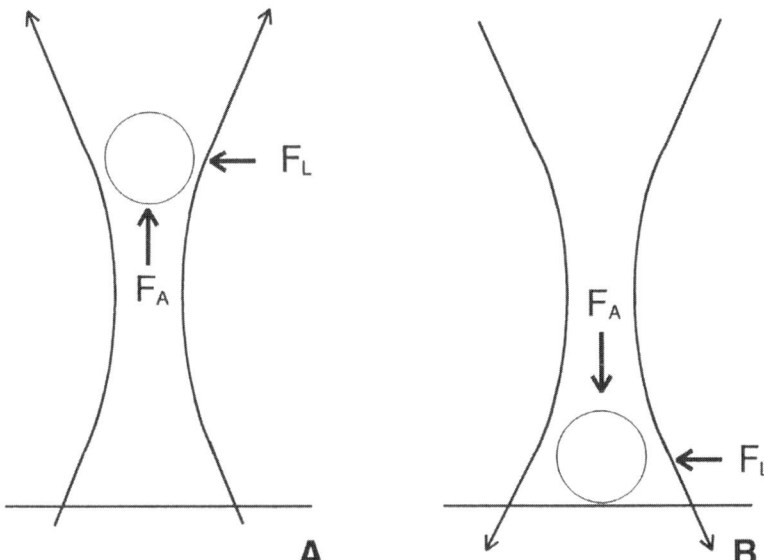

Fig. 7. Schematic diagram of forces exerted on a cell when using an inverted micro-scope with: (**A**) epi-illumination (i.e., laser focused through the objective) or (**B**) tran-sillumination (i.e., laser focused through the condenser). F_A is the axial force, and F_L is the lateral trapping force. Curved arrows represent the laser beam waist and point in the direction of light propagation.

the fluorescently tagged cells to a height greater than the thickness of the capil-lary tube wall. Then the stage was translated so that the capillary tube "slid" over the trapped cell. The capillary tube containing the selected cells could then be removed from the dish to transport the cells for further treatment. In the second part of the experiment, fluorescently tagged natural killer (NK) cells were opti-cally trapped and then placed in contact with target cells to observe the subsequent interaction. Using optical traps to place cells in close proximity has also been used in combination with a UV laser microbeam for laser-induced cell fusion *(25)*.

Because optical trapping does not require physical contact, cells can be manipulated in a completely enclosed environment in which physical and chemical parameters can be maintained over time. Specially designed cham-bers have been used that contain multiple compartments into which cells can be sorted automatically on the basis of optical characteristics not recogniz-able by conventional flow-sorting techniques *(26,27)*.

4.3. Micromanipulation of Cell-Surface Proteins

In another application, an optical trap was used to manipulate cell-surface proteins *(29)*. First, colloidal gold particles conjugated to monoclonal antibodies were bound to the cells. Then an optical trap was used to drag the particle across the surface of the cell and thereby drag the cell-surface protein through the membrane. When the cell-surface protein encountered a "boundary," the lateral force of the optical trap was overcome, and the particle was released. This approach was used to compare boundary limitations experienced by two different major histocompatability complex (MHC) class I glycoproteins. One was an integral membrane protein with a transmembrane region and cytoplasmic tail (H-2D^b), and the other (Qa2) was a cell-surface protein anchored by a glycosylphosphatidylinosital (GPI) linkage.

5. Concluding Remarks

The techniques discussed in this chapter are intended as an overview of how laser microbeams might be used in conjunction with antibodies to address various aspects of molecular and cellular biology. There may be other applications that were not covered, and there will likely be additional approaches developed as laser microbeams become increasingly available. Although microscopes are already a common tool of biological research, lasers are becoming so. With decreases in their size, complexity, and cost, lasers may become a standard accessory to the research microscope.

Acknowledgments

My thanks to Dr. Juanita Anders of the Uniformed Services University of the Health Sciences, Bethesda, MD, for two careful readings of the manuscript, which resulted in numerous clarifications and improvements. This chapter was written in my private capacity. No official support or endorsement by the Food and Drug Administration is intended or should be inferred.

References

1. Berns, M. W., Wright, W. H., and Wiegand Steubing, R. (1991) Laser microbeam as a tool in cell biology. *Int. Rev. Cyt.* **129,** 1–44.
2. Matyus, L. (1992) Fluorescence resonance energy transfer measurements on cell surfaces. A spectroscopic tool for determining protein interactions. *J. Photochem. Photobiol. B: Biol.* **12,** 323–337.
3. Szollosi, J., Damjanovich, S., Mulhern, S. A., and Tron, L. (1987) Fluorescence energy transfer and membrane potential measurements monitor dynamic properties of cell membranes: a critical review. *Prog. Biophys. Mol. Biol.* **49,** 65–87.
4. Szollosi, J., Matyus, L., Tron, L., Balazs, M., Ember, I., Fulwyler, M. J., and Damjanovich, S. (1987) Flow cytometric measurements of fluorescence energy transfer using single laser excitation. *Cytometry* **8,** 120–128.
5. Jovin, T. M. and Arndt-Jovin, D. J. (1989) FRET microscopy: digital imaging of fluorescence resonance energy transfer. Application in cell biology, in *Microspectrofluorometry of Single Living Cells* (Kohen, E., Ploem, J. S., and Hirschberg, J. G., eds.), Academic, Orlando, FL, pp. 99–117.

6. Szabo, G., Jr., Pine, P. S., Weaver, J. L., Kasari, M., and Aszalos, A. (1992) Epitope mapping by photobleaching fluorescence resonance energy transfer measurements using a laser scanning microscope system. *Biophys. J.* **61,** 661–670.

7. Szabo, G., Jr., Pine, P. S., Weaver, J. L., Rao, P. E., and Aszalos, A. (1992) CD4 changes conformation upon ligand binding. *J. Immunol.* **149,** 3596–3604.

8. Szabo, G., Jr., Pine, P. S., Weaver, J. L., Rao, P. E., and Aszalos, A. (1994) The L-selectin (Leu8) molecule is associated with the TcR/CD3 receptor; fluorescence energy transfer measurements on live cells. *Immunol. Cell Biol.* **72,** 319–325.

9. Wolf, D. E. and Edidin, M. (1981) Diffusion and mobility in surface membranes, in *Techniques in Cellular Physiology* (Baker, P., ed.), Elsevier, North Holland, pp. 1–14.

10. Anders, J. J. and Woolery, S. (1992) Microbeam laser-injured neurons increase in vitro astrocytic gap junctional communication as measured by fluorescence recovery after photobleaching. *Lasers Surg. Med.* **12,** 51–62.

11. Velez, M., Barald, K. F., and Axelrod, D. (1990) Rotational diffusion of acetylcholine receptors on cultured rat myotubes. *J. Cell Biol.* **110,** 2049–2059.

12. Hellen, E. H. and Axelrod, D. (1991) Kinetics of epidermal growth factor/receptor binding on cells measured by total internal reflection/fluorescence recovery after photobleaching. *J. Fluorescence* **1,** 113–128.

13. Edidin, M., Zagyansky, Y., and Lardner, T. J. (1976) Measurements of membrane protein lateral diffusion in single cells. *Science* **191,** 466–468.

14. Koppel, D. E. (1980) Lateral diffusion in biological membranes: a normal mode analysis of diffusion on a spherical surface. *Biophys. J.* **30,** 187–192.

15. Jay, D. G (1988) Selective destruction of protein function by chromophore-assisted laser inactivation. *Biochemistry* **85,** 5454–5458.

16. Linden, K. G., Liao, J. C., and Jay, D. G. (1992) Spatial specificity of chromophore assisted laser inactivation of protein function. *Biophys. J.* **61,** 956–962.

17. Jay, D. G. and Keshishian, H. (1990) Laser inactivation of fascilin I disrupts axon adhesion of grasshopper pioneer neurons. *Nature* **348,** 548–550.

18. Miller, J. P. and Selverston, A. I. (1979) Rapid killing of single neurons by irradiation of intracellularly injected dye. *Science* **206,** 702–704.

19. Schindler, M., Allen, M. L., Olinger, M. R., and Holland, J. F. (1985) Automated analysis and survival selection of anchorage-dependent cells under normal growth conditions. *Cytometry* **6,** 368–374.

20. Schindler, M., Jiang, L-W., Swaisgood, M., and Wade, M. H. (1989) Analysis, selection, and sorting of anchorage dependent cells under growth conditions. *Methods Cell Biol.* **32,** 423–446.

21. Jiwa, A. H. and Wilson, J. M. (1991) Selection of rare event cells expressing β-galactosidase. *Methods (San Diego, CA)* **2,** 272–280.

22. Weber, G. and Greulich, K. O. (1992) Manipulation of cells, organelles, and genomes by laser microbeam and optical trap. *Int. Rev. Cytol.* **133,** 1–41.

23. Ashkin, A., Dziedzic, J. M., and Yamane, T. (1987) Optical trapping and manipulation of single cells using infrared laser beams. *Nature* **330,** 769–771.

24. Seeger, S., Monajembashi, S., Hutter, K.-J., Futterman, G., Wolfrum, J., and Greulich, K. O. (1991) Application of laser optical tweezers in immunology and molecular genetics. *Cytometry* **12,** 497–504.

25. Wiegand Steubing, R., Cheng, S., Wright, W. H., Numajiri, Y., and Berns, M. W. (1991) Laser induced cell fusion in combination with optical tweezers: the laser cell fusion trap. *Cytometry* **12,** 505–510.

26. Buican, T. N., Smyth, M. J., Crissman, H. A., Salzman, G. C., Stewart, C. C., and Martin, J. C. (1987) Automated single-cell manipulation and sorting by light trapping. *Appl. Opt.* **26,** 5311–5316.

27. Buican, T. N., Neagley, D. L., Morrison, W. C., and Upham, B. D. (1989) Optical trapping, cell manipulation, and robotics. *SPIE Proc.* **1063,** 190–197.

28. Berns, M. W., Aist, J. R., Wright, W. H., and Liang, H. (1992) Optical trapping in animal and fungal cells using a tunable, near-infrared titanium-sapphire laser. *Exp. Cell Res.* **198,** 375–378.

29. Edidin, M., Kuo, S. C., and Sheetz, M. P. (1991) Lateral movements of membrane glycoproteins restricted by dynamic cytoplasmic barriers. *Science* **254,** 1379–1382.

22

Overview of Fluorescence Photomicrography

Lorette C. Javois and J. Michael Mullins

1. Introduction

The results of fluorescence labeling experiments may be photographed to produce a permanent record. One of the major hurdles associated with fluorescence photography is that most of the commercially available films have been designed to work at exposure times in the range of 0.01 to 0.1 s. However, the light levels emitted by fluorochromes are of low intensity, necessitating exposure times one to two magnitudes longer. Low-intensity light over a long period will have less effect on film than high-intensity light over a short period, a phenomenon known as reciprocity failure.

2. Practical Effects of Reciprocity Failure

Reciprocity failure of the film has two practical implications. The first is seen with the use of color-reversal film. Each of the three color-sensitive layers of the film has a different reciprocity failure factor resulting in incorrect color reproduction during long exposures. To correct for this problem, color-compensating filters can be placed in the light path to the camera.

All fluorochromes are susceptible to some degree of fading, bleaching, or quenching as a result of exposure to excitation light. Thus, long exposure times lead to bleaching, which in turn necessitates even longer exposure to insure an adequate image. This is a vicious circle that is hard to avoid. The key is to use as sensitive a film as possible, to optimize the emitted fluorescence, and to reduce exposure times.

2.1. Film Selection

400 ISO films such as Kodak Tri-X Pan or T-Max for black-and-white prints or Ektachrome for color slides are suitable for many applications. These films

From: *Methods in Molecular Biology, Vol. 115: Immunocytochemical Methods and Protocols*
Edited by: L. C. Javois © Humana Press Inc., Totowa, NJ

can be exposed at 800 or 1600 ISO and push-processed in high-contrast developer. A combination that has found favor in many laboratories is Kodak Tri-X with the Diafine two-bath developer *(1)*, which provides an ISO of 1600. In general, black-and-white film has a finer grain and is suitable in most situations, the exceptions being double-labeling procedures in which both fluorophores are to be photographed simultaneously or the preparation of color slides for presentations. An advantage of black-and-white film is that it allows considerable adjustment of image contrast in the printing process, a factor that is generally lost when doing color work. High-speed color films (400–1600 ISO) are now available, though one trades higher speed for a more grainy appearance.

2.2. Optimizing Fluorescence

Emitted fluorescence may be optimized in a number of ways. To start with, high-intensity excitation light at the absorption maximum of the fluorochrome should be used. Second, lenses with moderate magnification (to minimize bleaching and retain fluorescence intensity) and maximum numerical aperture (NA) are important. With epifluorescence illumination, the objective of the microscope serves as both the condensor and objective; therefore, fluorescence intensity is proportional to the fourth power of its NA. $NA = n(\sin \mu)$ where n is the refractive index of the medium and μ is the 1/2 angle of the margins of the light cone captured by the lens. Theoretically, the largest 1/2 angle of light that a lens can gather is 90° (half of 180°), the sine of which is 1.0. Therefore, the maximum NA that can be obtained is dependent on the refractive index of the medium between the objective and coverslip. The refractive index of air is 1.00, water is 1.33, glycerine is 1.47, and immersion oil is 1.52. Therefore, the highest practical numerical aperture is 1.52. Several microscope manufacturers sell oil-immersion objectives with numerical apertures approaching this theoretically maximum, including newly designed planapochromatic objectives that provide excellent light transmission for fluorescence combined with the flatness of field and superb optical correction characteristic of planapo lenses. In conjunction with their ability to correct for chromatic and spherical aberrations, these lenses currently retail in the range of $3000–$6000.

2.3. Reducing Exposure Times

Most microscopes are designed so that some or all of the light may be directed to either the binocular eyepieces for observation or to the camera for photography. To maximize the emitted fluorescence directed to the camera, a 0–100% reflector prism or mirror is a necessity. This arrangement allows one to direct 100% of the light to the camera when taking a picture.

Dimming of the fluorescence image resulting from photobleaching of fluorophores exposed to the intense excitation light produces further complications for observation and photography. Antifading (antibleaching) reagents, which are thought to work by scavenging free radicals, may be added to the mounting medium. As free radicals damage unexcited fluorochromes, reducing their presence reduces photobleaching of fluorochromes, especially fluorescein. Commonly used reagents are 1,4-diazobicyclo-[2.2.2]-octane (DABCO) *(2)*, *p*-phenylenediamine *(3,4)*, *n*-propyl gallate *(5)* and sodium azide *(6)*. Bacallao et al. *(7)* obtained successful retardation of photobleaching with a combination of 100 mg/mL DABCO and 0.1% sodium azide in a mounting medium of 50% glycerol in PBS. They note that in their hands, *n*-propyl gallate produced a generally dimmer fluorescence image. Also *p*-phenylenediamine has been reported to destroy stored samples over time *(2)*. In addition to these "home brews," several commercial antifading agents of unknown composition are available. For example, Molecular Probes (Eugene, OR) sells "Prolong," which reduces bleaching dramatically and also gels to provide the actual sealing of the coverslip to the slide.

Automatic exposure systems often present difficulties for determining the correct exposure for fluorescent images. Since the background in a fluorescent specimen is black or nearly black, an exposure meter that determines exposure based simply on the average intensity of the image will invariably produce overexposed slides or negatives. Metering systems that provide spot measurements of small areas of the image generally provide more accurate exposure, providing the spot area can be filled with regions of typical fluorescence intensity. Some photo systems (e.g., the Olympus PM30) provide metering modes in which the contribution of the dark, nonfluorescent background is discounted, allowing for accurate exposure assessment based only on fluorescence intensity. Expensive, automatic exposure systems, however, are not a necessity. Manual exposure times can be determined by exposing a test roll of film to determine an appropriate exposure range for a given type of specimen.

3. Counterstaining for Color Photography

When preparing color photographs or slides, a useful counterstain for fluorescein-labeled specimens is Evan's blue. Under the 490-nm excitation light used for fluorescein, Evan's blue fluoresces red, providing a good contrast for the yellow-green of fluorescein. For sectioned material, 5–10 min in 0.1% Evan's blue in distilled water followed by several rinses in distilled water before mounting is adequate. For whole mounts, staining time should be reduced to 1 min. It should be noted that overstaining with Evan's blue can mask some specific fluorescence, especially if the signal is weak.

Counterstaining in these circumstances also eliminates the problem that unlabeled regions of the specimen are not normally visible under epifluorescence. Without counterstaining, localization of the antigen requires superposition or comparison of a fluorescent image with a phase or bright field image of the specimen. With counterstaining, the entire specimen is present in the fluorescent image. Additionally, DAPI, which provides a nice nuclear/DNA stain, can also be useful for orientation of a fluorescence image. Vector Laboratories (Burlingame, CA) sells an antifade mounting medium that includes DAPI, making the counterstaining very easy.

4. Electronic Imaging

High-resolution video or solid-state cameras in conjunction with image intensifiers and/or image processors can be attached to the microscope in place of a film-based camera. This configuration allows one to capture images too dim to be seen clearly with the unaided eye or recorded on conventional photographic emulsions *(8)*. Of particular interest is the tracking of fluorophore-labeled native molecules and probes microinjected into living cells *(9,10)*. In addition, images captured in digital format from solid-state cameras or converted analog data from video cameras may be readily sharpened, enhanced, or quantified through computer processing. While heightened sensitivity is advantageous, it is accompanied by lower spatial resolution and increased image noise.

References

1. Osborn, M. and Weber, K. (1982) Immunofluorescence and immunocytochemical procedures with affinity purified antibodies: tubulin-containing structures. *Meth. Cell Biol.* **24,** 98–132.
2. Langanger, G., DeMay, J., and Adam, H. (1983) 1,4-Diazobizyklo-[2.2.2.] oktan (DABCO) verzogest das Ausbleichen von immunofluorenzpreparaten. *Mikroskopie* **40,** 237–241.
3. Johnson, G. D. and Nogueira Araujo, G. M. de C. (1981) A simple method of reducing the fading of immunofluorescence during microscopy. *J. Immunol. Meth.* **43,** 349–350.
4. Johnson, G. D., Davidson, R. S., McNamee, K. C., Russell, G., Goodwin, D., and Holborow, E. J. (1982) Fading of immunofluorescence during microscopy: a study of the phenomenon and its remedy. *J. Immunol. Meth.* **55,** 213–243.
5. Giloh, H. and Sadat, J. W. (1982) Fluorescence microscopy: reduced photobleaching of rhodamine and fluorescein protein conjugates by *n*-propyl gallate. *Science* **217,** 1252–1255.
6. Bock, G., Hilchenbach, M., Schauenstein, K., and Wick, G. (1985) Photometric analysis of anti-fading agents for immunofluorescence with laser and conventional illumination sources. *J. Histochem. Cytochem.* **33,** 699–705.

7. Bacallao, R., Bomsel, M., Stelzer, E. H. K., and DeMey, J. (1989) Guiding principles of specimen preservation for confocal fluorescence microscopy, in *Handbook of Biological Confocal Microscopy* (Pawley, J. B., ed.), Plenum, New York, pp. 197–205.
8. Inoué, S. (1986) *Video Microscopy*. Plenum, New York.
9. DeBiasio, R., Bright, G. R., Ernst, L. A., Waggoner, A. S., and Taylor, D. L. (1987) Five-parameter fluorescence imaging: wound healing of living Swiss 3T3 cells. *J. Cell Biol.* **105,** 1613–1622.
10. Taylor, D. L. and Wang, Y. L. (eds.) (1989) *Fluorescence Microscopy of Living Cells in Culture, Parts A and B. Methods in Cell Biology, vols. 29 and 30.* Academic, New York.

23

Overview of Antigen Detection
Through Enzymatic Activity*

Gary L. Bratthauer

1. Introduction

The use of enzymes together with immunoglobulins to identify specific substances emerged with the work by Nakane and Pierce, who labeled an immunoglobulin with the peroxidase enzyme rather than with a fluorescent compound *(1)*. The difficulty with this approach lies in attaching a relatively large molecule like an enzyme to another large molecule such as an immunoglobulin without either molecule losing the ability to function. In this case, labeling can occur without appreciable functional loss. The labeled antibody is still able to bind the antigen, and the attached enzyme is still able to catalyze the oxidative reaction. This direct-labeled technique was the forerunner of numerous other methods that bring enzymes and antibodies together to allow the enzyme action to identify the location of the antigen through the antibody intermediary.

Along with the direct-labeled methods, the indirect-labeled methods were developed, providing amplification and universality. By using a labeled secondary antibody, any number of primary antibodies can be used (to which the secondary antibody binds) *(2)*. This allows great freedom in selecting antigens to study and antibodies with which to study them. All that is needed is a secondary "antiantibody" reagent with an enzyme attached, a far less costly means of antigen detection. This makes the technique much more universal in application than the direct techniques are.

The other overriding and perhaps more important aspect of this technique is the increased amplification of the signal obtained.

---

*The opinions or assertions contained herein are the private views of the author and are not to be construed as official or as reflecting the views of the Department of the Army or the Department of Defense.

From: *Methods in Molecular Biology, Vol. 115: Immunocytochemical Methods and Protocols*
Edited by: L. C. Javois © Humana Press Inc., Totowa, NJ

By allowing multiple secondary antibodies, each with several enzyme molecules attached to bind to the primary antibody, the amount of enzyme at the site of the primary antibody–antigen interaction can be increased along with the resultant signal. This makes the reaction easier to observe but also increases the visibility of weak reactions, thereby increasing the overall sensitivity of the method. In addition, the background is reduced and the primary antibodies can be diluted even further with little loss in detectability. This produces a technique that is not only more cost-effective, but also is more specific resulting from a decrease in the concentration of any nonspecific antibodies in the primary antisera.

Increased amplification yielding more sensitivity and more specificity, along with a universal methodology providing ease of use and applicability, are the essential principles governing all further modifications of this technology as shown in Chapters 24 and 25.

2. Enzymes Used for Antigen Detection

There are a few common enzymes that have been employed in these types of assay systems over the years, the chief among them being the peroxidase enzyme *(3)*. Peroxidase has an oxidative function when in conjunction with a source of oxygen, transferring electrons to a molecule, which becomes oxidized. The peroxidase enzyme found in the horseradish plant has been used for its ability to carry out this function, for the fact that it is easily obtained, and for the antigenic differences from most mammalian forms of the enzyme. The oxidative function of this enzyme allows for the use of chromogens, which when oxidized, not only change color, but precipitate in such a manner as to render a permanent preparation.

Another common enzyme used in these procedures is alkaline phosphatase. Alkaline phosphatase will cleave phosphates off of a donor molecule, which then in turn acts as a mediator of a color change involving a third molecule. This system is often used because alkaline phosphatase can create more of the color-producing molecules per enzyme molecule than can peroxidase, resulting in better sensitivity. Alkaline phosphatase systems are especially sensitive for examining protein, or nucleic acid blots with enzyme labels. The problem when examining tissue is the presence of the endogenous enzyme in the tissues examined. Quenching the endogenous alkaline phosphatase activity can be difficult. The standard treatment is to use levamisole in a blocking step incubation, but the exact conditions for successful inactivation are often varied, depending on the tissue, and endogenous activity can persist *(4)*.

Other minor enzyme systems can be used such as betagalactosidase *(5)*. This system works well; however, it can lead to some false-positive problems resulting from endogenous enzymes that have a similar reactivity *(6)*. Another enzyme not commonly used is glucose oxidase. A glucose oxidase system can provide a sensitive and specific assay if other endogenous enzyme activity is a problem.

3. Chromogenic Substrates

In addition to the many enzyme systems available, there are with each a series of chromogenic substrate solutions that can be used to create different colors and locations of reaction products. For the peroxidase system, there are numerous oxidizable compounds that precipitate as a permanent color. The most common and still widely used is 3,3'diaminobenzidine tetrahydrochloride (DAB). This compound precipitates to a golden brown color when in solution with peroxidase and hydrogen peroxide. This brown color has many subtleties and readily stands out in a tissue section. With practice, it is possible to differentiate specific from nonspecific staining patterns just by examining the characteristics of the precipitated pigment. This material is also insoluble in alcohol and xylene, and therefore the tissue may be routinely dehydrated and cleared without loss of chromogen.

A drawback to the use of this chromogen is its close resemblance to some endogenous pigments like melanin. Most of the time, the characteristics and color subtleties of this compound are enough to distinguish it from melanin, lipofuchsin, hemosiderin, or formalin pigments. However, on occasion, depending on the tissue, the endogenous pigment may be startlingly close in appearance to DAB. In these situations, it may be necessary to try a different molecule that precipitates as an alternate color. One such molecule is 3-amino-9-ethylcarbazole (AEC).

AEC precipitates as an insoluble color substance when subjected to peroxidase and hydrogen peroxide. The color is a bright red and is often more intense than that achieved with DAB. It does not conflict with any endogenous pigments. The drawback to the use of this chromogen is that it is soluble in alcohol. This means that the tissue sections cannot be dehydrated and cleared as is commonly done. An aqueous mounting medium has to be used, and the sections often appear thickened under the microscope. Sometimes there is a loss of resolution when using high magnification. A new mounting medium used with some success is Crystal/Mount (Biomeda, Foster City, CA). It is baked to an insoluble plastic. Whereas the plastic coat protects the specimen and removes the refractility associated with wet mounts, the surface of the plastic is quite fragile and may be easily scratched. However, once the compound has dried and hardened, the section can be coverslipped using an organic mounting medium such as Permount, and the slide can be made more permanent. The resolution, though, is still not as good as with an alcohol-dehydrated, xylene-cleared specimen.

There are other chromogens that can be employed when using peroxidase enzymes. The compound 4-chloro-1-naphthol is one that is often used in immunoblotting and occasionally for tissue examination *(7)*. The signal achieved with this material is thought to be superior to that obtained with other chromogens. The blue color, though, which is the precipitant product, is harder to visualize under the microscope, since blue has traditionally been the color of histology counterstains such as hematoxylin. Many peroxidase chromogens used in other assays, such as *o*-phenylenediamine, change color when oxidized but are soluble in aqueous solutions and will not precipitate as a permanent pigment, thus being unsuitable for tissue analysis. Most of these oxidizable substrates are potential carcinogens and should be handled appropriately. There is, however, a compound that is thought to not be as hazardous as some, called tetramethylbenzidene (TMB) *(8)*. TMB precipitates to a dark blue-black and is insoluble in alcohol.

There are times when two color assays may be required. This can be accomplished by altering the enzyme system used for one of the assays or the chromogen used for one of the assays if the enzymes are not altered. Perhaps the easiest method is to alter the color product of the DAB chromogen for one of the assays. This can be done be adding a heavy metal compound such as cobalt or nickel to the dilute solution *(9)*. The resultant precipitate will be blue to black, not brown. It is helpful in this circumstance to use a Methyl Green counterstain, which will enhance the black precipitate and avoid the confusion of color seen with some other counterstains.

Alkaline phosphatase acts on many substrates as well, each precipitating as a different color. For example, a combination of 5-bromo-4-chloro-3-indolyl phosphate (BCIP) and nitro blue tetrazolium (NBT) results in a permanent blue precipitate at the site of alkaline phosphatase localization. There are other compounds that can also be tried, such as Fast Red TR/Naphthol AS-MX (Sigma, St. Louis, MO), which precipitates as a red color.

The use of multiple chromogens and multiple enzymes has emerged as a process whereby numerous antigens can be identified with increased sensitivity. The sensitivity is increased through more substrate precipitation or greater color resolution. Sometimes, though, it is better to keep to simple methods when starting to work with these systems. The peroxidase and alkaline phosphatase enzymes are most widely used because they are easily controlled, the assays work, and experience creates many avenues for problem solving. The peroxidase enzyme system is highlighted in Chapters 24 and 25 because it is easy to consume the endogenous enzyme activity, the substrates are varied and obtainable, and the products are familiar and recognizable.

4. Enzyme Conjugation to Immunoglobulins
4.1. Direct Label Technique

Using an enzyme-labeled primary antibody, a very quick direct assay is obtained. This assay may not be as sensitive as others and may involve more background, but a labeled antibody can still be used if detection by other means is problematic. One example would be using a human-derived antibody to analyze a human sample. With a directly labeled antibody, a simple antibody incubation step followed by the enzyme-substrate combination is all that is required. Labeling an immunoglobulin with an enzyme can be difficult. There can be problems with free enzyme sticking to the section, unlabeled antibody decreasing the sensitivity, and denaturation occurring, which creates labeled fragments that can interfere with the test.

In this technique, once the specimens have been prepared (*see* Chapters 8–13), the antibody solution at the appropriate dilution is applied. Following 30 min of incubation, extensive washing is done before incubation with the chromogen. Washes totaling 30 min should be performed. The antigen is then detected by incubation in the chromogenic substrate solution.

4.2. Indirect Label Technique

With this technique, the often numerous primary antibodies used do not all have to be labeled with enzyme, provided they are all of the same species that a secondary antibody raised against immunoglobulin will recognize. The secondary antibody, labeled with enzyme, provides for the detection of all of the primary antibodies without the need for them to be individually conjugated. The problems of background and denaturation associated with conjugation are still present and can be troublesome. However, an increment of sensitivity can be achieved greater than that obtained by the previous method.

This method is about 50 min longer then the direct label technique, since the primary antibody incubation is performed for 30 min with an unlabeled antibody directed against the antigen of interest. Following washing steps, the labeled antibody must then be applied for 30 additional min followed by a slightly reduced, 20-min wash. The antigen is once again detected with substrate incubation. This process only works if the secondary antibody is reactive against the species from which the primary antibody was obtained, inclusive of primary antibody immunoglobulin type.

Amplification can be attained with this type of method by employing two indirect techniques simultaneously. After the specimen is thoroughly rinsed, following the labeled secondary antibody incubation, a third antibody may be used, which is enzyme-labeled and reactive against the species immunoglobulin responsible for the secondary antibody.

5. Enzyme Incorporation into Large Antibody-Rich Polymers

In 1994, the Dako Corporation introduced its Envision™ system *(10)*. This immunoenzyme technique achieves both sensitivity and speed in a universal reagent. The secondary antibodies (derived from goats, in the case of the peroxidase-based system) are attached to a large inert polymer molecule labeled with horseradish peroxidase. These goat antibodies are directed against both rabbit and mouse immunoglobulins. This technique requires only two steps and often does not require a protein-blocking reagent. If needed though, a universal nonserum blocking solution may be used. By allowing any monoclonal or polyclonal reagent to be detected with this one compound, and by achieving high levels of sensitivity with numerous incorporated enzyme molecules, this type of system becomes ideal for rapid turnaround situations, while maintaining the sensitivity developed with the longer, more-involved techniques. This system is available as a kit, and the polymer/enzyme/antibody solution is ready to use.

6. Enzyme Incorporation in Immune Complexes

Other sandwich type techniques are entirely immunologic. The peroxidase-antiperoxidase (PAP) technique (see Chapter 24) was developed to alleviate the problems of antibody conjugation with enzyme and also to amplify the number of enzyme molecules that can be directed to a given site. By making use of the proteinaceous antigenicity of the peroxidase enzyme, Sternberger developed an immune complex of enzyme and antienzyme (11). This complex in solution in a 3:2, enzyme:antibody molar ratio allows enzyme incorporation through a linking antibody intermediate. The linking secondary antibody, produced in an alternate species, binds the primary antibody and the PAP complex, provided they are from the same species used as the immunoglobulin immunogen for the secondary species.

The advantage of this technique is an increase in sensitivity over the indirect techniques. A slight disadvantage occurs because of the increased time of a three-step procedure and a possible increase in background resulting from the increase in sensitivity. Time factors are not really an issue, though, because good reproducible results are desired regardless of perceived inconvenience. The main problem with this type of assay is the need for PAP made from many species, since quality primary antibodies are generated in rabbits, mice, guinea pigs, goats, rats, and sheep. Also, there are other methods that may provide more sensitivity than this method.

In addition to peroxidase-mediated immunologic assays, there are also alkaline phosphatase–antialkaline phosphatase (APAAP) reagents available (12). The general overall principle governing these immunologic assays remains the same.

7. Enzyme Conjugation to High-Affinity Molecules

The overriding problems seen with the direct and indirect techniques are the conjugation of a large enzyme to a large immunoglobulin, at the same time preventing denaturation, removing unconjugated species of both reagents, and preserving the normal function of both reagents. The more widespread means of labeling currently in use involves the vitamin biotin and the protein avidin. By biotinylating a compound, it can be linked virtually irreversibly to any other compound through an avidin intermediate. The use of the avidin D-biotin systems in the newer detection methods alleviates some of the problems cited above. These molecules are sufficiently small to prevent the problem of inactivation so common with large, cumbersome enzyme conjugations. Biotinylation is not a difficult process. There are numerous commercially available biotin-labeling kits and many procedures adapted for coupling *(13)*. Some allow for labeling to amino groups, others to the sulfhydryl groups available on the immunoglobulin molecule (*see* Chapter 7). Initially, biotin conjugation to antibody was detected by enzyme conjugation to avidin, or avidin was used as an intermediary in a bridged avidin-biotin technique. The avidin-biotin complex (ABC) molecule (*see* Chapter 25) was developed by Hsu and associates to increase the number of enzyme molecules available near any one antibody *(14)*. Essentially, by biotinylating an enzyme and reacting that enzyme-labeled biotin with avidin, a complex builds that consists of four molecules of biotin for every avidin and two molecules of avidin for every biotin. This complex, then, is able to bind the biotin attached to the secondary antibody.

Variations on this biotin-binding sandwich technique involve the use of enzyme-labeled streptavidin, which is a compound with a lower isoelectric point than avidin D. The lower isoelectric point helps to prevent some charge-mediated nonspecific binding. Also, long-arm spacer biotinylated secondary antibodies can be used that help alleviate the steric hindrance of these giant complexes so near the antibody. Labeled avidin techniques were developed that reduced the potential for steric hindrance. The idea that the more enzyme available at the antigen site, the more amplification and the better the result, is generally true. With the ABC technique, though, the complex can get too large for proper localization. It is sometimes better to make the complex smaller and allow more of it to reach the biotin molecules on the antibody. This is the theory of the labeled avidin-type systems. It is true that in some cases, a smaller enzyme complex can provide an increased signal through greater resolution in detection.

The above represent the past and present of the most common enzyme-mediated methods of antigen detection. There are alternate procedures available, involving such methods as antibiotin antibody steps that combine the avidin-biotin systems with a further antibiotin/antienzyme sandwich for still greater sensitivity. Also, there are methods that follow a PAP procedure with a biotinylated antibody to the PAP immunoglobulin followed by ABC detection *(15)*. The obvious problem created with this approach is the tremendous increase in steps and the ever more-increasing difficulty associated with troubleshooting and quality control with such a system. But, for an individual assay in a research setting where utmost amplification is required, there are many combinations of methods and reagents that may be appropriate.

8. Detection of Precipitated Marker Molecules

In the ongoing quest for signal amplification, a catalyzed reporter molecule process was developed for immunoassays *(16)*. This method was developed to accentuate the signal seen using an avidin-biotin type of detection. This process has application in formalin-fixed, paraffin-embedded tissue, to increase the sensitivity to the point where, in combination with antigen recovery methods, compounds can be identified that were previously negative *(17)*. The method uses a biotinylated tyramine, which, when exposed to the oxidative effects of peroxidase on hydrogen peroxide, becomes radicalized and attaches covalently to electron-rich amino acids in the vicinity of the peroxidase enzyme. The amount of biotin deposited at this site is tremendously enhanced as the enzyme action progresses. Following this, a marker-enzyme-labeled avidin molecule can bind to the numerous biotin molecules present. The chromogenic substrate can then be employed. By using, first, an enzyme substrate that doubles as a detection molecule, and second, the enzyme substrate molecule that is visualized, up to a 1000-fold increase in sensitivity may be obtained. These types of detection substrate chemicals are available from DuPont NEN (TSA) and from Dako (CSA) in kit form.

9. Conclusions

There are inherent problems associated with enzyme-mediated methods, regardless of the method used. The right conditions must be met, of course, for the enzyme action to take place. Unlike fluorochromes or gold particles (two other marker compounds), enzymes need to act chemically for the assay to work. Also, the enzyme action must only represent the marker molecule. Endogenous enzyme or enzyme-like activity can create problems only realized in systems that use enzymes. Also, the use of enzymes demands more attention to detail because of the increase in sensitivity that is often obtained. The problem of unwanted reactivity is enhanced in enzyme-mediated reactions more so than in others, in part because of the additional level of sensitivity brought about by the continuous action on a substrate.

There are certain specimen preparations that create problems for some of the sandwich methods. Frozen or fresh cell preparations may have inherent avidin-binding properties, as some fixed preparations can have. While this should not preclude the use of these methods, one should be aware of the potential limitations and have alternate methodologies available for confirmatory testing.

Taken as a whole, each enzyme method offers a unique feature, which, under the right conditions, is the ideal choice. The direct method is advantageous in situations in which an antiantibody step would be prohibitive. In this case, because of the problems of directly labeled enzyme, a better choice perhaps would be biotinylating the primary antibody and using labeled avidin or ABC for detection. If this is not possible because of the tissue being used, enzyme labels would be needed.

To quickly screen antibodies under consideration for use, an indirect method would be fine. However, once a reagent has been shown to be promising, a more powerful PAP or ABC system should be used to decrease potential background problems and boost sensitivity.

Among the sandwich techniques, the ABC is the most universal, with the PAP being the least prone to nonspecific staining. More sensitivity can be had with labeled avidin methods or the newer more-inventive amplified amplifications. A universally applied method with an alternate technique available, like the PAP for occasional problem specimens, will most often suffice.

Once a system is chosen, each facet of that system should be analyzed for potential problems. If the technology is understood, problem solving becomes easier. The simpler the system utilized, from the buffers and antibodies to the enzyme and chromogen, the easier it is to investigate when things go awry, which unfortunately is always a possibility. When selecting a method, choose a system that has familiar technology, performance test the system on several known positive specimens, and always perform positive and negative controls with every assay in order to monitor performance.

References

1. Nakane, P. and Pierce, G. (1966) Enzyme-labeled antibodies: preparation and application for the localization of antigens. *J. Histochem. Cytochem.* **14,** 929–931.
2. Farr, A. and Nakane, P. (1981) Immunohistochemistry with enzyme labeled antibodies: a brief review. *J. Immunol. Meth.* **47,** 129–144.
3. Swanson, P. (1988) Foundations of immunohistochemistry. *Am. J. Clin. Pathol.* **90,** 333–339.
4. Ayala, E., Martinez, E., Enghardt, M., Kim, S., and Murray, R. (1993) An improved cytomegalovirus immunostaining method. *Lab. Med.* **24,** 39–43.
5. Sakanaka, M., Magari, S., Shibasaki, T., Shinoda, K., and Kohno, J. (1988) A reliable method combining horseradish peroxidase histochemistry with immuno-β-Galactosidase staining. *J. Histochem. Cytochem.* **36,** 1091–1096.
6. Flugelman, M., Jakitsch, M., Newman, K., Casscells, S., Bratthauer, G., and Dichek, D. (1992) In vivo gene transfer into the arterial wall through a perforated balloon catheter. *Circulation* **85,** 1110–1117.
7. Musiani, M., Zerbini, M., Plazzi, M., Gentilomi, G., and LaPlaca, M. (1988) Immunocytochemical detection of antibodies to Epstein-Barr virus nuclear antigen by a streptavidin-biotin-complex assay. *J. Clin. Microbiol.* **26,** 1005–1008.

8. Bos, E., van der Doelan, A., van Rooy, N., and Schuurs, A. (1981) 3,3', 5,5'-tetra-methylbenzidine as an Ames test negative chromogen for horse-radish peroxidase in enzyme-immunoassay. *J. Immunoassay* **2,** 187–204.

9. Hsu, S. and Soban, E. (1982) Color modification of diaminobenzidine (DAB) precipitation by metallic ions and its application for double immunohistochemistry. *J. Histochem. Cytochem.* **30,** 1079.

10. Dako Envision System universal kit instructions. (1994) Dako Corporation, Santa Barbara, CA.

11. Sternberger, L. (1979) *Immunocytochemistry.* 2nd ed. Wiley, New York.

12. Hinglais, N., Kazatchkine, M., Mandet, C., Appay, M., and Bariety, J. (1989) Human liver Kupffer cells express CRT, CR3, and CR4 complement receptor antigens. *Lab Invest.* **61,** 509–513.

13. Tse, J. and Goldfarb, S. (1988) Immunohistochemical demonstration of estrophilin in mouse tissues using a biotinylated monoclonal antibody. *J. Histochem. Cytochem.* **36,** 1527–1531.

14. Hsu, S., Raine, L., and Fanger, H. (1981) Use of avidin-biotin-peroxidase complex (ABC) in immunoperoxidase techniques: a comparison between ABC and unlabeled antibody (PAP) procedures. *J. Histochem. Cytochem.* **29,** 577.

15. Swanson, P., Hagen, K., and Wick, M. (1987) peroxidase-antiperoxidase (ABPAP) Avidin-biotin-complex. *Am. J. Clin. Pathol.* **88,** 162–176.

16. Bobrow, M. N., Litt, G. J., Shaughnessy, K. J., Mayer, P. C., and Conlon, J. (1992) The use of catalyzed reporter deposition as a means of signal amplification in a variety of formats. *J. Immunol. Meth.* **150,** 145–149.

17. Merz, H., Malisius, R., Mannweiler, S., Zhou, R., Hartmann, W., Orscheschek, K., Moubayed, P., and Feller, A. C. (1995) A maximized Immunohistochemical method for the retrieval and enhancement of hidden antigens. *Lab. Invest.* **73,** 149–156.

24

The Peroxidase-Antiperoxidase (PAP) Method and Other All-Immunologic Detection Methods*

Gary L. Bratthauer

1. Introduction

These methods have, as an underlying unification, the sole use of antibody-antigen interaction to provide the binding of reagents, ultimately leading to the localization of enzyme at the site of cellular antigen. The earlier direct and indirect techniques, while still used today, were found to be less sensitive than the later-used peroxidase-antiperoxidase (PAP) techniques. The newer immune complex polymer type of detection system (Envision™; Dako Corporation, Santa Barbara, CA) increases the amplification with an indirect style of reaction while still keeping the binding entirely immunologic. These techniques have been supplanted mostly by the avidin–biotin binding schema (*see* Chapter 25). However, there are still times when a non-avidin–biotin system is preferred. The following procedure highlights the PAP type of reaction, the most commonly used immunologic reaction method, and additional versions—direct, indirect and polymer types of procedures—are discussed at the end of the chapter.

The PAP method was pioneered by Sternberger in 1979 *(1)*. The method uses an immunological sandwich amplification and the enzyme peroxidase to effect a signal. The unique feature of this procedure is the enzyme/antibody solution, the PAP immune complex. The horseradish peroxidase enzyme, itself an immunogenic protein, is used to inoculate a given species, and a polyclonal immune response is generated against the enzyme. This antiserum is harvested and placed in solution with the enzyme so that immune complexes form that

---

*The opinions or assertions contained herein are the private views of the author and are not to be construed as official or as reflecting the views of the Department of the Army or the Department of Defense.

From: *Methods in Molecular Biology, Vol. 115: Immunocytochemical Methods and Protocols*
Edited by: L. C. Javois © Humana Press Inc., Totowa, NJ

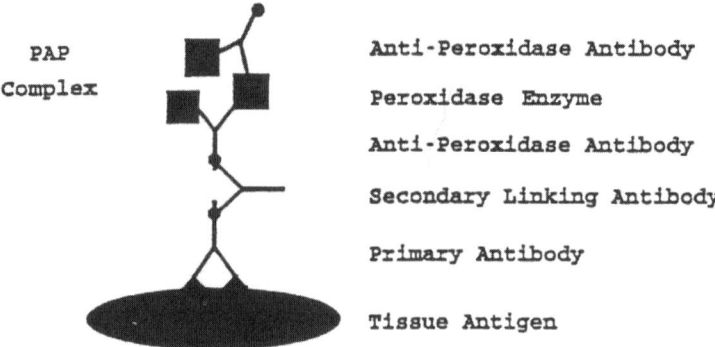

PAP
Complex

Anti-Peroxidase Antibody

Peroxidase Enzyme

Anti-Peroxidase Antibody

Secondary Linking Antibody

Primary Antibody

Tissue Antigen

Fig. 1. Diagram illustrating the molecular interactions of the PAP procedure. The PAP complex is comprised of horseradish peroxidase bound to an antiperoxidase antibody generated in the same animal species as the primary antibody which recognized the tissue antigen of interest. The primary antibody and the PAP complex are linked via a secondary antibody generated in a second animal species against immunoglobulin of the primary animal species (⋏, immunoglobulin; ∎, peroxidase enzyme).

remain soluble. These complexes form with a molar ratio of two molecules of IgG to three molecules of enzyme. Furthermore, not only does this complex remain soluble, but the enzymatic activity of the peroxidase is not affected by the attached immunoglobulins. The antiperoxidase antibodies are from the same species as that producing a primary antibody raised against a tissue antigen. These two antibodies, one directed against the tissue antigen and the other directed against the peroxidase enzyme, can be linked by another antibody raised in an alternate species against immunoglobulin from the first species (*see* **Fig. 1**).

The peroxidase enzyme most often used is that found in the horseradish plant. Horseradish peroxidase (HRP) is a 40-kDa enzyme capable of stimulating an immune response. The principle of the technique is the same as other immunolocalization techniques, which provide a means of getting a marker molecule (the enzyme) in close proximity to an antigen through the use of antibodies. The difference in this technique is that the three-step approach provides a further step in amplification, and this reaction is entirely based on immunologic binding. Without the need for conjugation of a marker molecule to an antibody, reactivity can be achieved with less background and more sensitivity *(2)*. The only background problems occur as a result of antibody binding nonspecifically, which can sometimes be corrected with the use of detergents and more extensive washing steps. In cases where endogenous biotin-binding capabilities preclude the use of the popular avidin–biotin complex (ABC) procedure (*see* Chapter 25), the PAP technique is a convenient alternative.

One disadvantage of the PAP method is that the primary antibody and the PAP complex must be from the same species. To have ready-made PAP complexes from multiple species to accommodate the vast library of primary antibodies available is costly and unwieldy. In fact, a particular PAP complex may not be available for all cases of primary antibody species. Also, the sensitivity is slightly less than that obtained by other technologies. However, for reducing background and ease in problem solving, this technique is still quite effective.

Finally, a PAP assay system is nice to have available for detecting more than one antibody on an individual specimen. In double-labeling experiments, having two completely different assay systems reduces the chances of cross-over reactivity. The first antigen can be detected using the standard ABC procedure (*see* Chapter 25), and the second antigen can be detected using the PAP system. This way, no harsh acid treatments need to be performed to remove the first series of reagents in preparation for the second. By using PAP instead of ABC for the second antigen detection, there is no danger of binding to the biotin present on the secondary antibody used in the first antigen detection.

2. Materials

1. Gloves, gowns, and masks.
2. Ultralow freezer (–70°C).
3. 45-L carboy.
4. Vortex.
5. Adjustable pipetman pipets with tips.
6. 4-mL glass Wheaton vials (Wheaton Scientific, Millville, NJ).
7. Humid chamber leveling tray with lid, for horizontal antibody application.
8. Absorbent paper towels.
9. Staining racks and dishes for solution incubation.
10. Stirring block with stir bars.
11. Light microscope.
12. Coverslips.
13. Phosphate-buffered saline (PBS): 0.01 M sodium phosphate, 0.89% sodium chloride, pH 7.40, ±0.05 (*see* **Note 1**).
14. 10 N Sodium hydroxide and concentrated hydrochloric acid for adjusting pH.
15. PAP immune complex reagent (Dako Corp., Carpinteria, CA).
16. Primary antibody reactive against the desired antigen from the same species as that used for the PAP immune complex (*see* **Note 2**).
17. A secondary linking antibody solution that is reactive against the immunoglobulin of the species responsible for the other two reagents.
18. 10% Normal serum from the species from which the secondary antibody was generated in PBS.
19. Chromogen solution: 3, 3' diaminobenzidine tetrahydrochloride (DAB) (Sigma; St. Louis, MO); 30% hydrogen peroxide (H_2O_2) (*see* **Note 3**).

20. 30% bleach in water.
21. Counterstain (e.g., Mayer's Hematoxylin).
22. 6 *M* ammonium hydroxide diluted 1:50 in deionized water.
23. 100% ethanol.
24. Xylene.
25. Permount mounting medium (Fisher Scientific, Pittsburgh, PA).

3. Methods
3.1. The Peroxidase Antiperoxidase Method

Depending on the starting material being used, preantibody incubation steps may vary and are outlined in Chapters 9–12. The following assay begins with removal of the slides from the overnight 10% normal serum incubation step. Since the PAP is generally produced from rabbits, a good secondary antibody to use is an antirabbit antibody produced from swine. Therefore, the overnight serum incubation in this case would have been with 10% normal swine serum in PBS.

1. Align a humid staining chamber, level the slide bars, and add water to the chamber (*see* **Note 4**).
2. Remove the slides from the dish of serum and place them on the bars in the chamber. Cover the specimens with 10% normal swine serum in PBS and replace the chamber cover.
3. Prepare the antibody solutions in PBS. If thawing out fresh-frozen reagent, allow plenty of time for aggregates to disperse. Also, allow sufficient time for reconstitution of lyophilized material (*see* **Note 2**).
4. Blot off the 10% swine serum from each slide by placing the end of the slide on absorbent paper towels.
5. Add the primary rabbit antibody directed against the antigen desired to the slide, making sure to cover the entire specimen. Work quickly to avoid any drying. Cover the chamber and incubate for 30 min (*see* **Note 5**).
6. Rinse the specimens using PBS with wash bottle force or a siphon stream. Rinse for several seconds, allowing the buffer to freely flow off the end of the slides sitting on the racks in the chamber. Repeat three times allowing a 1–2 min between washings to enable any nonspecifically adherent immunoglobulins to slowly diffuse away (*see* **Note 6**).
7. Blot off the excess PBS from each slide by placing the end of the slide on the absorbent paper towels.
8. Add more 10% normal swine serum in PBS, covering the specimens to further guard against nonimmunologic binding. Cover the chamber and incubate for 10 min (*see* **Note 7**).
9. Rinse with PBS briefly, then add more 10% normal swine serum in PBS and cover the chamber for another 10-min incubation.
10. Blot off the 10% swine serum from each slide by placing the end of the slide on the absorbent paper towels.

11. Add the swine antirabbit IgG secondary antibody to the slide, making sure to cover the entire specimen. Work quickly to avoid any drying. Cover the chamber and incubate for 30 min (*see* **Note 8**).
12. Repeat **step 6**.
13. Repeat **step 7**.
14. Add the PAP complex to the slide, making sure to cover the entire specimen. Work quickly to avoid any drying. Cover the chamber and incubate for 30 min.
15. Repeat **step 6**.
16. Place the slides in a staining slide rack and incubate for 10 min in a dish of PBS. Do not allow specimens to dry.
17. Prepare the chromogen solution (*see* **Note 3**).
18. Place the rack of slides into the chromogen solution, cover, and incubate for 15 min.
19. Remove the slide rack and wash in a dish with three changes of deionized water, for 2 min each.
20. Counterstain with Mayer's hematoxylin 1–5 min depending on the concentration and color intensity desired (*see* **Note 9**).
21. Rinse with deionized water, three changes, 2 min each.
22. Develop the nuclei blue with a 10-s incubation in ammonium hydroxide in water.
23. Rinse with deionized water, three changes, 2 min each.
24. Dehydrate the specimens with 100% ethanol, four changes for 2 min each.
25. Clear the specimens with xylene, four changes, for 2 min each.
26. Coverslip with Permount, dry, and observe. A positive reaction should be visible as a brown precipitate. The nuclei should be light blue (*see* **Notes 9–11**).

3.2. Additional Methods

3.2.1. Direct Assay

This test requires an antibody that is labeled with enzyme directed against the compound of interest. Though expensive, less sensitive, and occasionally plagued with background problems, this type of assay is still used—especially in cases where, for example, a particular patient has an antibody reactive with a virus or endogenous protein, and this antibody is enzyme-conjugated from serum and used on a human cellular substrate. Also, this type of test is performed as a detection method. A popular marker molecule in use today is digoxigenin. Nucleic acid probes labeled with digoxigenin are often detected in situ or in vitro using antibody-enzyme conjugates directed against digoxigenin *(3)*. In performing these types of assays, the blocking serum or protein solution could be either the normal serum of the species providing the antibody or one of the many universal blocking solutions available commercially.

For the procedure, use the conjugated antibody reagent in **Subheading 3.1., step 5**, and go directly from **step 7** to **step 16**. The time of incubation in DAB should be closely monitored, because this type of test can lead to false-positive results if there is heavy background present.

3.2.2. Indirect Assay

This test requires that an antibody be directed against another species immunoglobulin, conjugated with a marker molecule such as horseradish peroxidase. These techniques are also not used very often, but some newer technologies use affinity-purified reagents that decrease the problems associated with free antibody or enzyme. Moreover, these reagents can be designed for detecting more than one antibody species. In this manner, a universal type of secondary detection system can be employed. One must be cautious of the synergistic effects of pooled immunoglobulin, but overall, these reagents can provide a rapid, very effective nonbiotin type of reactivity by using a universal antibody solution. In performing these kinds of assays, a species-specific (if sole antibody solutions are used) or universal (if more than one species of antibody are used) blocking solution is utilized.

For the procedure, use the conjugated anti-immunoglobulin reagent in **Subheading 3.**, **step 11** and go directly from **step 13** to **step 16**. Again, monitoring the DAB development can be critical to a successful result.

3.2.3. Immune Polymer Assay

This newer type of indirect detection attaches the various immunoglobulins (reactive against the immunoglobulins of the most popular species used in these types of techniques) to a large inert polymer substance *(4)*. This polymer substance has, embedded within it, horseradish peroxidase molecules (*see* **Fig. 2**). In this way, this compound can react with any primary antibody and provide a large amount of enzyme molecules to act toward signal production. One problem with large-sized complexes, in general, is resolution. However, for antigens in small amounts, the amplification achieved is worth the loss in resolution, especially if the antigen is not in a large-enough quantity to create a problem of overreactive localization. The test is also relatively fast, omitting the final step, and is as sensitive as the three-step procedures. If this assay is needed, a universal blocking solution is required, since the antibodies used for detection are from multiple species. In some cases, because of the charge of the polymer, a blocking step may not be necessary. These types of tests are beneficial when time is an important factor.

For the procedure, use the polymer reagent (Envision™) in **Subheading 3.1.**, **step 11** and go directly from **step 13** to **step 16**. Remember, it may not be necessary to block for non-specific binding with this reagent. Therefore, following **Subheading 3.1., step 7**, one could go directly to **step 11**. Also, because this reagent is a commercially prepared one, no further dilution is required.

4. Notes

1. Prepare the PBS for large volume use. For 45 L of a simple buffer with low ionic strength.

Immunepolymer of Secondary
Antibody and Horseradish
Peroxidase

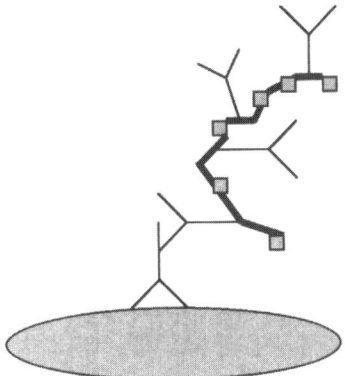

Primary Antibody

Tissue Antigen

Fig. 2. Diagram illustrating the immune polymer method, Dako's Envision™. This is a two step, fast method which allows the use of more dilute primary antibody. The enzyme-containing polymer reagent is "universal" in that it contains antirabbit/mouse immunoglobulins and will bind to rabbit or mouse primary antibodies (λ, immunoglobulin; ■, peroxidase enzyme).

    a. Weigh 50 g sodium phosphate and 150 g sodium chloride and dissolve in a 4-L flask, filled to within 100 mL of the 4-L mark.
    b. Bring the pH of the solution to 6.8 with 10 *N* sodium hydroxide (about 20 mL).
    c. Fill the remainder of the flask to 4 L and check the pH.
    d. Dilute the flask to 45 L in carboy.
    e. Mix well by shaking/rotating the carboy.
    f. Remove a beaker of solution and check the pH; it should be close to 7.4. If not, adjust with appropriate vol of hydrochloric acid or sodium hydroxide. The final pH of the 0.01 *M*, PBS should be 7.40 ±0.05.
2. Antibody solutions are made with PBS to a predefined dilution. These are experimentally determined on a known positive specimen. Usually, working antibody concentrations are in the range of 10–20 µg/mL. However, depending on the individual reagent, this concentration could vary considerably. For a beginning titration, start at 1:10 and do serial 1:10 dilutions resulting in 10-, 100-, 1000-, 10,000-fold dilutions of concentrated or "neat" antibody. Immunoassay a known positive specimen, following the procedure and examine the sections. Optimum results will occur within a range of the dilutions. Once a range is established, another titration can be performed within that range of 1:2 serial dilutions to better optimize the final dilution. The antibodies can be stored concentrated in aliquots in an ultralow freezer at –70 to –80°C. However, it is unwise to store the PAP reagent in the freezer as the freeze/thaw disrupts the immune enzyme. The other antibodies are stable indefinitely but should be thawed once and used, not refrozen. If desired, antibodies can be thawed and diluted as concentrated stock solutions from which more dilute working solutions can be prepared. These stock solutions can be kept at 4–8°C for 1 wk.

3. The chromogen, DAB, is a potential carcinogen, and should be handled accordingly. For small operations there exists commercially available tablets or prediluted concentrates designed to provide about 100 mL of reagent. In those instances where the volume of slides to examine is not great, these are preferred, because of the reduced contact this type of prepared chromogen offers. However, for large volumes, or in case alterations to the procedure are desired, the following is one method of preparing the DAB chromogen:
   a. DAB is purchased in 5-g bottles and stored desiccated in a –20°C freezer.
   b. Working in a fume hood wearing the appropriate gloves, gown, and mask, add PBS to the 5-g bottle and mix until dissolved.
   c. Add the contents of the 5-g DAB bottle to a 250-mL volumetric flask, fill to 250 mL with PBS and mix with stirring.
   d. Remove contents to a 500-mL beaker and pipet 4-mL vol to 4-mL glass Wheaton vials with constant stirring.
   e. The 4-mL vials should be frozen at –70°C until used.
   f. Prepare the working solution in the fume hood by adding one 4-mL vial of DAB to 500 mL PBS in a staining dish and then adding 400 μL of 30% H_2O_2. DAB should be prepared immediately before use and kept covered in the hood, because oxidation begins with the addition of H_2O_2 and will continue if exposed to the open air. This solution should be made fresh every day and should only be used for 3–4 h or for three racks of slides and discarded when having turned a dark brown color. The 30% H_2O_2 is extremely caustic and can cause burns. See the slight variation depending on conditions in **Note 6**.
   g. The contaminated glassware and any spills should be cleaned up with a 30% bleach solution or horseradish peroxidase and H_2O_2. Oxidation should be allowed to occur over 2–3 d before handling.
4. As with all immunological procedures, it is important to make sure the specimens stay hydrated throughout. Drying will result in nonspecific immunoglobulin binding. A level chamber rack is also important, because after 30 min, immunoglobulins may flow with gravity away from the specimen location.
5. Allow the antibody solutions and the slides to come to room temperature before applying. Also make sure that the antibodies to be used are fully dissolved or reconstituted. Sometimes the results are better if the antibody solutions are prepared the day before and left at 4°C overnight. Slight vortexing before application helps to ensure proper solution. To increase reactivity, incubations can be extended in length, or temperatures can be raised. This often increases background as well. Sometimes reaction intensity can be increased without background by incubation with primary antibody overnight at 4°C *(5)*. The inconvenience of using a refrigerator, especially with large volumes of slides, makes room temperature incubations more attractive, and the reduced background benefit that is sometimes possible with overnight incubations is slight enough to be discounted.

6. Vigorous rinsing is required for reduction of background. Often, when assaying tissue sections that have been proteolytically digested or subjected to microwave retrieval methods, too vigorous of a wash may cause the section to dislodge from the slide. Still, it is important to wash as thoroughly as possible. The wash stream should not directly splash the specimen surface as this may dislodge antibodies with low avidity. The stream should initiate at one end of the slide and the solution should flow rapidly across the slide surface. Washes are the most important aspect of background reduction and should therefore be extended if too much background is a problem. To reduce background further, 0.25% Triton X-100 can be added to the PBS *(6)*. In some cases, this will also reduce reactivity, because the charge interactions of antibody-antigen binding will be altered. However, if the antibody has a high avidity, and other unwanted clones are present to a lesser degree, the addition of detergent can reduce background staining. Also, if plus slides or poly-L-lysine coatings are used, it may be necessary to add detergent to the buffer because of the hydrophobic nature of these slides. The results, though, are generally better than with glue and are artifact-free. A compensation in chromogen concentration may offset the effects of the detergent on signal. The chromogen concentration may be doubled or increased up to five times (reducing time in solution accordingly) without much increase in background because of detergent washes. If performing the procedure in this fashion, PBS without detergent is needed for preparation of primary antibodies (some of which are not stable if stored in a detergent solution) and in the preparation of blocking sera and chromogen solutions. If there is substantial background, a preabsorption may be indicated (*see* Chapter 4). The antibody may be diluted in buffer containing 2% bovine serum albumin (BSA), secondary-species serum (swine in the aforementioned example), or even 0.5% normal human serum, if the contaminating clones are reactive against a serum-based constituent.

7. The 10% secondary-species serum incubation allows for protein to bind to charged sites on the specimen. The additional 10 min incubations are extra protection against unwanted antibody sticking nonspecifically. The concentration of the serum solution and the time of incubation may be increased if desired.

8. It is very important in PAP techniques for the secondary antibody to be applied in excess. This way, one arm of the divalent Fab portion of the immunoglobulin molecule can bind the primary antibody, and the other arm is free to bind the PAP complex. If not in excess, both arms of the molecule may bind the primary antibody, creating a prozone-type of effect. Without a free secondary antibody Fab site to capture the PAP complex, the assay will not work.

9. It is better if the counterstain is weak, because the principal reaction product may be masked by too much counterstain. Ideally, just enough hematoxylin to identify structure is all that is necessary.

10. These slides are permanent and should not fade with time. If the presence of endogenous pigment is a problem with a particular specimen, or a color other than brown is desired as an indicator, different chromogenic compounds can be

used. The compound 3-amino-9-ethylcarbazole (AEC) can be used, for example, to create a red color product at the site of enzyme deposition. This compound can be purchased commercially in kit form, and is also thought to be potentially carcinogenic. Using the AEC substrate system, begin at **Subheading 3., step 17**, and prepare the AEC chromogen as follows:

a. Place 2 mL of acetate buffer in a Wheaton vial.
b. Add 1 drop of AEC to the vial and mix.
c. Add 1 drop of H_2O_2 to the vial and mix.
d. Filter if necessary. A precipitate may develop that will not affect the results in any way, but it can be filtered out if desired.

Continue with **Subheading 3., step 18** in the method. The positive reaction product with this chromogen will be red, with the nuclei a light blue. Rather than dehydrating the specimens in ethanol and xylene, allow them to dry, add 1 drop of Crystal/Mount (Biomeda Corp, Foster City, CA) to the specimen and bake them in a 60°C oven for 30 min. This preparation is permanent and can be coverslipped with Permount if needed. The Crystal/Mount will form a hard plastic coating on the slide, but it can be damaged by smudging.

As another alternative, 1 mL of 1% cobalt or nickel chloride can be added to the DAB solution prior to slide incubation, and the resultant precipitate will be dark blue to black, not brown *(7)*. This can increase the overall sensitivity of the reaction, but it is not popular because of the different color counterstain that is usually required. A 5% solution of Methyl Green used as a counterstain for 5 min provides enough contrast to the blue to be a good background for interpreting assays with a nuclear location.

11. To use a PAP technique with a monoclonal antibody (generally made in mice), the PAP complex must contain mouse immunoglobulin. A good secondary reagent is a rabbit antimouse IgG, which necessitates the use of normal rabbit serum for the nonspecific binding-blocking solution. To further enhance reactivity, the assay may be repeated following the PAP step, from the secondary antibody step, essentially reacting the PAP complex with a new secondary antibody incubation, followed by a localization reaction *(8)*.

References

1. Sternberger, L. (1979) *Immunocytochemistry.* 2nd ed. Wiley, New York.
2. Cerio, R. and MacDonald, D. (1988) Routine diagnostic immunohistochemical labeling of extracellular antigens in formol saline solution—fixed, paraffin-embedded cutaneous tissue. *J. Am. Acad. Dermatol.* **19,** 747–753.
3. Boehringer Mannheim product insert for anti-digoxigenin-POD, Fab fragments. (1995) Boehringer Mannheim, Hanover, Germany.
4. Dako Envision System universal kit instructions. (1994) Dako Corporation, Santa Barbara, CA.
5. Clements, J. and Beitz, A. (1985) The effects of different pretreatment conditions and fixation regimes on serotonin immunoreactivity: a quantitative light microscopic study. *J. Histochem. Cytochem.* **33,** 778–784.

6. Laitinen, L., Laitinen, A., Panula, P., Partanen, M., Tervo, K., and Tervo, T. (1983) Immunohistochemical demonstration of substance P in the lower respiratory tract of the rabbit and not of man. *Thorax* **38,** 531–536.
7. Hsu, S. and Soban, E. (1982) Color modification of diaminobenzidine (DAB) precipitation by metallic ions and its application for double immunohistochemistry. *J. Histochem. Cytochem.* **30,** 1079.
8. Ordronneau, P., Lindstrom, P., and Petrusz, P. (1981) Four unlabeled antibody bridge techniques: A comparison. *J. Histochem. Cytochem.* **29,** 1397–1404.

25

The Avidin-Biotin Complex (ABC) Method and Other Avidin-Biotin Binding Methods*

Gary L. Bratthauer

1. Introduction

These methods involve, at their core, the vitamin biotin and the protein avidin, which bind together irreversibly. By establishing a biotin link, through avidin, between the horseradish peroxidase enzyme and a secondary antibody reagent, enzyme localization can be achieved at the site of primary antibody interaction with the specimen. These procedures are more universal than the purely immunologic techniques (*see* Chapter 24), requiring only a biotinylated secondary antibody for each species of primary antibody used. The biotin molecule is small and can be easily conjugated to immunoglobulin by amino substitution at alkaline pH without the loss of immunoglobulin activity. These secondary antibodies are quite inexpensive, readily available commercially, and reactive against immunoglobulin from a wide variety of species. The fact that the binding in this technique is not entirely immunologic and almost irreversible because of the high affinity of avidin for biotin, makes this technique less prone to errors related to assay conditions than other purely immunologic techniques. While the sensitivity provided by these methods is generally better than that obtained with the peroxidase-antiperoxidase (PAP) technique (*see* Chapter 24), sometimes resulting from fixation or for other reasons, the sample may not be suitable for avidin-biotin methodology (*1*). Tissues may have endogenous biotin or possess biotin-like binding capabilities, which may result in nonspecific binding. The irreversible nature of the binding also allows for the procedure to be undertaken with the most stringent of conditions. The

*The opinions or assertions contained herein are the private views of the author and are not to be construed as official or as reflecting the views of the Department of the Army or the Department of Defense.

From: *Methods in Molecular Biology, Vol. 115: Immunocytochemical Methods and Protocols*
Edited by: L. C. Javois © Humana Press Inc., Totowa, NJ

conditions still need to be conducive to antigen–antibody binding, but the strong avidin–biotin interaction makes this procedure slightly more forgiving. The advantages of these techniques lie in the ease of finding suitable secondary reagents. If a primary antibody has been found to be reliable, regardless of the species, there probably exists a biotinylated antibody reactive against it, or one could be easily engineered (*see* Chapter 7). The need for different PAP reagents appropriate to the primary species is removed (*see* Chapter 24).

The first method to incorporate avidin and biotin was the avidin–biotin complex (ABC) method. This method, described in 1981 by Hsu and associates, makes fundamental use of the covalent and irreversible binding seen between avidin, an egg white protein, and biotin, a vitamin *(2)*. The horseradish peroxidase enzyme can incorporate many biotin molecules without the loss of enzymatic activity. Biotin can also be conjugated to immunoglobulin, which accepts the molecules, as did the enzyme, without the loss of apparent activity *(3)*. The two biotin molecules can be joined via an avidin molecule by creating a complex of avidin and biotinylated enzyme and attaching it to the biotinylated secondary antibody. There are four binding sites for biotin on each avidin molecule, and two binding sites for avidin on each biotin molecule. Together, the biotinylated enzyme and avidin molecules form a lattice complex and remain in solution. The ratio is such that there is always an available biotin-binding site on the avidin–biotin complex for binding of the biotinylated secondary antibody (*see* **Fig. 1**). This technology allows for more enzyme to be located at the antigen site through these avidin–biotin complexes, thus increasing the sensitivity. The ABC complex can also become so large that overall binding is decreased through steric hindrance. This can, in effect, decrease the resolution of the ABC technique.

Variations on the ABC technique can also be used to incorporate different enzymes that result in different chromogenic products. Alkaline phosphatase ABC is one example. The difficulty with this system is the consumption of endogenous alkaline phosphatase, which is more prevalent than peroxidase and is harder to remove. However, the alkaline phosphatase enzyme does provide more product per unit than does peroxidase and is therefore a slightly more sensitive means of detection. Endogenous alkaline phosphatase can be blocked by an incubation in 3 mM levamisole for 15 min, but some enzyme may escape consumption. Alkaline phosphatase has many substrates, too, the most popular being BCIP/NBT, which precipitates to a dark blue (*see* Chapter 25).

Finally, an ABC assay system is nice to have available for detecting more than one antibody on an individual specimen. In double-labeling experiments, having two completely different assay systems reduces the chances of crossover reactivity. The first antigen can be detected with the standard ABC procedure, and the second antigen is then detected using the PAP system (*see*

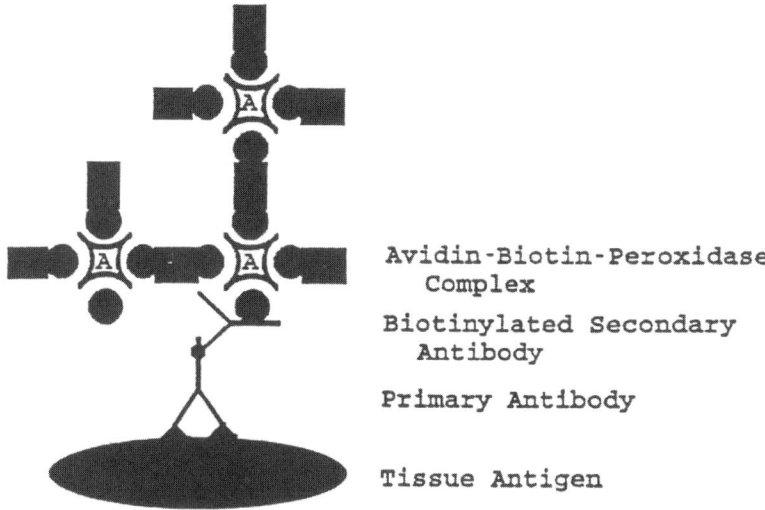

Avidin-Biotin-Peroxidase
    Complex

Biotinylated Secondary
    Antibody

Primary Antibody

Tissue Antigen

Fig. 1. Diagram illustrating the molecular interactions of the ABC procedure. The primary antibody against the antigen of interest is linked to the avidin-biotinylated peroxidase complex via a biotinylated secondary antibody raised against immunoglobulin of the animal species used to generated the primary antibody (λ, immunoglobulin; ●, biotin; **A**, avidin; ■, peroxidase).

Chapter 24). Using the two technologies provides less chance of crossover detection, especially if alkaline phosphatase is employed along with the peroxidase enzyme *(4)*.

This method, either standard or elite (increased molar ratio), remains the forefront of much of the immunocytochemistry being performed today, and will be the main subject of this chapter. There are new methods, though, that are being used with increased frequency, such as the labeled-avidin binding method, sometimes called the streptavidin-binding method, and a newer catalyzed amplification method that uses avidin, biotin, peroxidase, and a biotinyl tyramide to achieve even more sensitivity. These methods will be discussed at the end of this chapter.

2. Materials

1. Gloves, gowns, and masks.
2. Ultralow freezer (–70°C).
3. 45-L carboy.
4. Vortex.
5. Adjustable pipetman pipets with tips.
6. 4-mL glass Wheaton vials (Wheaton Scientific, Millville, NJ).
7. Humid chamber leveling tray with lid, for horizontal antibody application.

8. Absorbent paper towels.
9. Staining racks and dishes for solution incubation.
10. Stirring block with stir bars.
11. Light microscope.
12. Coverslips.
13. Phosphate buffered saline (PBS): 0.01 M sodium phosphate, 0.89% sodium chloride, pH 7.40, ±0.05 (*see* **Note 1**).
14. 10 N Sodium hydroxide and concentrated hydrochloric acid for adjusting pH.
15. Primary antibody reactive against the desired antigen (*see* **Note 2**).
16. A secondary antibody labeled with biotin, which is reactive against the species of immunoglobulin used for the primary reagent.
17. 10% normal serum in PBS from the species from which the secondary antibody was generated (For this example: horse serum).
18. Avidin–biotin complex horseradish peroxidase reagent (Vectastain ABC HRP Kit, Vector Laboratories, Inc., Burlingame, CA).
19. Chromogen: 3,3'diaminobenzidine tetrahydrochloride (DAB) (Sigma, St. Louis, MO); 30% hydrogen peroxide (H_2O_2) (*see* **Note 3**).
20. 30% bleach in water.
21. Counterstain (e.g., Mayer's Hematoxylin).
22. 6 M Ammonium hydroxide diluted 1:50 in deionized water.
23. 100% ethanol.
24. Xylene.
25. Permount mounting medium (Fisher Scientific, Pittsburgh, PA).

3. Method

3.1. The Avidin–Biotin Complex Method

Depending on the starting material, preantibody incubation steps may vary and are outlined in Chapters 9–12. The following assay begins with the removal of the slides from the overnight 10% normal serum incubation step. Since the ABC technique is universal in application, and a biotinylated secondary antibody exists for virtually any primary antibody, this protocol will assume a monoclonal assay and a primary antibody from a mouse hybridoma. A good secondary antibody to use in this situation is a biotinylated antimouse antibody made in horse. Therefore, the overnight serum incubation in this case would have been with 10% normal horse serum in PBS (*see* **Note 4**).

1. Align a humid staining chamber, leveling the slide bars, and adding water to the chamber (*see* **Note 5**).
2. Remove the slides from the dish of serum and place them on the bars in the chamber. Cover the specimens with 10% normal horse serum in PBS and replace the chamber cover.
3. Prepare the antibody solutions in PBS. If thawing out fresh frozen reagent, allow plenty of time for aggregates to disperse; likewise for any reconstitution of lyophilized material (*see* **Note 2**).

4. Blot off the 10% horse serum from each slide by placing the end of the slide on absorbent paper towels.

5. Add the primary monoclonal antibody directed against the antigen desired to the slide making sure to cover the entire specimen. Work quickly to avoid any drying. Cover the chamber and incubate for 30 min (*see* **Note 6**).

6. Rinse the specimens using PBS with wash bottle force or a siphon stream. Rinse for several seconds, allowing the buffer to freely flow off the end of the slides sitting on the racks in the chamber. Repeat three times allowing a minute or two between subsequent washings to enable any nonspecifically adherent immunoglobulins to slowly diffuse away (*see* **Note 7**).

7. Blot off the excess PBS from each slide by placing the end of the slide on the absorbent paper towels.

8. Add more 10% normal horse serum in PBS covering the specimens to further guard against nonimmunologic binding. Cover the chamber and incubate for 10 min (*see* **Note 8**).

9. Rinse with PBS briefly, then add more 10% normal horse serum in PBS and cover the chamber for another 10-min incubation.

10. Blot off the 10% horse serum from each slide by placing the end of the slide on the absorbent paper towels (*see* **Note 9**).

11. Add the biotinylated horse antimouse IgG secondary antibody to the slide, making sure to cover the entire specimen. Work quickly to avoid any drying. Cover the chamber and incubate for 30 min (*see* **Notes 4** and **10**).

12. Prepare the ABC reagent according to the manufacturer's instructions. For Vector Laboratories' Vectastain Elite kit, add 2 drops of reagent A to 5 mL of PBS, then add 2 drops of reagent B. Mix and incubate for 30 min at room temperature. Additional volume may be prepared keeping the ratio of reagents constant.

13. Repeat **step 6**.

14. Blot off the excess PBS from each slide by placing the end of the slide on the absorbent paper towels.

15. Add the avidin–biotin complex to the slide, making sure to cover the entire specimen. Work quickly to avoid any drying. Cover the chamber and incubate for 30 min.

16. Repeat **step 6**.

17. Place the slides in a staining slide rack and incubate for 10 min in a dish of PBS. Do not allow specimens to dry.

18. Prepare the DAB chromogen solution (*see* **Note 3**).

19. Place the rack of slides into the DAB dish, cover, and incubate for 15 min.

20. Remove the slide rack and wash in a dish with three changes of deionized water, for 2 min each.

21. Counterstain with Mayer's hematoxylin 1–5 min depending on the concentration and color intensity desired (*see* **Note 11**).

22. Rinse with deionized water, three changes, 2 min each.

23. Develop the nuclei blue with a 10-s incubation in ammonium hydroxide in water.

24. Rinse with deionized water, three changes, 2 min each.

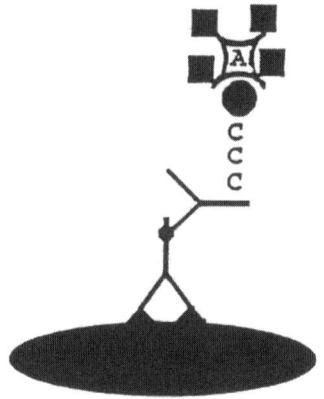

Peroxidase Labeled
Avidin

Biotinylated Secondary
Antibody

Primary Antibody

Tissue Antigen

Fig. 2. Diagram illustrating the molecular interactions of the LAB procedure. Horse-radish peroxidase is covalently linked to avidin. The primary antibody against the antigen of interest is linked to the enzyme labeled avidin complex (LAB) via a biotinylated secondary antibody raised against immunoglobulin of the animal species used to generate the primary antibody (λ, immunoglobulin; **CCC**, long carbon arm extension; ●, biotin; **A**, avidin; ■, peroxidase).

25. Dehydrate specimens with 100% ethanol, four changes for 2 min each.
26. Clear specimens with xylene, four changes, for 2 min each.
27. Cover slip with Permount, dry, and observe (*see* **Note 12**). Positive reaction should be visible as a brown precipitate. The nuclei should be light blue.

3.2. Additional Methods

3.2.1. Labeled Avidin Binding Assay

The labeled avidin binding (LAB) method provides even more sensitivity than the other popular immunohistochemistry methods *(5)*. The technological change seen with this method is that the enzyme peroxidase is covalently linked directly to the avidin molecule. This enhances sensitivity, since the avidin molecule can be labeled extensively with enzyme, and the complex remains relatively small. The small size enables many avidin molecules to bind to the biotinylated secondary antibody *(6)*. In addition, the biotin is more accessible, because it is positioned on a long carbon arm extension from the secondary antibody, reducing steric hindrance (*see* **Fig. 2**). With the numerous enzyme-labeled avidin molecules able to attach to any biotin on the secondary antibod-ies, the sensitivity and the resolution of this technique are superior *(7)*. The avidin molecule in use in this system is generally obtained from *Streptomyces avidinii* bacteria and is called streptavidin, a molecule with a more neutral pH and a reduced tendency toward inappropriate binding *(8)*. There is also another form of avidin developed by Belovo Chemicals called Neutralite Avidin (available from Accurate Chemical and Scientific, Westbury, NY). This avidin molecule has been engineered without sugar residues and modified to have a neutral isoelectric point. When properly labeled with an enzyme marker, this molecule should have less tendency to bind nonspecifically to charged sites on the specimen. In addition, like the ABC method, there are alkaline-phosphatase-labeled streptavidin molecules that can be used to provide different chromogenic products.

For this procedure, a secondary antibody labeled with a long-arm-spacer biotin, reactive against the species of immunoglobulin used for the primary reagent (Kirkegaard & Perry Laboratories, Gaithersburg, MD), is needed. This reagent is used in **Subheading 3.1., step 11** of the above method.

Then, the peroxidase-labeled streptavidin (Kirkegaard & Perry) reagent is used in **step 15**. The labeled streptavidin is added at the appropriate dilution, which must be experimentally determined (*see* **Note 2**). It does not need to be prepared a minimum of 30 min in advance like the ABC reagent.

3.2.2. Amplified Biotin Substrate Assay

The amplified biotin methods involve the use of biotinylated tyramine, which, when subjected to peroxidase enzyme, deposits biotin at the enzyme site *(9)*. This reagent can be utilized in any detection method incorporating peroxidase. Following the biotin deposition (which occurs when this reagent is used in place of DAB or some other chromogen), an additional avidin-enzyme reagent would be employed (*see* **Fig. 3**). This, reagent would bind to all of the deposited biotin, and detection with a subsequent chromogen could increase the sensitivity 1000-fold. The tremendous increase in sensitivity may require considerable manipulation of an already-established technique. Also, the resolution will be far worse than with the ABC-type procedure. However, if used for instances in which the sought-after antigen is in very small quantity, the decreased resolution will not be a factor, and the increase in sensitivity may allow for the detection of otherwise undetectable substances.

For the procedure, add the biotinyl tyramide (Dupont NEN, Boston, MA, or Dako, Carpinteria, CA) solution in place of the DAB solution in **Subheading 3.1., step 19** and incubate for 15 min. Rinse as in **Subheading 3.1., step 16**.

Then, return to **Subheading 3.1., step 15** and add more ABC solution or a labeled streptavidin compound. Continue on with the procedure from there.

The added steps will increase the overall turn-around time for these procedures, but the potential identification of previously undetectable substances is worth the extra hour or so.

There are other alterations as well to the basic procedures outlined in these chapters that involve combinations of all of these methods, allowing further

Peroxidase-Labeled Avidin

Biotinyl Tryamide
Oxidized by Peroxidase
to Biotin

Avidin-Biotin-Peroxidase
Complex
Biotinylated Secondary
Antibody
Primary Antibody
Tissue Antigen

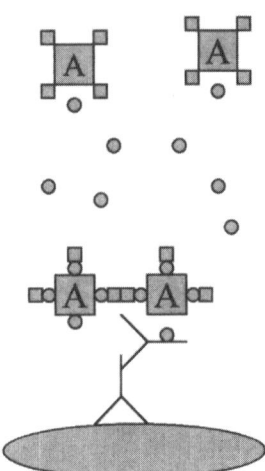

Fig. 3. Diagram illustrating the amplified biotin substrate method. The biotinyl tyramide compound TSA forms biotin when oxidized by peroxidase. The deposition of biotin in the vicinity of the original ABC reagent allows for further amplification by the addition of peroxidase-labeled avidin or more ABC, followed by substrate (⅄, immunoglobulin; ●, biotin; A, avidin; ■, peroxidase).

amplification through the use of antibodies to biotin, avidin, or both. For example, a biotinylated antibody can be recognized by an antibiotin antibody, and a linking reagent can be used to bind the antibiotin antibody to a PAP type of complex; or, a secondary biotinylated antibody can be employed, following the antibiotin step, which could then result in an ABC means of detection. Alternatively, an antiavidin may be used to further amplify a LAB technique. With each successive step, a greater level of amplification can be achieved. However, the possibility of higher background and poorer resolution from severe steric hindrance is also increased.

4. Notes

1. Prepare the PBS for large volume use. For 45 L of a simple buffer with low ionic strength:
   a. Weigh 50 g sodium phosphate and 150 g sodium chloride and dissolve in a 4-L flask, filled to within 100 mL of the 4-L mark.
   b. Bring the pH of the solution to 6.8 with 10 N sodium hydroxide (about 20 mL).
   c. Fill the remainder of the flask to 4 L and check the pH.
   d. Dilute the flask to 45 L in carboy.
   e. Mix well by shaking/rotating the carboy.
   f. Remove a beaker of solution and check the pH. It should be close to 7.4. If not, adjust with appropriate volumes of hydrochloric acid or sodium hydroxide. The final pH of the 0.01 M PBS should be 7.40 ±0.05.

2. Antibody solutions are made with PBS to a predefined dilution. These are experimentally determined on a known positive specimen. Usually, working antibody concentrations are in the range of 10 to 20 µg/mL. However, depending on the individual reagent, this concentration could vary considerably. For a beginning titration, start at 1:10 and do serial 1:10 dilutions, resulting in 10-, 100-, 1000-, and 10,000-fold dilutions of concentrated or "neat" antibody. Immunoassay a known positive specimen following the procedure, and examine the sections. Optimum results will occur within a range of the dilutions. Once a range is established, another titration can be performed within that range of 1:2 serial dilutions to better optimize the final dilution. The antibodies can be stored concentrated in aliquots in an ultralow temperature freezer at –70 to –80°C. Antibodies are stable in the freezer indefinitely but should be thawed once and used, not refrozen. However, the manufacturer recommends that the ABC reagent components be stored in the refrigerator. If desired, antibodies can be thawed and diluted to a concentrated stock solution from which more dilute working solutions can be prepared. These stock solutions can be kept at 4–8°C for 1 wk.

3. The chromogen DAB is a potential carcinogen and should be handled accordingly. For small operations, there exists commercially available tablets or prediluted concentrates designed to provide approx 100 mL of reagent. In those instances where the volume of slides to prepare is not great, these are preferred because of the reduced contact that this type of prepared chromogen offers. However, for large volumes, or in case alterations to the procedure are desired, the following is one method of preparing the DAB chromogen:

   a. DAB is purchased in 5-g bottles and stored desiccated in a –20°C freezer.
   b. Working in a fume hood and wearing the appropriate gloves, gown, and mask, add PBS to the 5-g bottle and mix until dissolved.
   c. Add the contents of the 5-g DAB bottle to a 250-mL volumetric flask, bring the volume to 250 mL with PBS and mix by stirring.
   d. Remove the contents to a 500-mL beaker, and pipet 4-mL vol to 4-mL glass Wheaton vials, with constant stirring.
   e. The 4-mL vials should be frozen at –70°C until used.
   f. In the fume hood, add one 4-mL vial of DAB to 500 mL of PBS in a staining dish, and then add 400 µL of 30% H_2O_2. DAB should be prepared directly before use and kept covered in the hood, because oxidation begins with the addition of H_2O_2 and will continue if exposed to the open air. This solution should be made fresh every day and should only be used for 3–4 h or for three racks of slides and discarded when it turns a dark brown color. DAB is a potential carcinogen and should be treated as hazardous waste, and 30% H_2O_2 is extremely caustic and can cause burns (*see* slight variation on conditions in **Note 7**).
   g. The contaminated glassware and any spills should be cleaned up with a 30% bleach solution, or with peroxidase enzyme and H_2O_2.

4. Many biotinylated secondary antibodies from various species that are reactive against various species' immunoglobulin can now be pooled. Some manufacturers provide pooled or universal secondary reagents, which can be used regardless

of the primary antibody species. The only requirement is that a universal type of blocking serum also be used so crossreactivity will not occur. The various manufacturers also provide blocking agents that are protein-, milk-, or casein-based solutions. While this provides convenience, it does not always produce better results.

5. As with all immunological procedures, it is important to make sure the specimens stay hydrated throughout. Drying will result in nonspecific immunoglobulin binding. A level chamber rack is also important, because in approx 30 min, immunoglobulins may flow with gravity away from the specimen location.

6. Allow the antibody solutions and the slides to come to room temperature before applying. Also make sure that the antibodies used are fully dissolved or reconstituted. Sometimes the results are better if the antibody solutions are prepared the day before and left at 4°C overnight. Slight vortexing before application helps to ensure proper solution. To increase reactivity, incubations can be extended in length, or temperatures can be raised. This often increases background as well. Sometimes reaction intensity can be increased without background by incubation with primary antibody overnight at 4°C *(10)*. The inconvenience of using a refrigerator, especially with large volumes of slides, makes room-temperature incubations more attractive, and the reduced background benefit sometimes possible with overnight incubations is small enough to be discounted.

7. Vigorous rinsing is required for reduction of background. Often, when assaying tissue sections that have been proteolytically digested or subjected to microwave retrieval methods, too vigorous of a wash may cause the section to dislodge from the slide. Still, it is important to wash as thoroughly as possible. The wash stream should not directly splash the specimen surface, as this may dislodge antibodies with low avidity. The stream should initiate at one end of the slide, and the solution should flow rapidly across the slide surface. Washes are the most important aspect of background reduction and should therefore be extended if too much background is a problem. To further reduce background, 0.25% Triton X-100 can be added to the PBS *(11)*. This will also, in some cases, reduce reactivity as the charge interactions of antibody-antigen binding will be altered. However, if the antibody has a high avidity and other unwanted clones are present to a lesser degree, the addition of detergent can reduce background staining. Also, if plus slides or poly-L-lysine coatings are used, it may be necessary to add this detergent to the buffer because of the hydrophobic nature of these slides. The results, though, are generally better than with glue and are artifact-free. A compensation in chromogen concentration may offset the effects of the detergent on signal. The chromogen concentration may be doubled or increased up to five times (reducing time in solution accordingly) without much increase in background because of detergent washes. If the procedure is performed in this fashion, PBS without detergent is needed for preparation of primary antibodies (some of which are not stable if stored in a detergent solution) and in the preparation of blocking sera and chromogen solutions. If there is substantial background, a preabsorption may be indicated (*see* Chapter 4). The antibody may be diluted in buffer containing 2% bovine serum albumin (BSA), secondary species serum (horse in the above

example), or even 0.5% normal human serum if the contaminating clones are reactive against a serum-based constituent.

8. The 10% secondary species serum incubation allows for protein to bind to charged sites on the specimen. The additional 10-min incubations are extra protection against unwanted antibody sticking nonspecifically. The concentration of the serum solution and the time of incubation may be increased if desired.

9. Some tissue samples, especially fresh or frozen ones, can have intrinsic biotin-binding capabilities that must first be blocked before using an ABC type of detection. The simple Blocking Kit from Vector can be used to add avidin to the specimen to bind biotin, followed by biotin to bind avidin. Incubations of 15 min with brief PBS rinses should suffice.

10. The ABC complex requires 30 min to form. Make sure the complex is prepared right after the secondary antibody addition so that it has time to build. The complex is stable in the refrigerator for 72 h.

11. It is better if the counterstain is weak, because the principal reaction product may be masked by too much counterstain. Ideally, just enough hematoxylin to identify structure is all that is necessary.

12. These slides are permanent and should not fade with time. If the presence of endogenous pigment is a problem with a particular specimen, or a color other than brown is desired as an indicator, different chromogenic compounds can be used. The compound 3-amino-9-ethylcarbazole (AEC) can be used, for example, to create a red color product at the site of enzyme deposition. This compound can be purchased commercially and in kit form. It is also thought to be potentially carcinogenic. Using the AEC substrate system, begin at **Subheading 3.1., step 18** and prepare the AEC chromogen as follows:

   a. Place 2 mL of acetate buffer in a Wheaton vial.
   b. Add 1 drop of AEC to the vial and mix.
   c. Add 1 drop of H_2O_2 to the vial and mix.
   d. Filter if necessary.

   A precipitate may develop that will not affect the results in any way, but can be filtered out if desired. Apply to the specimen, making sure that it is covered incubate for 15 min, and continue with the protocol. Instead of dehydrating in alcohol and xylene, allow specimens to dry, add 1 drop of Crystal/Mount to each and bake in a 60°C oven for 30 min. This preparation is permanent and can be cover-slipped with Permount if needed. The Crystal/Mount will form a hard plastic coating on the slide, but can be damaged by smudging. The positive reaction product with this chromogen will be red, with the nuclei a light blue.

   As an alternative, 1 mL of 1% cobalt or nickel chloride can be added to the DAB solution prior to incubation, and the resultant precipitate will be dark blue to black, not brown *(12)*. This can increase the overall sensitivity of the reaction, but it is not popular because of the different color counterstain that is usually required. A 5% solution of Methyl Green used as a counterstain for 5 min provides enough contrast to the blue to be a good background for interpreting assays with a nuclear location.

References

1. Hsu, S. and Raine, L. (1981) Protein A, avidin, and biotin in immunohistochemistry. *J. Histochem. Cytochem.* **29,** 1349–1353.
2. Hsu, S., Raine, L., and Fanger, H. (1981) The use of avidin-biotin-peroxidase complex (ABC) in immunoperoxidase technique—a comparison between ABC and unlabeled antibody (PAP) procedures. *J. Histochem. Cytochem.* **29,** 577.
3. Guesdon, J., Ternynck, T., and Avrameas, S. (1979) The use of techniques avidin-biotin interaction in immunoenzymatic techniques. *J. Histochem. Cytochem.* **27,** 1131–1139.
4. Gillitzer, R., Berger, R., and Moll, H. (1990) A reliable method for simultaneous demonstration of two antigens using a novel combination of immunogold-silver staining and immunoenzymantic labeling. *J. Histochem. Cytochem.* **38,** 307–313.
5. Elias, J., Margiotta, M., and Gabore, D. (1989) Sensitivity and detection efficiency of the peroxidase antiperoxidase (PAP), avidin-biotin complex (ABC), and the peroxidase-labeled avidin-biotin (LAB) methods. *Am. J. Clin. Pathol.* **92,** 62.
6. Leary, J., Brigati, D., and Ward, D. (1983) Rapid and sensitive colormetric method for visualizing biotin-labeled DNA probes hybridized to DNA or RNA immobilized on nitrocellulose. *Proc. Natl. Acad. Sci. USA* **80,** 4045–4049.
7. Milde, P., Merke, J., Ritz, E., Haussler, M., and Rauterberg, E. (1989) Immunohistochemical detection of 1,25-dihydroxyvitamin D_3 receptors and estrogen receptors by monoclonal antibodies: comparison of four immunoperoxidase methods. *J. Histochem. Cytochem.* **37,** 1609–1617.
8. Ayala, E., Matinez, E., Enghardt, M., Kim, S., and Murray, R. (1993) An improved cytomegalovirus immunostaining method. *Lab. Med.* **24,** 39–43.
9. Bobrow, M. N., Litt, G. J., Shaughnessy, K. J., Mayer, P. C., and Conlon, J. (1992) The use of catalyzed reporter deposition as a means of signal amplification in a variety of formats. *J. Immunol. Meth.* **150,** 145–149.
10. Clements, J. and Beitz, A. (1985) The effects of different pretreatment conditions and fixation regimes on serotonin immunoreactivity: a quantitative light microscopic study. *J. Histochem. Cytochem.* **33,** 778–784.
11. Laitinen, L., Laitinen, A., Panula, P., Partanen, M., Tervo, K., and Tervo, T. (1983) Immunohistochemical demonstration of substance P in the lower respiratory tract of rabbit and not of man. *Thorax* **38,** 531–536.
12. Hsu, S. and Soban, E. (1982) Color modification of diaminobenzidine (DAB) precipitation by metallic ions and its application for double immunohistochemistry. *J. Histochem. Cytochem.* **30,** 1079.

26

Avidin–Biotin Labeling
of Cellular Antigens in Cryostat-Sectioned Tissue

Mark Raffeld and Elaine S. Jaffe

1. Introduction

Immunohistochemical staining techniques are widely used for the identification of a variety of diverse antigens in paraffin-embedded tissues (*1*). Nonetheless, a major limitation in the use of paraffin-embedded fixed tissue is that many potentially interesting antigens are denatured and their antigenicity destroyed by the process of tissue fixation. To overcome this problem, many laboratories have employed the use of cryostat cut frozen sections (*2–4*). The principles of the staining reactions are identical to those used for paraffin sections, and the antigenicity of the majority of cellular proteins and carbohydrate moieties are preserved. The frozen section technique, however, has its own drawbacks. Special equipment is needed for freezing, cutting, and storing the frozen tissue blocks. Second, good morphologic detail requires greater skill in cutting the tissue sections and careful attention to particular steps in the procedure (*see* **Subheading 3.**). Even under optimal conditions, the morphology of the immunostained frozen section will be inferior to the morphology obtained from paraffin-embedded fixed tissue.

As previously mentioned, the principles of staining are identical to those used in paraffin sections. The techniques used may be direct or indirect as described in Chapters 15–18. Indirect techniques are generally more sensitive and, therefore, preferable. Indirect techniques can be broken down into three steps. In the first step, an antibody directed against the antigen of interest is applied to the tissue section. In the second step, a labeled secondary antibody directed against the first antibody is applied. The last step consists of a detection step that is composed of linking the secondary antibody to a detection system and

From: *Methods in Molecular Biology, Vol. 115: Immunocytochemical Methods and Protocols*
Edited by: L. C. Javois © Humana Press Inc., Totowa, NJ

an enzymatic reaction that converts a substrate to a colored product (*see* **Notes 1** and **2**).

The technique described here is for use with monoclonal primary antibodies of mouse origin, but can easily be adapted for use with polyclonal antibodies from other species (i.e., rabbit). This method uses a secondary biotin-labeled antibody and a detection system that employs a biotin–avidin horseradish peroxidase complex linker step, the so-called ABC (avidin–biotin complex) detection system *(5)* (*see* Chapter 25). In this detection system, avidin acts as a bridge between the biotinylated secondary antibody and a biotin-labeled peroxidase enzyme. The anchored enzyme, in the presence of H_2O_2 can then convert the substrate, diaminobenzidine, to a brown or black reaction product that is easily identifiable in the tissue section.

The ABC detection system has been shown to be more sensitive than most other detection system *(5,6)*, primarily because of the large size of the preformed ABC complexes, which result in amplification of the signals. Alternative detection systems for immunohistochemical analysis include the peroxidase–antiperoxidase (PAP) *(1)* and the alkaline phosphatase–antialkaline phosphatase (APAAP) systems *(7)* (*see* Chapter 24). These approaches are conceptually and technically similar, and will not be discussed here.

2. Materials

1. Freezing solution: 2-methylbutane and dry ice or liquid nitrogen.
2. OCT embedding compound (Miles Laboratories, Naperville, IL).
3. Tris-buffered saline (TBS): 0.05 *M* Tris-HCl and 0.15 *M* NaCl, pH 7.6.
4. Acetone.
5. Normal goat serum: 1–3% in TBS.
6. Primary monoclonal antibody (of mouse origin), at appropriate dilution.
7. Secondary antibody: biotin-labeled F(ab')^2 fragment of goat antimouse immunoglobulin (TAGO, Burlingame, CA) or biotin-labeled goat antimouse immunoglobulin (Vector Laboratories, Burlingame, CA), at appropriate dilution.
8. Avidin–biotin complex (Vector ABC Elite kit) (Vector Laboratories): ABC solution should be made fresh 10 min before use, so that complexes have time to form according to instructions of manufacturer.
9. Diaminobenzedine (DAB) solution: 100 mg DAB (Sigma, St. Louis, MO) and 0.01% H_2O_2 in 200 mL TBS (optional, 1 mL of 8% nickel chloride). DAB is a suspected carcinogen and should be handled with caution. DAB solution should be made fresh, immediately before use (*see* **Notes 3** and **4**).
10. Dewar flask for snap-freezing tissues.
11. Cryostat for preparing frozen sections; freezing chucks.
12. Clean glass slides previously treated with 3-aminopropyltriethoxysilane (Sigma) or other adhesives, such as gelatin, poly-L-lysine, or glue adhesive.
13. Humid chamber for incubations.
14. Suitable low-temperature freezer or Dewer flask for tissue storage.

15. Absolute alcohol, 95% ethanol, and xylene.
16. Glass coverslips.
17. Permount.

3. Methods

3.1. Preparing the Tissue

1. Place the tissue in normal saline or other media to prevent drying. Do not place tissue on gauze or paper towel since desiccation will occur. Slice the tissue with a sharp scalpel blade or razor. Sections should be 2–3 mm in thickness.
2. Chill the freezing chuck by dipping it briefly into freezing solution, either 2-methyl-butane with added dry ice or liquid nitrogen. Use a clamp or other device to hold chuck.
3. Apply a thin layer of OCT embedding compound to the cold chuck. Place the tissue section on the embedding compound, which by now should have begun to form a solid base. Do not allow the embedding compound to solidify fully. Apply additional OCT compound to cover entire tissue.
4. Immerse the chuck with the tissue slice into the freezing solution. Freezing is indicated by a change in color of the embedding solution from clear to white and will be completed in 5–10 s. After 10 s, remove the chuck and tissue from the freezing solution, and place them on dry ice (*see* **Note 5**).
5. Store the tissue on the chuck if sections are to be cut in the immediate future; if not, wrap the tissue in aluminum foil, and place it in a small sealable plastic bag. Store at –70°C or in the vapor phase of a liquid nitrogen freezer (*see* **Note 6**).

3.2. Preparing Frozen Section Slides

1. When staining is to be performed, place the chuck containing the frozen tissue slice in the cryostat for preparation of frozen sections.
2. Clean slides, pretreated with 3-aminopropyltriethoxysilane to assist in retaining frozen sections to slide, should be used in the following steps. In addition to sialinated slides, other adhesive solutions can be used including gelatin, amino poly-L-lysine, Histostik (Accurate Chemical & Scientific, Westbury, NY), or glue (10–15% aqueous solution of Elmers Glue-All, Borden, Inc., Columbus, OH). Alternatively, one can purchase charged slides from one of several suppliers of glass microscope slides (Fisher Scientific, Pittsburgh, PA; Baxter Laboratories, McGaw Park, IL).
3. Cryostat sections should be prepared as thin as possible, 4–8 μm in thickness. Air-dry slides for 1–3 h at room temperature (*see* **Note 7**).
4. Fix sections in acetone for 5 min at room temperature. Place slides in a plastic box, and store at –70°C with desiccant if staining is not to be performed the same day. Alternatively, sections may be kept at 4°C for several days (*see* **Notes 8** and **9**).

3.3. Staining Procedure (see **Note 10**)

1. Quickly transfer the slides to TBS and wash, dipping in and out of the solution 10 times. Transfer to another Coplin jar containing TBS for 5 min Transfer to

another Coplin jar for a third wash in TBS plus 1–3% normal goat serum for 5 min (*see* **Notes 11** and **12**).

2. Apply primary antibody at the appropriate dilution, and incubate for 2 h at room temperature in a humid chamber. To keep the chamber humid, a paper towel soaked with H_2O placed at the bottom of the chamber will suffice (*see* **Notes 13–18**).

3. After incubation with the primary antibody, wash slides two times in TBS for 3 min. Then wash in TBS containing 1–3% goat serum for a third wash for 5 min.

4. Apply secondary biotinylated antibody, and incubate for another 30 min at room temperature.

5. Repeat washing as in **step 3**: two washes in TBS for 3 min each, and a third wash in TBS containing 1–3% goat serum for 5 min.

6. Application of ABC solution: Preform the ABC complex by mixing 10 µL of avidin and 10 µL of biotin-peroxidase (Reagents A and B from Vector ABC Elite kit) in 1 mL of TBS and incubating for 10 min. Apply 100 µL of ABC solution to each tissue section for 30 min at room temperature.

7. Wash slides three times in TBS for 3 min. Do not use the goat serum in the last wash at this step.

8. Incubate slides in DAB solution. The DAB solution should be made up fresh for each staining procedure just before use. Mix 100 µg DAB into 200 mL of TBS, and add H_2O_2 to 0.01%. The DAB development reaction should be monitored under the microscope. The usual development reaction takes 10–20 min. Using a known positive control, monitor development under the microscope to desired level of intensity (*see* **Note 19**). The brown (or black) reaction product should be specific to the cells bearing the antigen of interest, with little or no background staining, as illustrated in **Fig. 1**.

9. Wash slides in TBS.

10. Counterstain in hematoxylin. Other counterstains, such as Methyl Green or eosin, may be used (*see* **Note 20**).

11. Dehydrate the slides in three changes of 95% alcohol and two changes of absolute alcohol, and clear in two changes of xylene.

12. Mount with Permount, and apply a coverslip (*see* **Notes 21** and **22**).

4. Notes

1. There are many minor variations of the basic method, but the constant features of all indirect immunostaining procedures are:
   a. A primary antibody incubation.
   b. A secondary antibody incubation.
   c. A detection step with color development.

2. If the antigen of interest is abundant and additional sensitivity is not required, then a two-step direct method, rather than a three-step indirect method, can be employed. With the two-step method, the primary antibody is a biotin-labeled mouse monoclonal antibody, which is followed directly by the ABC incubation step. Because of the abundance of human immunoglobulins in tissue sections, methods for staining immunoglobulin often employ biotin-labeled mouse antihuman immunoglobulin reagents.

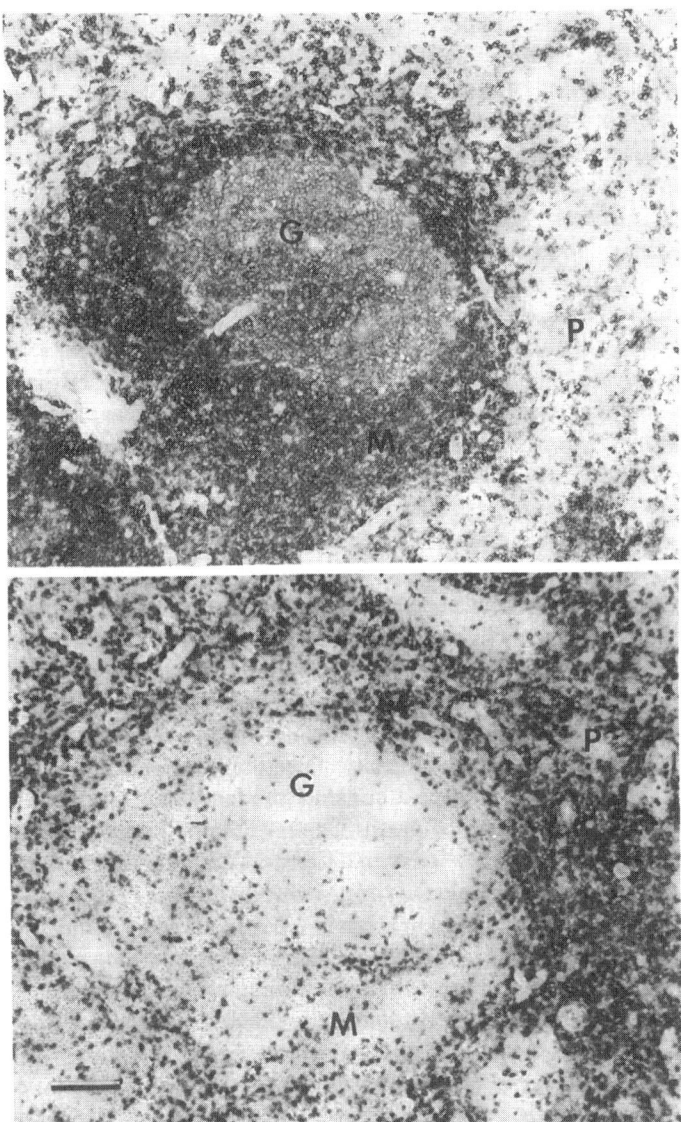

Fig. 1. Photomicrographs of a reactive lymph node follicle with germinal center (G), mantle (M), and surrounding paracortex (P) immunostained using the avidin–biotin complex technique. (Top) Follicle stained with an antibody specific to B-cells, B1 (CD20) (Coulter Immunology, Hialeah, FL), counterstained with Methyl Green. (Bottom) Parallel section of the same follicle stained with an antibody specific to T-cells, Leu 4 (CD3) (Becton-Dickinson, Mountain View, CA). Scale bar = 100 μ.

3. DAB is a carcinogen and should be handled with care. The powder should be weighed in a chemical hood. Some companies (e.g., Sigma) sell preweighed tablets ready to be dissolved in H_2O. Waste from the DAB solution should be collected and disposed of as hazardous waste.

4. Alternate chromogens to DAB can be employed. One such chromogen is aminoethylcarbizol (AEC). AEC is stated to be less carcinogenic than DAB. However, the reaction product is soluble in organic solvents, such as Permount. Thus, an alternate gelatin-based mounting medium must be employed. These media take longer to dry, and there is often fading with time. Fading of a reaction product is not observed with DAB.

5. If the tissues have not been appropriately snap frozen, ice crystals will be present in the tissue, identifiable as slits or fracture lines through the tissue. Artifactual staining may occur along such fracture lines. Therefore, ice crystal damage should be avoided. The tissues should always be snap frozen. If freezing is performed slowly over dry ice, ice crystals will accumulate in the tissue.

6. Tissue embedded in OCT should also always be maintained in the frozen state. If tissues are allowed to thaw, they cannot be refrozen without resulting in extensive damage to the tissue and often loss of antigenicity.

7. It is often helpful to mark a circle around the tissue section with a diamond glass marking pencil. This allows one to identify the area where the section is located, ensuring that reagents will be applied evenly on the tissue section.

8. Although acetone is a very good general-purpose fixative for preserving antigenicity in frozen sections, other fixatives may be preferable for preserving specific antigens. Alcohol-based fixatives are a widely used alternative to acetone. The fixative of choice for terminal deoxynucleotide transferase (TdT) and certain other nuclear antigens is 4% paraformaldehyde for 30 min (*see* Chapter 8).

9. Frozen sections can be stored dessicated at 4°C for at least 7 d without loss of reactivity of most antigens. For some antigens, slides can be kept for up to 30 d. It is always preferable to stain the slides within a few days of cutting the sections.

10. The entire staining protocol takes about 6 h. Because it is desirable to monitor the final incubation step with DAB, the number of sections stained at any one time should be limited, so that appropriate monitoring can be performed. Most experienced technologists can easily stain up to 50 slides in one run.

11. An optional step is the blocking of endogenous peroxidase activity. This procedure may result in some loss of antigenicity and decreased sensitivity. We recommend omitting this step unless absolutely necessary. Blocking endogenous peroxidase may be required for tissues with high endogenous peroxidase activity, such as spleen or bone marrow. To block endogenous peroxidase, place sections in methanol containing 3% H_2O_2 for 30 min. Blocking should be performed after **step 1** of **Subheading 3.3.** and should be followed by an additional wash step.

12. Control slides of known reactivity should be run with each set of slides immunostained. Appropriate controls should include an irrelevant antibody of the same immunoglobulin class at an equivalent concentration.

13. Most primary commercial monoclonal antibodies are used at titers ranging from 1:40 to 1:100. Antibodies should be diluted in TBS. Concentrated monoclonal antibodies should be stored with 0.1% sodium azide to prevent contamination. Most antibodies may be frozen down in concentrated form. Rare antibody preparations will lose activity on freezing.

14. Freezing and thawing of the monoclonal antibodies should be avoided. Most monoclonal antibodies are stable for months at 4°C if stored at an adequate protein concentration. Dilute antibodies in small aliquots, and use the diluted aliquot for a month or so, rather than diluting a large amount of the antibody at once. Use sterile technique when handling expensive antibodies.

15. Each primary antibody should be titered before its use in the laboratory against a known tissue expressing the antigen of interest.

16. If the reagents are not evenly applied, artifacts of staining will occur. The most common artifact is an "edge" effect or a rim of false positivity around the edge of the tissue section. This usually results from some drying of the antibodies at the edge of the tissue section. If any portion of the tissue has not been coated with reagents at any step in the procedure, this area will appear unstained. Usually a sharp line demarcates such artifactually negative areas. **Once the primary antibody has been applied to the test slide, the section should never be allowed to dry.**

17. Alternate staining protocols provide for incubation of the tissue sections with the primary antibody at 4°C overnight. This sometimes results in enhanced staining without an increase in background.

18. The method presented may be easily modified to accommodate primary polyclonal heteroantisera prepared in goats, rabbits, or other species. The advantages of monoclonal antibodies are the high specificity of these reagents and their standardization from laboratory to laboratory. Monoclonal antibodies will react only with specific and crossreacting epitopes. Serum-derived antibodies (including some sold as "affinity-purified" antibodies) may be reactive with many irrelevant antigens.

19. Nickel chloride (1 mL, 8% solution) may be added to the DAB solution. Without nickel chloride, the reaction product is brown. Nickel chloride produces a black reaction product that may be preferable for black and white photography of stained slides.

20. Methyl Green is the preferable counterstain if black and white photography is to be performed because the positive black reaction product will be clearly evident against the pale green background. If photography is not anticipated, hematoxylin may be desirable, since it provides better nuclear detail. Eosin may be preferable if one is staining nuclear antigen, such as TdT, Ki-67, or p53.

21. The pattern of anticipated results will vary with the tissue antigen identified. It is important to know the cellular distribution of the antigen of interest. Some staining patterns will be membranous. Others will be cytoplasmic or concentrate in the Golgi region. Yet others will be nuclear in their distribution. Knowledge of the cellular localization of the antigen of interest will help one to judge whether an observed staining pattern is appropriate for that antigen or whether the observed staining pattern may be the result of some artifact.

22. Some antigens are restricted to cells, whereas others are also found in interstitial spaces, appearing to produce a high background. For example, antibodies against immunoglobulins will not only stain B cells; they will also stain interstitial tissues because of free immunoglobulins present in serum and interstitial spaces.

References

1. Sternberger, L. A. (1978) *Immunocytochemistry.* Prentice-Hall, Englewood Cliffs, NJ.
2. Hsu, S. M., Cossman, J., and Jaffe, E. S. (1983) Lymphocyte subsets in normal human lymphoid tissues. *Am. J. Clin. Pathol.* **80,** 21–30.
3. Sheibani, K. and Tubbs, R. R. (1984) Enzyme immunohistochemistry: technical aspects. *Semin. Diagn. Pathol.* **1,** 235–251.
4. Warnke, R. A. and Rouse, R. V. (1985) Limitations encountered in the application of tissue section immunodiagnosis to the study of lymphomas and related disorders. *Hum. Pathol.* **16,** 326–331.
5. Hsu, S. M., Raine, L., and Fanger, H. (1981) The use of avidin–biotin–peroxidase complex (ABC) in immunoperoxidase technique: a comparison between ABC and unlabeled antibody PAP procedures. *J. Histochem. Cytochem.* **29,** 577–580.
6. Hsu, S. M., Cossman, J. C., and Jaffe, E. S. (1983) A comparison of ABC, unlabeled antibody and conjugated antibody methods with monoclonal and polyclonal antibodies—an examination of germinal center of tonsils. *Am. J. Clin. Pathol.* **80,** 429–435.
7. Cordell, J. L., Falini, B., Erber, W. N., Ghosh, A. K., Abdulaziz, Z., MacDonald, S., Pulford, K. A. F., Stein, H., and Mason, D. Y. (1984) Immunoenzymatic labeling of monoclonal antibodies using immune complexes of alkaline phosphatase and monoclonal anti-alkaline–phosphatase (APAAP complexes). *J. Histochem. Cytochem.* **32,** 219–229.

27

Multiple Antigen Immunostaining Procedures

Tibor Krenacs, Laszlo Krenacs, and Mark Raffeld

1. Introduction

The immunohistological demonstration of more than one antigen in the same tissue section may be used for studying the topographic relationships of cell populations and for correlating cell phenotype with functional or prognostic markers or with microbial infection *in situ* (*1*). In addition, single- or double-label immunohistochemistry may also be combined with the detection of nucleic acid sequences by *in situ* hybridization for revealing messenger RNA and translated protein side by side. The combination of these techniques is primarily dependent on the compatibility of the antigen retrieval technique needed and the stability of the RNA and DNA sequences in question (*2*) (*see* Chapter 46).

Two basic strategies have evolved for the detection of multiple antigens in tissue section. One strategy is to combine immunoenzymatic methods that use different chromogenic-substrate reactions (*3,4*) (*see* Chapter 23) and/or with the silver-enhanced immunogold technique (IGSS) (*5–7*) (*see* Chapter 29). The different colored reaction products can easily be studied against the recognizable background structure with a conventional light microscope, particularly when traditional histological stains such as hematoxylin for nuclei or PAS for basement membranes are additionally applied. With careful balancing of the subsequent chromogenic reactions, up to four antigens situated in separate cell populations or within different compartments of the same cell (nucleus, cytoplasm, or cell membrane) can be distinguished in a single tissue section (*6,7*).

The second strategy uses combinations of different antibodies coupled to fluorochromes with distinct emission maxima (*8,9*). The most relevant fluorochromes for combined antigen detection are fluorescein isothiocyanate (FITC; abs. max. 494 nm, emiss. max. 517 nm), rhodamine isothiocyanate (TRITC;

From: *Methods in Molecular Biology, Vol. 115: Immunocytochemical Methods and Protocols*
Edited by: L. C. Javois © Humana Press Inc., Totowa, NJ

540 and 550 nm), 7-amino-4-methylcoumarin-3-acetic acid (AMCA; 350 and 450 nm), Texas Red (359 and 615 nm) and cyanin 5.1 (Cy5; 652 and 667 nm) (*see* Chapter 14). The visualization of fluorochromes requires the use of a fluorescence microscope with special filters. With proper filter combinations, up to four antigens may be selectively stained side by side *(8)*. With appropriate filters, overlapping reactions representing antigen coexpression in the same tissue/cell compartment can be clearly demonstrated. Significant improvement in microscopic resolution can be achieved when the fluorescence signal is digitally magnified and studied with confocal laser scanning microscopy (*see* Chapter 20). In this way, objects as small as 0.2–0.5 µm, such as desmosomes and gap junctions, can be analyzed in known tissue volumes beyond the resolution of a light microscope *(9)*.

There are two basic variations for multiple antigen staining protocols, depending on whether the reagents of one immunoreaction can interact with the reagents from the second immunoreaction. When such "cross-talk" between the two staining reactions is excluded, the parallel reagents from the individual immunoreactions (e.g., primary antibodies and/or secondary antibodies) may be mixed and used simultaneously. If cross-talk cannot be avoided in the experimental design (i.e., when two mouse monoclonal primary antibodies must be used), the reaction steps for each antigen have to be completed consecutively, including some of the blocking steps. In this chapter, we present guidelines and describe protocols suitable for performing immunohistochemical double- and triple-antigen detection. Although we have not detailed specific methods for double staining using immunofluorescence, the principles of staining are identical, and suggestions for immunofluorescent detection are provided.

1.1. Principles of Reagent Combination

The principles followed for multiple antigen staining are identical to those in single immunostaining except that there is the possibility that unwanted reactions (cross-talk) between the different staining sequences can occur. This possibility must be considered when selecting an appropriate staining protocol. Cross-talk is not expected when two primary antibodies from noncrossreactive species, such as mouse and rabbit, are combined. The same is true for combinations consisting of antibodies of different immunoglobulin isotypes that are detected with isotype-specific secondary antibodies, or if the primary antibodies are of identical species or isotype, but are directly coupled to different enzymes or fluorochromes *(10)*. These combinations of reagents may be applied simultaneously in the same incubation steps. Only the chromogenic-enzyme reactions need to be carried out consecutively.

Noncrossreactive antibodies, however, are not always available in the laboratory. This has forced researchers to develop ways of avoiding cross-talk when combining autologous or crossreactive primary antibodies. The silver shell of IGSS and the compact 3,3'-diaminobenzidine tetrachloride (DAB) precipitate can completely block the immunological sequences of the associated antigen/antibody complex *(6)*. Therefore, when the antigens of interest are expected in different cell populations or in different cellular compartments, primary antibodies of the same animal species may be used in consecutive immunoreactions. IGSS or peroxidase-labeled immunoglobulin (IPO)/DAB (brown) can be used to develop the first immunoreaction, followed by the application of either IPO/3-amino-9-ethylcarbazole (AEC) or alkaline phosphatase-labeled immunoglobulin (IAP)/Fast Red, Blue, New Fuchsin, or nitroblue tetrozolium (NBT)-X-phosphate to develop the second immunoreaction (*see* Chapter 23). For visualization of some antigens, one can combine two IPO/DAB reactions, as long as one is developed in the presence of nickel chloride, which will generate a black precipitate to contrast with the brown precipitate (*see* **Fig. 1**). We prefer this relatively simple solution (*see* **Subheading 3.1.**); however, several alternative protocols are available for consideration. The simplest, but least reliable method is to attempt to disrupt the first antibody–antigen reaction after color development has occurred by lowering the pH with hydrochloric acid treatment *(11)*. This should release the interfering primary antibody from the first immunoreaction to prevent it from reacting in the immunodetection step of the second reaction. In another approach, the first monoclonal primary antibody is detected with a streptavidin method using a biotinylated antimouse secondary. Prior to the second immunoreaction, the free binding sites of the biotinylated antimouse Ig are saturated with normal mouse serum. This is followed by the application of the second primary mouse monoclonal antibody, which is FITC-labeled and detected using peroxidase-(PO) or alkaline phosphatase-(AP) conjugated anti-FITC reagents *(12)*. Yet another procedure uses preformed complexes of autologous primary antibodies and their cognate enzyme-labeled secondaries in combination. This variation has a low risk of unwanted crossreactions *(13)*; however, it is very laborious to determine the optimal saturation for the complexes. One can also take advantage of significant differences in the detection sensitivities of two different immunostaining procedures. For example, the streptavidin-biotin method with biotinylated-tyramide amplification can detect antigens by using such low concentrations of primary antibody that it can be coupled to almost any indirect system used for detection of the second immunoreaction *(14)* (*see* Chapter 25).

2. Materials

With the assistance of wet-heat mediated antigen retrieval, most antigens of interest to a general histopathology laboratory can be detected in archival tis-

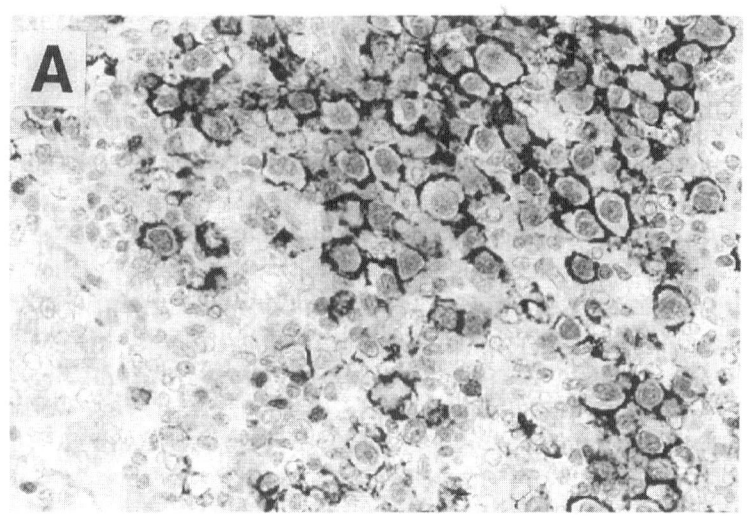

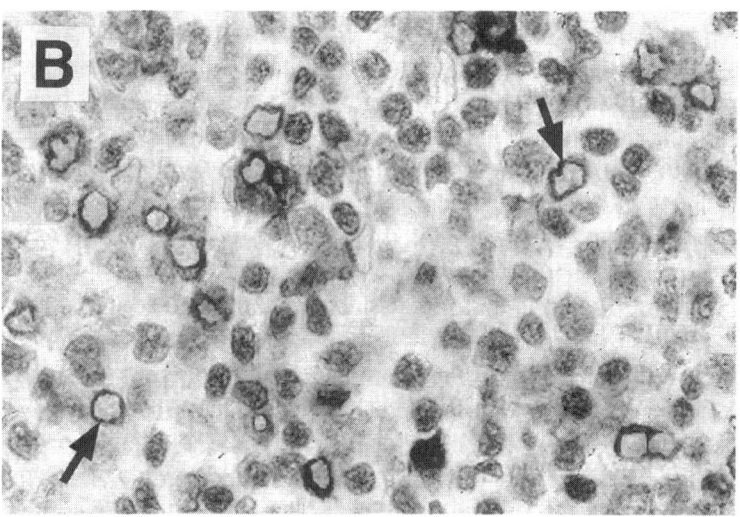

Fig. 1. Double staining of membrane and nuclear antigens using consecutive immunoreactions. (A) depicts a large B-cell lymphoma (LBCL) doubly stained for the B-cell-associated membrane antigen CD20 and the B-cell-associated nuclear antigen B-cell specific activator protein (BSAP). Note the colocalization of BSAP and CD20 in the tumor cells. (B) depicts the same LBCL doubly stained for BSAP and the membrane-associated T-cell-associated antigen CD3 (arrows). Note the divergence of nuclear BSAP staining of tumor cells and membrane CD3 staining of reactive T-cells. Double-staining method used: IPO/DAB (brown) (first antibody) and IPO/DAB + NiCl (black) (second antibody). (Abbreviations are as in text.)

sues fixed either in 10% buffered formaldehyde or B5 and embedded in paraffin wax (*see* Chapter 13). If it is not possible to detect one or more antigens of interest in paraffin-embedded tissues, or if a necessary retrieval procedure is incompatible with the detection of one antigen, frozen tissue may be used instead (*see* **Notes 1–5**). Antibodies, immunoreagents, chemicals, and equipment needed are identical to those used for single antigen detection (*see* Chapters 15–18 and 23–25).

1. Lugol's iodine: 1% iodine and 2% potassium iodide in distilled water.
2. 2.5% and 5% sodium thiosulfate in distilled water.
3. Methanol with 0.5% hydrogen peroxide.
4. Acetone.
5. TBS: 0.05 M Tris-HCl, 0.15 M NaCl, pH 7.6.
6. Blocking solution (BS): TBS with 1% bovine serum albumin (BSA), 0.1% sodium azide, and 5% normal goat serum (NGS).
7. TBS with 1% Tween-20; TBS with 0.05% Tween-20.
8. Mouse monoclonal or rabbit polyclonal primary antibodies.
9. Secondary antibodies: goat antimouse or goat antirabbit antibodies conjugated with colloidal gold (1.4 nm gold; 1:30 dilution in TBS with 0.05% Tween-20; Nanogold, Stony Brook, NY or 5 nm gold; 1:30 dilution in TBS with 0.05% Tween-20; Sigma, St. Louis, MO); conjugated with horseradish peroxidase or alkaline phosphatase (1:50 dilution in BS without azide; Dako, Carpinteria, CA); or conjugated with separate fluorescent fluorophores; for example, FITC, TRITC, Texas Red, Cy-5, or AMAC (1:50 dilution in BS; Vector Labs, Burlingame, CA or Dako).
10. Glycerol-gel mounting medium containing an antifading agent (Vectashield; Vector).
11. Double-distilled water.
12. Silver acetate developer: part a: hydroquinone; 0.1 M citrate buffer, pH 3.5 (*see* **Note 6**); and part b: silver acetate (*see* **Note 7**).
13. DAB substrate solution: DAB; 0.05 M Tris-HCl buffer, pH 7.6; and hydrogen peroxide.
14. AEC substrate solution: AEC; dimethylformamide; 0.1 M acetate buffer, pH 4.6; and hydrogen peroxide.
15. Fast Blue substrate solution: naphthol AS-MX (Sigma: N,N-dimethylformamide; 0.1 M Tris-HCl buffer, pH 9.0; and 1 M levamisole (*see* **Note 8**).
16. Fast Red substrate solution (code: K699; Dako).
17. New Fuschin substrate solution (code: K698; Dako).

3. Methods

3.1. Sequential Antigen Detection

In sequential detection protocols (*see* **Note 9**), the detection of each antigen is followed sequentially by the detection of the following antigen. As discussed in **Subheading 1.1.**, the precipitated reaction products of IGSS, IPO, or IAP can mask the immunological sequences of their associated antigen–antibody complex, thereby preventing secondary antibodies used in a separate sequence from binding to autologous primary antibodies in the preceding sequence. Primary antibodies of the same animal species or crossreacting antibodies can be combined, provided that the antigens are localized in different cell types or in different areas of the same cell. Immunofluorescence methods with crossreacting antibodies should not be used in this way.

For the following protocol, use 5-μm-thick paraffin or frozen sections. Paraffin sections and frozen sections should be mounted on charged or appropriately coated glass slides (*see* **Notes 1** and **2**) (*see* Chapters 9 and 12).

For paraffin sections, begin with **step 1** and skip **steps 5** and **6**; for frozen sections begin with **step 5** (*see* **Note 10**).

1. Dewax paraffin sections in three changes of xylene and ethanol for 2 min each.
2. Apply wet-heat mediated antigen retrieval, pressure cooking or microwaving when appropriate (*see* Chapter 13).
3. Place sections in Lugol's iodine for 5 min, then in 2.5% sodium thiosulfate for 30 s if IGSS development is used.
4. Block endogenous peroxidase in methanol containing 0.5% hydrogen peroxide for 15 min.
5. Allow freshly cut frozen sections to dry for 15 min at room temperature.
6. Fix in acetone for 5–10 min and let sections dry again for at least 2 h at room temperature before use.
7. Wash dewaxed or frozen sections in TBS for 3 min and incubate in BS for 5–10 min. Use this diluent throughout the protocol except where otherwise noted.
8. Apply primary antibody (mouse monoclonal or rabbit polyclonal) for 2 h for paraffin sections and for 30 min for frozen sections at room temperature.
9. Wash slides for 3×3 min in TBS containing 0.1% Tween-20.
10. Apply goat antimouse or antirabbit secondary antibody labeled either with colloidal gold or with horseradish peroxidase for 30 min (*see* **Notes 5** and **11**).
11. Wash for 2×2 min in TBS containing 0.1% Tween-20.
12. Before silver amplification of gold particles, rinse the sections for 3×30 s and then wash for 3×2 min in double-distilled water.
13. Use silver development for IGSS (*see* **Notes 12** and **13**) or DAB chromogenic-substrate reaction for IPO as detailed in **Subheadings 3.3.1.** and **3.3.2.**
14. Rinse slides in TBS and apply the primary antibody of the second immunological sequence for 2 h on paraffin sections or for 30 min on frozen sections.
15. Wash 3×2 min with TBS and apply the appropriate secondary antibody; that is, goat antirabbit or antimouse Ig coupled to peroxidase or to alkaline phosphatase, for 30 min (*see* **Note 14**).
16. Wash slides 3×2 min in TBS.
17. Develop the chromogenic reaction of the second sequence using AEC/H_2O_2 development for IPO or naphthol phosphate/Fast Blue, Fast Red, or New Fuchsin for IAP development as detailed in **Subheading 3.3.** (*see* **Note 15** for triple labeling).

3.2. Simultaneous Antigen Detection

In this method, the primary antibodies are combined into a single cocktail and applied simultaneously in a single-reaction step. The same is done for the labeled secondary antibodies. Antibodies of different animal species or immunoglobulin isotypes (noncrossreactive) or autologous antibodies that are directly coupled to different enzymes/fluorochromes are required. Immuno-fluorescent methods are particularly suited for this method.

Steps 1–6 are identical to those described in **Subheading 3.1.**

7. Mix primary antibodies (e.g., a rabbit polyclonal with a mouse monoclonal) in BS. Prepare 2X concentrations of each primary antibody, mix equal volumes, and apply the 1X primary antibody mixture onto the sections. Incubate paraffin sections for at least 2 h and frozen sections for 30 min.
8. Wash slides for 3 × 3 min in TBS containing 0.05% Tween-20.
9. Apply appropriate, labeled secondary antibodies together (e.g., goat antimouse and antirabbit immunoglobulins), labeled with either colloidal gold, horseradish peroxidase, or alkaline phosphatase, respectively. Alternatively, use secondary antibodies labeled with separate fluorescent dyes. Incubate paraffin sections for 1–2 h and frozen sections for less than 30 min.
10. Wash for 3 × 2 min in TBS.
11. Use the relevant visualization steps described in **Subheading 3.3.** for IGSS, IPO, or IAP, in sequential reactions. Sections stained for immunofluorescence should be mounted with a glycerol-gel solution containing an antifading agent such as Vectashield (Vector).
12. Rinse slides stained with immunoenzyme or IGSS methods and mount them with glycerol-gel.

3.3. Visualization of Reaction Products for Light Microscopy

The wide range of chromogenic-substrate systems available allows one to obtain excellent color contrast for double/multiple antigen detection. Chromo-genic reactions resulting in black (IGSS), yellow-brown (IPO/DAB), red-brown (IPO/AEC), blue (IAP/Fast Blue, or 5-bromo-4-chloro-3-indolyl phosphate/nitroblue tetrazolium [BCIP-NBT]), and purple (IAP/Fast Red or New Fuchsin) can be used in different combinations (*see* Chapter 23). For development of these products, the following protocols are provided (*see* **Note 16**).

3.3.1. IGSS

1. Prepare solution a (silver acetate developer) *(15)* by dissolving 250 mg hydro-quinone in 50 mL of 0.1 *M* citrate buffer, pH 3.5. Prepare solution b by dissolving 100 mg silver acetate in 50 mL double-distilled water (*see* **Notes 6** and **7**).
2. Mix solutions a and b just before use.
3. Immerse sections in developer and place the jar in a dark place. After 5–8 min, monitor the intensity of the gray-black product under the microscope.

4. When desired intensity is achieved, wash slides in distilled water for 2 min, then in 5% sodium thiosulfate for 2 min.

3.3.2. IPO/DAB

1. Dissolve 20 mg DAB in 100 mL of 0.05 M Tris-HCl buffer, pH 7.6, and add 100 μL hydrogen peroxide to it just before use.
2. Place slides into DAB solution in Coplin jar and develop for 5–15 min under microscopic control to obtain a yellow-brown product.

3.3.3. IPO/AEC

1. Mix 2.5 mL of 1% AEC dissolved in dimethylformamide into 90 mL of 0.1 M acetate buffer, pH 4.6. Add 100 μL hydrogen peroxide to the AEC solution before use.
2. Place slides into AEC solution in Coplin jar and develop for 5–15 min under microscopic control to obtain a red-brown product.

3.3.4. IAP/Fast Blue (see **Note 17**)

1. Dissolve 2 mg naphthol ASMX phosphate in 200 μL N,N-dimethyl-formamide, add 9.8 mL of 0.1 M TrisHCl buffer, pH 9.0, and 1 drop of 1 M levamisole (*see* **Notes 17** and **18**).
2. Before use, add 10 mg Fast Blue BB salt, shake to dissolve, filter, and drop the mixture onto the sections; a blue product will form.

3.3.5. IAP/Fast Red (see **Note 17**)

1. Using a commercial kit available from Dako (or other comparable kit), dissolve one tablet that contains the substrate, napthol phosphate, the chromogen, Fast Red, and levamisol in 2 mL of 0.1 M Tris-HCl buffer, pH 8.2, provided in the kit.
2. Filter the solution through the filter device provided and drop onto the section. A purple-red product will form in 5–15 min.

3.3.6. IAP/New Fuchsin (see **Note 17**)

1. Using the commercial kit available from Dako (or another comparable kit), add 1 drop Tris-buffer concentrate (vial a) and 1 drop of substrate concentrate (vial b) to 2 mL distilled water.
2. Mix 1 drop New Fuchsin (vial c) with 1 drop sodium nitrite (vial d). After 2 min, mix this chromogen solution to the buffer-substrate solution, shake, and cover the sections with this solution. A purple product will form in 5–15 min.

4. Notes

1. 3-amino-propyltriethoxysilane (APES) coating: Clean glass slides by immersing in ethanol for 5 min. After drying, immerse the slides for 5 min in 2% APES dissolved in acetone. Rinse the slides briefly in distilled water, and keep them at 56°C overnight before use. Heat activation of slides for 2 h with the sections

mounted is also important to avoid detachment of tissue sections during antigen retrieval.

2. When IGSS is used for color development, sections should be mounted on high-purity gelatin or poly-L-lysine-coated slides instead of APES-coated slides. APES may take part in silver reduction causing background in the reaction. This background can be reduced by careful washing and by the addition of detergent to the wash buffer and diluent. APES is the preferable gluing agent for sections when harsh antigen retrieval is used.

3. Frozen sections fixed in acetone are vulnerable to washing and soaking. Therefore, the immunostaining protocol should be kept as short as possible to avoid structural damage. Overnight drying of frozen sections at RT following acetone fixation is recommended to help protect section integrity during the longer double/triple staining procedure. Most antigens will survive the overnight drying step.

4. Epoxy resin-embedded semithin sections of 1–2 μm may also be used following extraction of the resin with a sodium meth(et)hoxide treatment for 5–8 min. Sodium meth(et)hoxide is prepared by saturating methanol or ethanol with sodium hydroxide pellets.

5. Sodium azide inhibits peroxidase enzyme, therefore it should be left out from the diluents used with PO-immunoglobulin conjugates. Add 0.05% Tween-20 when diluting gold reagents.

6. Prepare citrate buffer as follows: Dissolve 2.56 g citric acid and 2.36 g trisodium citrate in 50 mL double-distilled water; stir with a magnetic stirrer for 15 min. The buffer may be stored at 4°C for approx 2 mo.

7. Use a magnetic stirrer for 15 min to dissolve the silver acetate.

8. This solution can be stored at –20°C for 2 mo.

9. In general, the accompanying two-step indirect methods provide sufficient sensitivity to successfully complete a double/triple immunolabeling procedure in 1 d.

10. F(ab) fragments of both conjugated and unconjugated antibodies are preferred in frozen section immunohistochemistry of lymphoid tissues in order to avoid nonspecific binding of antibodies through their Fc fragment to Fc receptors (e.g., on B lymphocytes).

11. As a negative control, the crossreactivities of the secondary antibodies may be checked by exchanging them in single-staining methods.

12. The silver development step of IGSS is very sensitive to the ambient temperature in the laboratory. There can be significant differences in the reaction speed in summer and winter depending on air conditioning and heating.

13. A further advantage of the IGSS method is that its reaction product can clearly be differentiated from immunoenzyme products with epipolarization microscopy.

14. When increased sensitivity is needed, Dako's Envision (indirect) reagents are recommended. These products consist of long dextran polymers coupled to tens of enzyme and immunoglobulin molecules, which provide increased sensitivity. Monospecific EnVision+ reagents labeled with PO or AP are most useful, but dual-specificity reagents (i.e., antimouse and antirabbit) can also be used following the blocking of the irrelevant antibody with normal serum (*see* Chapter 24).

15. Double labeling with IGSS and IPO may be followed by the detection of a third antigen consecutively with an indirect IAP method as detailed above. Recommended combinations for sequential triple antigen detection are:
    a. IGSS (black) + IPO (AEC, red-brown) + IAP (Fast Blue, blue).
    b. IGSS (black) + IPO (DAB, brown) + IAP (Fast Red or New Fuchsin, purple).
    c. IPO (DAB, brown) + IPO (AEC, red) + IAP (Fast Blue, blue; or X-phosphate, dark-blue).
16. Preconditioning of the sections in buffer for 3–5 min in the substrate-chromogen buffers is important for avoiding unexpected precipitates of the chromogens resulting from sudden changes in pH or salt molarity.
17. Development of IAP products are very hard to standardize; therefore, the use of commercial kits is highly recommended.
18. This solution can be stored at –20°C for 2 mo.

References

1. Krenacs, T. and Krenacs, L. (1994) Immunogold-silver staining (IGSS) for immunoelectron microscopy and in multiple detection affinity cytochemistry, in *Modern methods in analytical morphology* (Gu, J. and Hacker, G. W., eds.), Plenum, New York, pp. 225–251.
2. Brahic, M. and Ozden, S. (1992) Simultaneous detection of cellular RNA and proteins, in *In situ hybridization*, Oxford University Press, Oxford, UK, pp. 85–102.
3. Mason, D. Y. and Sammons, R. (1978) Alkaline phosphatase and peroxidase for double immunoenzymatic labeling of cellular constituents. *J. Clin. Pathol.* **31,** 454–460.
4. Krenacs, T., DobÛ, E., and Laszik, Z. (1990) Characteristics of endocrine pancreas in chronic pancreatitis as revealed by simultaneous immunocytochemical demonstration of hormone production. *J. Histotechnol.* **13,** 213–218.
5. Krenacs, T., Krenacs, L., BozÛky, B., and Ivanyi, B. (1990) Double and triple immunocytochemical labeling at a light microscopic level in histopathology. *Histochem. J.* **22,** 530–536.
6. Krenacs, T., Laszik, Z., and DobÛ, E. (1989) Application of immunogold-silver staining and immunoenzymatic methods in multiple labeling of human pancreatic Langerhans islet cells. *Acta Histochem.* **85,** 79–85.
7. Van Noorden, S., Stuart, M. C., Cheung, A., Adams, E. F., and Polak, J. M. (1986) Localization of pituitary hormones by multiple immunoenzyme staining procedures using monoclonal and polyclonal antibodies. *J. Histochem. Cytochem.* **34,** 287–292.
8. Ferri, G.-L., Gaudio, R. M., Castello, I. F., Berger, P., and Giro, G. (1997) Quadruple Immunofluorescence: a direct visualization method. *J. Histochem. Cytochem.* **45,** 155–158.
9. Krenacs, T. and Rosendaal, M. (1995) Immunohistological detection of gap junctions in human lymphoid tissue: Connexin 43 in follicular dendritic and lympho-endothelial cells. *J. Histochem. Cytochem.* **43,** 1125–2237.

10. Chaubert, P., Bertholet, M.-M., Correvon, M., Laurini, S., and Bosman, F. T. (1997) Simultaneous double immunoenzymatic labeling: a new procedure for the histopathologic routine. *Mod. Pathol.* **10,** 585–591.

11. Nakane, P. K. (1968) Simultaneous localization of multiple tissue antigens using the peroxidase-labeled antibody method: a study on pituitary glands of the rat. *J. Histochem. Cytochem.* **16,** 557–560.

12. Gillitzer, R., Berger, R., and Moll, H. (1990) A reliable method for simultaneous demonstration of two antigens using a novel combination of immunogold-silver staining and immunoenzymatic labeling *J. Histochem. Cytochem.* **38,** 307–313.

13. Krenacs, T., Uda, H., and Tanaka, S. (1991) One-step double immunolabeling of mouse interdigitating reticular cells: simultaneous applications of preformed complexes of monoclonal rat antibody M18 with horseradish peroxidase-linked anti-rat immunoglobulin and of monoclonal mouse antiIa antibody with alkaline phosphatase coupled antimouse immunoglobulins. *J. Histochem. Cytochem.* **39,** 1719–1723.

14. Hunyady, B., Krempels, K., Harta, G. Y., and Mezey, E. (1996) Immunohistochemical signal amplification by catalyzed reporter deposition and its application in double immunostaining. *J. Histochem. Cytochem.* **44,** 1353–1362.

15. Hacker, G. W., Grimelius, L., Danscher, G., Bernatzky, G., Muss, W., Adam, H., and Thurner, J. (1988) Silver acetate autometallography: an alternative enhancement technique for immunogold-silver staining (IGSS) and silver amplification of gold, silver, mercury and zinc in tissues. *J. Histotechnol.* **11,** 213–221.

28

ELISA on Attachment-Dependent or Suspension Grown Cells

Julie Ann Brent

1. Introduction

Enzyme-linked immunosorbant assay (ELISA) has been a powerful technique in immunology since it was first developed in 1971 by Engvall and Perlman (1). The technique is based on the fact that proteins can be absorbed nearly irreversibly to plastic (2). The antigen (or antibody) is immobilized in the well of a 96-well plate. An enzyme-conjugated antibody (or antigen) is then added to the well to bind to a specific target. Any unbound antibody (antigen) is rinsed away. Substrate is added to react with the conjugated enzyme to allow a color reaction to occur. The intensity of the color is directly related to the amount of labeled antibody (antigen) bound to the target (3). This technique allows for specific and quantitative results. The assay is significant because it is straightforward and the materials readily available.

There are several modifications to the procedure that allow the target to be identified and quantified. Since the topic is immunocytochemistry, the procedures described deal with ELISAs performed on intact cells fixed onto the wells of 96-well plates. There are two forms of the procedure: direct and indirect. The direct method calls for linking the enzyme directly to the antibody of interest. Depending on the availability of the antibody and its activity, direct conjugation of the enzyme to the antibody may interfere with specificity or success of binding with the target. The preferred method is the indirect ELISA. The antigen or cell of interest is immobilized onto the well. The primary antibody (often a monoclonal) from a known host animal is allowed to bind to the immobilized target. A secondary antibody, directed against the primary host animal's immunoglobulin (usually an IgG), is conjugated to a reporter enzyme and allowed to bind to the primary antibody. Then, the complex is visualized by addi-

From: *Methods in Molecular Biology, Vol. 115: Immunocytochemical Methods and Protocols*
Edited by: L. C. Javois © Humana Press Inc., Totowa, NJ

tion of a substrate and formation of a colored reaction product. This procedure allows specificity to be maintained and often allows the signal to be amplified *(4)*.

The two most common enzyme labels for secondary antibodies are alkaline phosphatase (AP) used in conjunction with the substrate *p*-nitrophenyl phosphate, which results in a yellow reaction product, and horseradish peroxidase (HRP) used in conjunction with the substrate ABTS and H_2O_2, which results in blue-green reaction product (*see* Chapter 23).

An overview of the procedure for an ELISA is as follows: Fix cells onto the well of the plate; block the cells and well surface to prevent nonspecific binding; incubate the cells in the presence of the primary antibody; rinse away any unbound antibody; incubate the cells in the presence of the enzyme-linked secondary antibody; rinse away unbound antibody; add enzyme substrate; read the color reaction with a plate reader.

2. Materials

1. 96-Well tissue-culture plates (Corning, Corning, NY; Costar, Cambridge, MA; Falcon, Los Angeles, CA).
2. Phosphate buffered saline (PBS): 137 mM NaCl, 2.7 mM KCl, 4.3 mM Na$_2$HPO$_4$, 1.4 mM KH$_2$PO$_4$, pH 7.2–7.4.
3. Poly-L-lysine: 0.01 mg/mL in PBS (Sigma, St. Louis, MO).
4. Cell suspension to be tested.
5. 0.5% Glutaraldehyde in PBS (Sigma) (*see* **Note 1**).
6. 0.05% NP 40 in PBS.
7. 100 mM Glycine in PBS with 0.1% bovine serum albumin (BSA).
8. Blocking solution: 1% BSA and 0.1% Tween-20 in PBS.
9. Primary antibody (*see* **Note 2**).
10. Enzyme-conjugated secondary antibody (*see* **Note 3**) (Kirkegaard and Perry Laboratories, Gaithersburg, MD).
11. Detection reagent (substrate) for enzyme-labeled antibody: 2,2-azino-di(3-ethyl-benzthiazoline sulfonate-6) (ABTS) 0.3 g/L in 0.15 M sodium citrate with 0.1% H_2O_2 for the detection of horseradish peroxidase-labeled secondary antibody or *p*-nitrophenyl phosphate (pNPP) 1 g/L in 1 M diethanolamine in water for the detection of an alkaline phosphatase-labeled antibody (Kirkegaard and Perry Laboratories).
12. 1% Sodium dodecyl sulfate (SDS).
13. 5% Ethylene-diaminetetraacetic acid (EDTA).
14. Multiwell plate reader to measure color at 405–410 nm wavelength.
15. Low-speed centrifuge with microplate carriers.

3. Methods

3.1. Preparation of ELISA Plates with Cells

3.1.1. Attachment-Dependent Cells

1. Prepare a cell suspension in growth medium from a healthy, log-phase culture.
2. Dilute the cells to a final concentration 4–5 × 10^4 cells/mL.

3. Inoculate the wells with 100 µL of cell suspension. *See* **Note 4** for how many wells need to be filled.

4. Allow the cells to incubate at 37°C in a CO_2 incubator for 24–48 h (depending on how fast the cells divide) so that the wells are 90–100% confluent on the day of the assay.

5. Remove growth medium from the wells.

6. Rinse the wells two times by adding 100 µL of PBS, and flicking the plate into the sink or onto paper towels to remove excess PBS (*see* **Note 5**).

7. Fix the cells with 50 µL of 0.5% glutaraldehyde in PBS for 15 min at room temperature.

8. Remove the glutaraldehyde by flicking, and then lyse the cells with 50 µL of 0.05% NP 40 in PBS for 15 min at room temperature.

9. Wash the wells two times with PBS by immersion (*see* **Note 6**).

10. Add 200 µL of 100 m*M* glycine in PBS with 1% BSA to block excess glutaraldehyde. Incubate at room temperature for 30 min.

11. Remove glycine, and wash the wells two times with PBS by immersion.

12. Add 100 µL of blocking solution to prevent nonspecific binding. Incubate at room temperature for 2 h or at 37°C for 45 min (*see* **Note 7**).

13. Remove blocking solution by flicking, and wash the wells two times with PBS by immersion. Leave PBS on the cells until ready to start the assay (*see* **Note 8**).

3.1.2. Preparation of ELISA Plates with Suspension Grown Cells

1. Pretreat the wells of the 96-well plate with 100 µL of poly-L-lysine in PBS for 1–16 h at room temperature.

2. Remove the excess poly-L-lysine from the wells by flicking.

3. Wash the suspension cells with PBS, and dilute the cells to a final concentration of 1×10^6 cells/mL.

4. Add 50 µL of cell suspension (50,000 cells) to each well needed for the assay (*see* **Note 4**).

5. Centrifuge the plates at 500*g* for 5 min in microplate carriers centrifuge (*see* **Note 9** if microplate carriers are not available).

6. Add 100 µL of 0.5% glutaraldehyde in PBS. Incubate at room temperature for 1 h (*see* **Note 10**).

7. Remove glutaraldehyde by flicking the plate.

8. Rinse the wells two times by adding 50 µL of PBS and flicking the plate.

9. Lyse the cells with 50 µL of 0.05% NP 40 in PBS for 15 min at room temperature.

10. Remove NP 40 by flicking and wash the plate with PBS by immersion.

Proceed with **steps 10–13** as described in **Subheading 3.1.1.**

3.2. ELISA on Fixed Cells

1. Dilute the primary antibody in blocking solution. Dilutions must be determined empirically for the end point of your particular assay (*see* **Note 4**).

2. Add 50 µL of each dilution to the wells (*see* **Notes 4, 11**, and **12**).

3. Incubate for 1 h at room temperature or 30 min at 37°C.

4. Wash the wells three times with PBS by immersion (*see* **Note 13**).
5. Dilute secondary antibody (enzyme-conjugate antibody) in blocking solution. Use dilutions recommended by manufacturer (*see* **Note 4**).
6. Add 50 μL of the diluted secondary antibody to each well (*see* **Notes 4, 11**, and **12**).
7. Incubate the plate for 1 h at room temperature or for 30 min at 37°C.
8. Wash the wells three times with PBS by immersion.
9. Add 100 μL of substrate solution/well including the blank wells. For HRP-labeled antibodies, add ABTS with 0.1% H_2O_2. For AP-labeled antibodies, add p-NPP in $1M$ diethanolamine buffer.
10. Incubate at room temperature for 10–15 min. If the reaction occurs slowly, warm the plate to 37°C.
11. Stop the reaction with equal volumes of stop solution added to each well before reading the plates: 1% SDS is the stop solution for the ABTS substrate, and 5% EDTA is the stop solution for p-NPP substrate.
12. Read the plate using the plate reader set at 405–410 nm wavelength.

4. Notes

1. Glutaraldehyde should be prepared fresh in PBS and stored at 4°C. It can be used for up to 1 m after being diluted.
2. Primary antibodies to specific antigens are available commercially from a number of sources, including Enzo Diagnostics (Syosset, NY) and Sigma from the host animal of choice.
3. Secondary antibodies are enzyme-linked antibodies directed against the primary antibody host animal's immunoglobulin (usually an IgG). If the primary antibody is mouse anticytokeratin the secondary antibody would be horseradish peroxidase-labeled or alkaline phosphatase-labeled antimouse IgG.
4. The number of wells to be filled with cells depends on the number of different concentrations of primary and secondary antibodies being used. A possible template for setting up a reaction could look like this:

	1	2	3	4	5	6
A	Cells	1:50	1:100	1:200	1:400	PBS	1° antibody
	Only	1:100	1:100	1:100	1:100	1:100	2° antibody
B	Duplicate of A						
C	Cells	1:100	1:100	1:100	1:100	1:100	1° antibody
	Only	1:100	1:200	1:500	1:1000	PBS	2° antibody
D	Duplicate of C						

   The wells filled with cells only are referred to as blanks. No primary or secondary antibody will be added to these wells. Instead, these wells will only get PBS. This column will be used to "zero" the microtiter plate reader. Wells that have PBS instead of the primary or secondary antibody are used as negative controls to assess background. The dilutions in this template are recommendations only.

5. Flicking refers to inverting the plate to empty all the wells at once. A fast movement of the wrist helps "flick" the liquid out of the wells.

6. Washing by immersion refers to the wells being filled with a squeeze bottle containing PBS. This allows for quick and thorough washing of each well.

7. Blocking may need to take place for longer than 2 h if nonspecific binding of the antibody takes place. This can be determined by examining the negative control wells *(5)*.

8. It is important that the wells do not dry out during the assay. It can cause nonspecific binding of the antibodies and high background.

9. When microplate carriers are not available, allow the cells to settle to bottoms of the wells for 4 h at 37°C. Add the fixative directly to the wells without removing the PBS.

10. Longer fixation times may be necessary to fix the suspension grown cells adequately to the plate.

11. Leave the first column of wells blank to "zero" the microplate reader. Fill the wells with PBS only.

12. Always use a fresh pipet tip for each dilution.

13. If high background occurs, wash the wells with 0.05% Tween-20 in PBS.

References

1. Engvall, E. and Perlman, P. (1971) Enzyme-linked immunosorbant assay (ELISA): quantitative assay of Immunoglobulin G. *Immunocytochemistry* **8,** 871–879.

2. Avrameas, S. and Ternynck, T. (1969) The cross linking of proteins with glutaraldehyde and its use for the preparation of immunoadsorbants. *Immunochemistry* **6,** 53–66.

3. Engvall, E. and Perlman, P. (1972) Enzyme-linked immunosorbant assay, ELISA. Quantitation of specific antibodies by enzyme-labeled anti-immunoglobulin in antigen-coated tubes. *J. Immunol.* **109,** 129.

4. Avrameas, S. and Guilbert, B. (1972) Enzyme-immunoassay for the measurement of antigens using peroxidase conjugates. *Biochemie* **54,** 837.

5. Trivers, G. E., Harris, C. C., Rougeot, C., and Dary, F. (1983) Development and use of ultrasensitive enzyme immunoassays. *Methods Enzymol.* **103,** 409.

29

Use of Immunogold with Silver Enhancement

Constance Oliver

1. Introduction

Although gold particles are readily detectable by transmission electron microscopy, they can be difficult to visualize by bright-field light microscopy. If the particle size is large enough and the labeling dense enough, the gold particles will stain tissue red (1,2). However, unless the gold is silver-enhanced to help visualize it, the sensitivity of the staining is fairly low. During silver-enhancement, the colloidal gold serves as a nucleation site for the deposition of metallic silver. The silver layer increases the size of the gold and imparts a black color to the stained tissue when viewed by bright-field microscopy (**Fig. 1A**). The silver-enhanced gold particles can also be visualized using epipolarization, where they appear bright against a dark background (**Fig. 1B**). The silver enhancement method has its basis in 19th-century photographic techniques. The enhancing solutions are physical developers that contain both silver ions and a reducing agent, buffered to an acid pH. The developers most commonly used contain silver lactate as the source of silver ions (3). The silver lactate has a low dissociation coefficient that allows for more control of the reduction. Hydroquinone (1,4-dihydroxybenzene) is the only reducing agent that has been used in silver-enhancement techniques. A protective colloid, such as gum arabic, bovine serum albumin, dextran, polyethylene glycol (PEG), or polyvinylpyrrolidone (PVP), is frequently added in order to inhibit the autocatalytic reaction between the silver salt and the reducing agent. The protective colloid also helps in providing even distribution of the components during the development. The developing solution is very unstable and must be protected from light. The samples should also be protected from light during silver-enhancement. Recently, commercial silver-enhancing kits have become available. These kits have the advantage that their components are stable and may be stored in the

From: *Methods in Molecular Biology, Vol. 115: Immunocytochemical Methods and Protocols*
Edited by: L. C. Javois © Humana Press Inc., Totowa, NJ

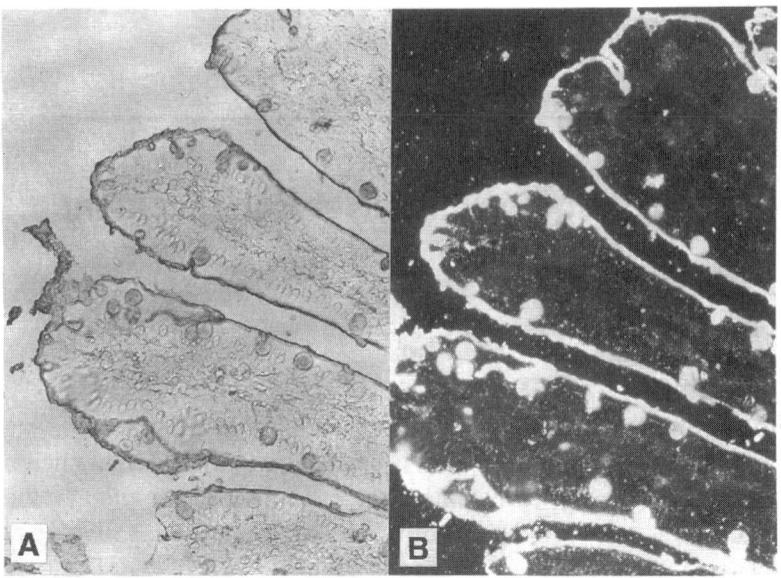

Fig. 1. Small intestine: formaldehyde-fixed, paraffin-embedded. Section was stained with Concanavilin A conjugated to 5-nm colloidal gold. The gold was then silver-enhanced. Staining of mucous on the cell surface and in goblet cells is seen. **(A)** Bright-field microscopy. The silver-enhanced gold appears as a dark stain. **(B)** Epipolarization microscopy. The silver-enhanced gold appears bright against a dark background.

refrigerator for months. They are also insensitive to light, so that the enhancement procedure can be monitored by light microscopy.

The use of silver-enhancement for detecting gold in tissue sections was introduced in 1935 by W. J. Roberts, who published a photochemical method for detecting injected gold salts in animal tissues *(4)*. By the 1980s, immunogold staining was firmly established for electron microscopy. Danscher's *(3,5)* evaluation of photochemical methods to visualize gold for light and electron microscopy, and the application of these methods by Holgate et al. *(6,7)* demonstrated the feasibility of applying physical development methods to localize immunogold-stained sites. The silver-enhancement method has also been used successfully to enlarge small-diameter gold particles for visualization by scanning electron microscopy *(8,9)*. Although the use of small gold particles will increase the efficiency of the reaction, particles smaller than 20 nm are too small to be easily seen using standard scanning electron microscopic methods. The use of silver enhancement to enlarge the smaller gold particles overcomes this problem. When combined with backscattered imaging, the ability to resolve immunogold by scanning electron microscopy is further increased. Silver-enhancement has been applied to a wide variety of tissues and antigens for both light and scanning electron microscopy *(10–12)*.

2. Materials

1. Sections cut onto aminosilane-coated slides and rehydrated to water (*see* **Note 1**).
2. Tris-buffered saline (TBS): 2.4 g Tris and 8.76 g NaCl. Adjust the pH to 7.4 with HCl then bring vol to 1 L with deionized glass-distilled water.
3. 1% Bovine serum albumin (BSA): 1 g BSA and 100 mL TBS, pH 7.4. Add BSA to TBS with stirring.
4. Primary antibody diluted in 1% BSA in TBS.
5. Silver-enhancing solution *(11)*: 60 mL protective colloid (25% gum arabic or 50% PEG [20,000 mol wt] or PVP), 10 mL 2 *M* citric acid or sodium citrate, and 850 mg hydroquinone dissolved in 15 mL deionized glass-distilled water; mix thoroughly; adjust pH to 3.8; immediately before use, add 110 mg silver lactate dissolved in 15 mL deionized glass-distilled water (*see* **Note 2**).
6. Commercially available silver enhancement kit (*see* **Note 3**).
7. Colloidal gold conjugate (*see* **Note 4**).

3. Methods

3.1. Silver Enhancement for Light Microscopy

Immunogold staining can be used successfully at the light microscopic level if the gold is silver-enhanced. Enhancing solutions may be made up in the laboratory, but because of their instability and light sensitivity, the commercially available silver-enhancing kits are preferable.

1. Block tissue for 15 min in TBS containing 1–5% BSA.
2. Rinse five times in TBS containing 1% BSA.
3. Incubate slides in a moist chamber in primary antibody diluted in TBS plus 1% BSA for 1–2 h at room temperature. Plastic-embedded sections may need to be incubated for 2 h at 37°C.
4. Rinse five times in TBS containing 1% BSA.
5. If required, incubate 30 min at room temperature in bridging antibody diluted in TBS plus 1% BSA, and rinse five times in TBS plus 1% BSA.
6. Incubate slides for 30 min at room temperature in colloidal gold conjugate diluted in TBS plus 1% BSA.
7. Rinse the sections five times in TBS, and then rinse five times in deionized glass-distilled water (*see* **Note 5**).
8. Prepare silver-enhancing solution immediately prior to use, and cover the sections with the solution (*see* **Note 6**).
9. Rinse the sections five times in deionized glass-distilled water (*see* **Note 7**).
10. Counterstain and mount. All commonly used counterstains and mounting media can be used.

3.2. Silver Enhancement for Scanning Electron Microscopy (SEM)

The procedure given above may be used successfully to surface-label cells or tissue for examination by SEM.

1. Immunolabel samples either before or after fixation. Any size gold can be used for scanning electron microscopy. The size of the gold particles is limited by the resolution of the instrument.
2. After fixation and labeling, silver-enhance the gold using a commercially available kit. Generally, the time needed for enhancement for SEM is around 5 min.
3. Dehydrate the samples, and critically point-dry.
4. The silver-enhanced gold can be detected in the scanning electron microscope using backscattered electron imagining.

4. Notes

1. Do not use coatings, such as poly L-lysine or chrom-alum, since the charge will interfere with the silver intensification reaction. For paraffin sections, remove paraffin with xylene and rehydrate to water. Penetration of reagents into plastic-embedded sections may be improved by treating the sections for 15 min with xylene and rehydrating to water. Following rehydration, the sections should not be allowed to dry.
2. All components, except the protective colloid, should be prepared immediately before use. Protect hydroquinone, silver lactate, and complete developer from light.
3. Commercially available kits are more stable and not as light-sensitive as enhancing solutions prepared in the laboratory.
4. The colloidal gold should be conjugated to an antibody raised against the species of the primary antibody or to Protein A, such that it will bind to the primary antibody (*see* Chapter 41). For light microscopy, 1- and 5-nm colloidal gold conjugates are used most frequently. The 1-nm gold may be somewhat more difficult to silver-enhance. For scanning electron microscopy, the size of the gold depends on the resolution of the instrument: 10–20 nm gold is the size most frequently used for scanning electron microscopy.
5. The ions from the buffer must be removed, or they will serve as nucleation sites for the silver and increase the background. If rinsing the samples in distilled water will damage the tissue or if the acid pH of the enhancing solution will remove antibodies with low affinities, the samples may be briefly fixed and quenched before rinsing in water and proceeding to the next steps.
6. For commercial enhancing solutions, incubate the sections 5–15 min. With these solutions, the optimal incubation time may be determined by monitoring the deposition of the silver by light microscopy. For developing solutions prepared in the laboratory, development should be done in a dark room with a safety light. The samples are immersed in the developer for 10–45 min. For longer development times, the solution may have to be changed.
7. After development, the slides must be thoroughly washed in deionized glass-distilled water to stop development.

References

1. DeMey, J., Moeremans, M., Geuens, G., Nuydens, R., and DeBrander, M. (1981) High resolution light and electron microscopic localization of tubulin with the IGS (immuno-gold staining) method. *Cell Biol. Int. Rep.* **5,** 889–899.
2. Geoghegan, W. D., Scillian, J. J., and Ackerman, G. A. (1978) The detection of human B lymphocytes by both light and electron microscopy utilizing colloidal gold labeled anti-immunoglobulin. *Immunol. Commun.* **7,** 1–12.
3. Danscher, G. (1981) Histochemical demonstration of heavy metals. A revised version of the sulphide silver method suitable for both light and electron microscopy. *Histochemistry* **71,** 1–16.
4. Roberts, W. J. (1935) A new procedure for the detection of gold in animal tissues: Physical development. *Proc. R. Acad. Sci. Amsterdam* **38,** 540–544.
5. Danscher, G. (1981) Localization of gold in biological tissue. A photochemical method for light and electron microscopy. *Histochemistry* **71,** 81–88.
6. Holgate, C. S., Jackson, P. I., Cowen, P. N., and Bird, C. C. (1983) Immunogold-silver staining: new method of immunostaining with enhanced sensitivity. *J. Histochem. Cytochem.* **31,** 938–944.
7. Holgate, C. S., Jackson, P., Lauder, I., Cowen, P., and Bird, C. C. (1983) Surface membrane staining of immunoglobulins in paraffin sections of non-Hodgkin's lymphomas using immunogold-silver staining technique. *J. Clin. Pathol.* **36,** 742–746.
8. Goode, D. and Maugel, T. K. (1987) Backscattered electron imaging of immuno-gold labeled and silver-enhanced microtubules in cultured mammalian cells. *J. Electron Microsc. Tech.* **5,** 263–273.
9. de Harven, E. (1989) Backscattered electron imaging of the colloidal gold marker on cell surfaces, in *Colloidal Gold*, vol. 1 (Hayat, M. A., ed.), Academic, New York, pp. 229–249.
10. Larsson, L.-I. (1988) Immunocytochemical detection systems, in *Immunocytochemistry: Theory and Practice* (Larsson, L.-I., ed.), CRC, Boca Raton, FL, pp. 77–146.
11. Scopsi, L. (1989) Silver-enhanced colloidal gold method, in *Colloidal Gold*, vol. 1 (Hayat, M. A., ed.), Academic, New York, pp. 251–295.
12. Hacker, G. W. (1989) Silver-enhanced colloidal gold for light microscopy, in *Colloidal Gold*, vol. 1 (Hayat, M. A., ed), Academic, New York, pp. 297–321.

IV

Fluorescence Activated
Cell Sorter (FACS) Analyses

30

Overview of Flow Cytometry and Fluorescent Probes for Cytometry*

Robert E. Cunningham

1. Introduction

It was shown by Creech and Jones *(1)* in 1940 that proteins, including antibodies, could be labeled with a fluorescent dye (phenylisocyanate) without biological or immunological effects to the intended target. In theory, fluorescent reporters (tracers, probes, antibodies, stains, and so on) can be used to detect or measure any cell constituent, provided that the tag reacts specifically and stoichiometrically with the cellular constituent in question *(2)*. Today, the repertoire of fluorescent probes is expanding almost daily (*see* Chapter 14). One area that has benefited from the ever-increasing number of fluorescent probes is flow cytometry.

1.1. Overview and Advantages of Flow Cyotometry

Flow cytometry is a high-precision technique for rapid analysis and sorting of cells and particles. The principles underlying flow cytometry are discussed in many comprehensive review articles *(3–5)* and will only be touched on here. This fluorescent detection technique provides statistical accuracy, reproducibility, and sensitivity, and allows the simultaneous measurement of multiple parameters on a cell-to-cell basis. Information is derived from the optical responses of the fluorescent probes and the light-scattering capabilities of the labeled, individual cells as they stream single file through an excitation laser light beam. Each flash of fluorescence or scattered light is collected, filtered, converted to an electronic signal, amplified, digitized, and stored for later

*The opinions or assertions herein represent the personal views of the author and are not to be construed as official or as representing the views of the Department of the Army or the Department of Defense.

From: *Methods in Molecular Biology, Vol. 115: Immunocytochemical Methods and Protocols*
Edited by: L. C. Javois © Humana Press Inc., Totowa, NJ

analysis. Cells may be fluorescently labeled with probes that bind directly to cellular constituents (*see* Chapter 15) or indirectly, using an unlabeled first probe followed by a second, labeled probe (e.g., fluorescently labeled antibody; *see* Chapters 32,33). Extrinsic and intrinsic fluorescent probes allow selective examination of both functional and structural components of cells. Many commercial flow cytometers are now equipped with up to six detectors and two lasers for two- or three-color fluorescence analysis. Combined with computer processing and display procedures, up to eight one-dimensional fully cross-correlated data parameters may be analyzed. Cytometers are also capable of sorting cells by deflecting electrostatically charged droplets containing single cells with given parameters into collection tubes for subsequent manipulation.

As a technical discipline, flow cytometry is playing an increasingly significant role in both the research and clinical laboratories. When the cells or organelles being investigated for a particular property are all homogeneous, flow cytometry offers no advantages over conventional biochemical techniques. However, when the population is heterogeneous for a property or properties, flow cytometry allows the analysis and isolation of these discrete populations. Some applications include chromosome analysis and sorting *(6,7)*, studies of DNA proliferation *(8,9)*, studies of cellular components other than DNA *(2)*, analysis of semen in the evaluation of human infertility *(10)*, and studies of the hemopoietic system *(11–14)*, to name a few. Additional applications can be found in articles published in *Cytometry, Analytical and Quantitative Cytology* and in journals related to hematology, immunology, oncology, and cytogenetics. Of major clinical relevance is the ability of flow cytometry to detect and characterize rare events. For example, by using combinations of specific antibodies, each labeled with a different fluorochrome, panels can be designed to follow residual disease with greater precision than with any other method. When flow cytometric results are combined with histologic results, the confidence in pathologic interpretation is far greater than either can provide alone (*see* Chapter 51). It is likely that the future application of these approaches will significantly improve diagnosis and patient care.

The development of nucleic acid probes, other reporters, and techniques with molecular biology has been of revolutionary significance in the biological and medical sciences *(15)*. These probes offer the promise of disease diagnosis through direct genomic analysis and the detection of genomic products at an unprecedented level of sensitivity. cDNA probes have been used for the diagnosis of infectious diseases, genetic diseases, and neoplasia. The combination of flow cytometry and fluorescent molecular biology techniques has unprecedented power in the biological sciences. It combines the ability of the flow cytometer to identify, quantify, and separate rare cell populations with the unique specificity of cDNA probes for genetic sequences in cells or microbial organisms *(16)*. One technique that was exclusively used for slide-based assays, fluorescence *in situ* hybridization (FISH), has recently been available to flow cytometry *(see* Chapter 48). FISH techniques demonstrate great promise for detecting specific sequences of nucleic acid by flow cytometry, particularly in instances where multiple copies of the nucleic acid sequence are present. Detection of a single gene copy remains difficult due to poor signal-to-noise ratios and limits the application of flow cytometric assays for direct examination of individual nucleic acid sequences, such as point mutations of oncogenes *(17)*. It is quite possible that this technical obstacle will be overcome in the near future.

One of the most eloquent techniques in the realm of molecular biology is a microtechnique known as the polymerase chain reaction (PCR), and combined with flow cytometry, it has further expanded cell sorting *(18)* *(see* Chapter 49). Nuclei can be sorted on the basis of cell cycle, digested to single-stranded DNA, and the desired segment of DNA amplified. Further, sorting does not necessarily require live cells for postsorting analysis and/or PCR, thereby decreasing the complexity of the sorting process. The requirement for live cells can be replaced by the retrospective analysis of paraffin-embedded tissue *(see* Chapter 35). Another molecular technique that utilizes flow cytometry and sorting is cytogenetics, which is the study of karyotype anomalies through the loss or gain of chromosomal material and structural changes. Molecular biology gives a means of recognizing chromosome losses and especially of studying oncogenic or antioncogenic mutations *(19)*. Sorting allows for the separation of individual chromosomes. Studying these alterations will allow better prediction of high-risk subjects in cancer families *(20)*. The integration of FISH and flow cytometric analysis provides more information on the chromosomal abnormalities of these neoplasms *(21)*. A large number of the methods for staining and measuring properties of individual cells are available *(22)*. New protein and DNA dyes have made it possible to individually analyze large numbers of cells for multiple properties. These techniques have had a great impact on cellular immunology and the study of cell proliferation. New fluorescent molecules that report on intracellular conditions are increasingly used to study cell physiology. Chromosome analysis and sorting by flow cytometry is becoming a valuable tool, and refinements in the techniques for manipulating small quantities of DNA will increase the application of chromosome sorting in molecular biology. However, the analysis of rare cell populations is still hampered by shortcomings in the present generation of commercial instruments. Flow cytometry can be used for the study of multidrug resistance (MDR), which is the study of the capacity of modulating agents to result in overexpression of the P-glycoprotein as well as the functional aspect of MDR in expulsion of the cytotoxic agents *(23)*.

1.2. Disadvantages Associated with Flow Cytometric Analysis

Successful flow cytometric analysis depends on adequate sample preparation (*see* Chapters 30–31), appropriate selection of probes or markers (*see* **Subheading 2.**), instrumentation, and data display and analysis. Each of these areas is interrelated and requires adequate attention to avoid the introduction of artifacts and misinterpretation of results. Flow cytometers tend to be excessively complicated and require a skilled operator for alignment and calibration, though manufacturers are introducing more compact, user-friendly data acquisition and image processing systems.

One major drawback of flow cytometry is the lack of visual control and structural information. Morphological details of cell shape, size, or texture are not obtained. Detection of intracellular proteins is less well developed and has some possible problems because of nonspecific antibody binding to dead cellular components. This results in considerable "biological noise" and reduces the sensitivity for detecting the desired protein within the cell. Finally, natural components of cells "autofluoresce" at shorter wavelengths, so ideal fluorochromes would be those that emit in the orange-to-red color range (575–800 nm), thus avoiding overlap with background autofluorescence, which is stronger in the blue-to-green range (450–535 nm).

2. Fluorescent Probes

2.1. Characteristics of Ideal Probes

In order to identify complex patterns of cell antigens, it is necessary to be able to visualize several antigens at one time. This can be accomplished through the use of fluorochromes, which are excited by the same wavelength of light but have distinct emissions spectra. Through the appropriate combination of optical filters, each fluorochrome's emission can be separated and distinguished. The number of different fluorochromes that can be detected on a single cell is limited only by the ability of the optical filters to separate emitted light. Ideally, the fluorochromes should also have high absorption coefficients, high quantum yields, and a large Stokes shift (*see* Chapter 14).

Table 1 lists some of the more popular probes used in flow cytometry. This is by no means a complete listing of probes, but it is a good point from which to begin, with particular attention to the excitation capability of the argon laser, which is the most widely used light source on flow cytometers. One of the most interesting new dyes is green fluorescent protein (GFP). This is a green protein fluorophore that is used to study genetically modified (transgenic) cells and essentially forms a fusion product with any protein and then retains its fluorescent properties *(24)*. This allows for protein localization in living cells without the concomitant problems of injection and protein handling. This

Table 1
Selected Fluorescent Probes for Flow Cytometry

| | Wavelength, nm | |
Fluorescent probe	Excitation	Emission
Adriamycin	472	570
7-Aminoactinomycin D	550	650
Acridine orange[a]	480	520, 640
Acriflavine[a]	480	550–600
Allophycocyanin	650	660
CY3	550	565
CY5	652	667
4',6-diamido-2-phenylindole (DAPI)	345	470
Dansyl chloride (DANS)	340	578
Indocarbocyanine (DiL-Cn-(3))	485	505
Indodicarbocyanine (DiL-Cn-(5))	548	567
Oxacarbocyanine (DiO-Cn-(3))	646	668
Eosins[a]	527	550
Erythrosin[a]	530	540
Ethidium bromide	518	610
Fluorescein diacetate (FDA)	475	530
Fluorescein isothiocyanate (FITC)[a]	494	517
Fura-2[a]	335	515
Green fluorescent protein (GFP)	395±	509±
Hoechst 33258[a]	365	480
Hoechst 33342	355	465
Image-orange	545	585
Indo-1	330	400
Mithramycin	421	575
Nile red (acetone)	530	605
Phycocyanin-C	620	640
Phycoerythrin-B	546	575
Phycoerythrin-R	566	575
Propidium Iodide (PI)	520	610
Pyronin Y(G)	555	571
Quinacrine[a]	440	510
Rhodamine isothiocyanate (RITC)[a]	540	550
Red dye 613	488[b]	613
Red dye 670	488[b]	670
Rhodamine 123	511	534
4-Acetamido-4'-isothiocyanate-stilbene-2,2'-disulfonic acid (SITS)	350	420
SNARF-1 (pH 10.0)	574	636
Tetramethylrhodamine isothiocyanate (TRITC)	540	550
Texas red	595	615
Thiazole orange	453	480
XRITC	580	605

[a]MEDLINE mesh subheadings.
[b]These dyes are specifically designed to preferentially absorb with the 488 nm wavelength of an argon laser.

becomes an elegant way to track a constructed genetic expression vector and intracellular modification. Also, mutants of GFP allow for varied spectral properties and differing emission and excitation frequencies. Undoubtedly, more dyes of this type will be developed in the future.

To find additional fluorescent probes or to enhance the rendering of existing probes through careful matching of emission spectra, journal articles, books, and manufacturers' data sheets are valuable entities *(25)* (*see* **Notes 1** and **2**).

2.2. Staining Techniques

Fluorescence staining for flow cytometric analysis falls into three categories: methods in which a fluorescent ligand accumulates on or within the cell (*see* Chapters 36,38); methods that require the ligand to interact with a cellular component to release the fluorophore or result in light emission (*see* Chapters 34,39); and methods that rely on fluorophore-coupled antibody binding (*see* Chapters 32,33).

Cellular DNA content is one of the most widely measured parameters using flow cytometry (*see* Chapter 34). Nuclei can be extracted from archival material for ploidy studies (*see* Chapter 35). DNA- or nucleic acid–specific ligands (DAPI, Hoechst, ethidium bromide, propidium iodide) as well as polyanion dyes (acridine orange, acriflavine) are commonly employed. Acridine orange has the unusual property of fluorescing green (520 nm) when bound to double-stranded nucleic acids and red (640 nm) when bound to single strands (*see* **Table 1**). Newly synthesized DNA that has incorporated bromodeoxyuridine (BrdU) can be detected using the commercially available anti-BrdU (*see* **Note 3**).

3. Notes

1. The *Handbook of Fluorescent Probes and Research Chemicals* by R. P. Haugland is a useful resource for over 1800 fluorescent probes and their applications. It is published yearly by Molecular Probes, 4849 Pitchford Ave., Eugene, OR 97402. *See also* http://www.probes.com.
2. Another source of information and protocols is the Internet. It supplies not only sources, but also valuable protocols and hints. A particularly useful site is the Purdue University Cytometry group, http://www.cyto.purdue.edu, which has links to many other flow cytometry websites.
3. There are currently two sources of monoclonal antibodies to bromodeoxyuridine: Becton Dickinson Immunocytochemistry Systems and Boehringer Mannheim Biochemicals. Both antibodies have worked equally well for the author, and both are directed against single-stranded DNA, therefore necessitating the removal of DNA-associated histones and denaturation of the double-stranded DNA.

References

1. Creech, H. J. and Jones, R. N. (1940) The conjugation of horse serum albumin with 1,2-benzathryl isothiocyanate. *J. Am. Chem. Soc.* **62**, 1970–1975.

2. van Dam, P. A., Watson, J. V., Lowe, D. G., and Shepherd, J. H. (1992) Flow cytometric measurement of cell components other than DNA: virtues, limitations, and applications in gynecological oncology. *Obstet. Gynecol.* **79,** 616–621.

3. Melaned, M. R., Mullaney, P. F., and Mendelsohn, M. L. (1979) *Flow Cytometry and Sorting.* John Willey and Sons, New York.

4. Kamentsky, L. A., Melamed, M. R., and Derman, H. (1965) Spectrophotometer: new instrument to ultrarapid cell analysis. *Science* **150,** 630,631.

5. Shapiro, H. M. (1988) *Practical Flow Cytometry,* 2nd ed. Liss, New York.

6. Hashimoto, K. (1992) Flow karyotyping and chromosome sorting. *Nippon Rinsho.* **50,** 2484–2488.

7. Gray, J. W., Kuo, W. L., and Pinkel, D. (1991) Molecular cytometry applied to detection and characterization of disease-linked chromosome aberrations. *Baillieres Clin. Haematol.* **4,** 683–693.

8. Wakita, A. and Kaneda, T. (1992) Detection of proliferative cells by DNA polymerase as a proliferation associated marker. *Nippon Rinsho.* **50,** 2338–2342.

9. Videl, L. L. and Christensen, I. J. (1990) A review of techniques and results obtained in one laboratory by an integrated system of methods designed for routine clinical flow cytometric DNA analysis. *Cytometry* **11,** 753–770.

10. Morrell, J. M. (1991) Applications of flow cytometry to artificial insemination: a review. *Vet. Rec.* **129,** 375–378.

11. Loken, M. R., Brosnan, J. M., Bach, B. A., and Ault, K. A. (1990) Establishing optimal lymphocyte gates for immunophenotyping by flow cytometry. *Cytometry* **11,** 453–459.

12. Garratty, G. (1990) Flow cytometry: its applications to immunohaematology. *Baillieres Clin. Haematol.* **3,** 267–287.

13. Packman, C. H. and Lichtman, M. A. (1990) Activation of neutrophils: measurement of actin conformational changes by flow cytometry. *Blood Cells* **16,** 193–207.

14. Zola, H., Flego, L., and Sheldon, A. (1992) Detection of cytokine receptors by high-sensitivity immunofluorescence/flow cytometry. *Immunobiol.* **185,** 350–365.

15. Keren, D. F. (1989) Clinical molecular cytometry: merging flow cytometry with molecular biology in laboratory medicine, in *Flow Cytometry in Clinical Diagnosis* (Keren, D. F., Hanson, C. A., and Hurtubise, P. E., eds.), Chicago, ASCP, pp. 614–634.

16. Tim, E. A. and Stewart, C. C. (1992) Fluorescence in situ hybridization in suspension (FISHES), using digoxigenin-labeled probes and flow cytometry. *Biotechniques* **12,** 362–367.

17. Bauman, J., Bentvelzen, P., and van Bekkum, D. (1987) Fluorescent in situ hybridization of MRNA in bone marrow and leukemic cells measured by flow cytometry. *Cytometry* (Suppl. 1):4.

18. Gyllensten, U. B. (1989) PCR and DNA sequencing. *Biotechniques* **7,** 700–708.

19. Milan, D., Yerle, M., Schmitz, A., Chaput, B., Vaiman, M., Frelat, G., and Gellin, J. (1993) A PCR-based method to amplify DNA with random primers: determining the chromosomal content of porcine flow-karyotype peaks by chromosome painting. *Cytogenet. Cell Genet.* **62,** 139–141.

20. Gray, J. W. and Cram, L. S. (1990) Flow karyotyping and chromosome sorting, in *Flow Cytometry and Sorting* (Melamed, M. R., Lindmo, T., and Mendelsohn, M., eds.), Wiley-Liss, New York, pp. 503–530.

21. Carrano, A. V., Gray, J. W., Langlois, R. G., Burkhart, S. K., and Van Dilla, D. (1979) Measurement and purification of human chromosomes by flow cytometry and sorting. *Proc. Natl. Acad. Sci. USA* **76,** 1382–1384.

22. Cram, L. S., Bartholdi, M. F., Ray, F. A., Meyne, J., Moyzis, R. K., Schwarzacher-Robinson, T., and Kraemer, P. M. (1988) Overview of flow cytogenetics for clinical applications. *Cytometry* (Suppl. 3):94–100.

23. Herzog, C. E. and Bates, S. E. (1994) Molecular diagnosis of multidrug resistance. *Cancer Treat Res.* **73,** 129–147.

24. Chalfie, M., Tu, Y,. Euskirchen, G., Ward, W. W., and Prasher, D. C. (1994) Green fluorescent protein as a marker for gene expression. *Science* **263,** 802–805.

25. Sasaki, K. and Kurose, A. (1992) Cell staining for flow cytometry. *Nippon Rinsho.* **50,** 2307–2311.

31

Tissue Disaggregation*

Robert E. Cunningham

1. Introduction

The extracellular matrix of mammalian tissue is composed of a complex mix of constitutive proteins. This matrix must be broken down to recover single cells effectively for culture and/or staining (1). Tissue dissociation and its affiliated problems were described and defined over 80 yr ago by Rous and Jones (2). More recent reviews (3,4) have revealed newer methods for creating single-cell suspensions. Numerous procedures exist for dissociating solid tumors. They are usually multistep procedures involving one or a combination of mechanical, enzymatic, or chemical manipulations. Ideally, the dissociation protocol is individualized for the tissue of interest and evaluated relative to both optimal and representative cell yield.

In the laboratory, I employ a modified mechanical/enzymatic method to isolate cells. Mechanical dissociation of tissue may involve repeated mincing with scissors or sharp blades, scrapping the tissue surface, homogenization, filtration through a nylon or steel mesh, vortexing, repeated aspiration through pipets or small-gage needles, abnormal osmolality stress, or any combination of these techniques. These methods result in variable cell yields and cell viability. There are various enzymes that can be used, alone or in combination to digest desmosomes, stromal elements, and extracellular and/or intercellular adhesions. Enzymes commonly used include: trypsin, pepsin, papain, collagenase, elastase, hyaluronidase, pronase, chymotrypsin, catalase, and dispase. The most routinely used enzymes are collagenase and dispase. DNase is used with these proteolytic

---

*The opinions or assertions contained herein are the private views of the author and are not to be construed as official or as reflecting the views of the Department of the Army or the Department of Defense.

From: *Methods in Molecular Biology, Vol. 115: Immunocytochemical Methods and Protocols*
Edited by: L. C. Javois © Humana Press Inc., Totowa, NJ

enzymes to hydrolyze the DNA–protein complexes, which often entrap cells and can lead to reaggregation of suspended cells. The different specificities of these enzymes for intercellular components allow one to design a dissociation protocol to a specific tumor and for specific purposes. Many enzymes are crude extracts that contain varying amounts of contaminating proteolytic enzymes.

The enzymatic method *(5,6)* is probably the method of choice as a starting point for most tissue types because of its ability not only to release a large number of cells, but also to preserve cellular integrity and viability *(7)*. Last, chemical dissociation is commonly used in conjunction with mechanical or enzymatic procedures. Chemical methods are designed to omit or sequester the Ca^{2+} and Mg^{2+} ions needed for maintenance of the intercellular matrix and cell-surface integrity. Ethylenediaminoacetate (EDTA) or citrate ion is commonly used to remove these cations, but does not adequately dissociate all types of tissue.

2. Materials

1. Tissue, fresh.
2. Enzyme cocktail: 0.5 mg/mL collagenase, 0.25% w/v trypsin or dispase, and 0.002% w/v DNase.
3. Dulbecco's phosphate-buffered saline (PBSG), pH 7.0–7.2 without Ca^{2+} and Mg^{2+}, but with 1.0 g/L glucose: 2.7 mM KCl, 1.2 mM KH_2PO_4, 138 mM NaCl, 8.1 mM Na_2HPO_4, and 5.6 mM D-glucose.
4. "Complete" medium: Tissue-culture basic salt medium (e.g., RPMI, EMEM, DMEM), with serum (e.g., 10% fetal calf v/v final concentration).
5. Hot plate/stirrer.
6. Forceps.
7. Razor blades and/or no. 11 surgical blades.
8. Ice bath.
9. Low-speed centrifuge.
10. Collection tubes: 100 × 13-mm polypropylene tubes.
11. Cell counter.
12. Trypan blue: 0.4% w/v in distilled water.
13. Erlenmeyer flasks, 25 mL and 250 mL.
14. Glass Petri dish.

3. Method

1. In a glass Petri dish with 15 mL of PBSG, chop the tissue into 2- to 3-mm diameter pieces with the use of a no. 11 scalpel or equivalent (*see* **Note 1**).
2. Transfer the tissue pieces into a 250-mL Erlenmeyer flask.
3. Add 100 mL of enzyme cocktail, and stir at 200 rpm for 30 min at 37°C. This is done on a hot plate-stirrer (*see* **Note 2**).
4. Allow the fragments to settle, and then collect the supernatant.
5. Centrifuge the supernatant at 500*g* for 5 min.

6. Resuspend the pellet in complete medium with serum.
7. Store the cell suspension on ice for up to 15 min (*see* **Note 3**).
8. Add 100 mL of fresh enzyme cocktail to the flask.
9. Repeat **steps 3–7** until disaggregation is complete; this is usually evident when no visible clumps are seen and the enzyme cocktail changes from a milky, opaque solution to a clearer solution without the cloudy background.
10. Collect and pool the cells by centrifugation; count the cells (*see* **Notes 4** and **5**).
11. The cell suspension is now ready to introduce into tissue culture, immunostain, or deposit onto a microslide with a Cytospin.

4. Notes

1. This protocol can be used as a first-run attempt at tissue disaggregation. Further experience with the tissue will probably yield insights as to the best enzyme types, enzyme concentration, digestion time, and digestion temperature.
2. This technique uses enzymes to disrupt the tissue. Alternatively, mechanical methods, such as "squashing" tissue through metal mesh (50–100 μm opening) or disrupting the tissue by passing small pieces through sequentially smaller needles (e.g., 16, 20, 23 gage), can be employed. Chemical methods consist of such procedures as changing the pH, use of chelators, such as EDTA or EGTA, or the increase or decrease in the salt concentration of the digestion. The most suitable disaggregation procedure usually involves the combination of enzymatic and mechanical techniques, although permutations of all three techniques are possible.
3. The time that the cells remain on ice is dependent upon the final use of the cell suspension. If the cells are to be returned to tissue culture for possible establishment of a cell line, then the time should be held to a maximum of 15 min. Alternatively, if the cells are to be used for staining, they may remain on ice for up to 1 h before additional manipulation.
4. If the yield of viable cells is low try the following:
   a. Be sure that the enzymes are stored cold and dry, and that aliquots of enzyme are stored frozen.
   b. Use more collagenase.
   c. Depending on tissue type, additional enzymes may be needed, e.g., elastase, protease.
5. If the yield of cells has decreased viability, try the following:
   a. Reduce the exposure time to enzyme; try half the original time, and decrease by this factor if necessary.
   b. Reduce the amount of mechanical disruption.
   c. Readjust the pH of the digestion cocktail often.
   d. Increase digestion cocktail concentration thereby decreasing the exposure time to the enzymes; try doubling the initial enzyme concentration and halving the time. If this is unsuccessful, increase the time in 10-min increments.
   e. Add albumin or serum to the digestion cocktail up to 10% v/v in PBSG.
   f. Be gentle in all aspects of the disaggregation process.
   g. Remove separated cells from the digestion cocktail more frequently.

References

1. Berwick, L. and Corman, D. R. (1962) Some chemical factors in cellular adhesion and stickiness. *Cancer Res.* **22,** 982–986.
2. Rous, P. and Jones, F. S. (1916) A method for obtaining suspensions of living cells from the fixed tissues and for the plating of individual cells. *J. Exp. Med.* **23,** 549–555.
3. Waymouth, C. (1974) To disaggregate or not to disaggregate. Injury and cell disaggregation, transient or permanent? *In Vitro* **10,** 97–111.
4. Freshney, R. I. (1983) *Culture of Animal Cells. A Manual of Basic Technique.* Liss, New York.
5. Lewin, M. J. M. and Cheret, A. M. (1989) Cell isolation techniques: use of enzymes and chelators. *Methods Enzymol.* **171,** 444–461.
6. Cerra, R., Zarbo, R. J., and Crissman, J. D. (1990) Dissociation of cells from solid tumors. *Methods Cell Biol.* **33,** 1–12.
7. Costa, A., Silvestrini, R., Del Bino, G., and Motta, R. (1987) Implications of disaggregation procedures on biological representation of human solid tumors. *Cell Tissue Kinet.* **20,** 171–180.

32

Indirect Immunofluorescent Labeling of Viable Cells*

Robert E. Cunningham

1. Introduction

The combination of the specificity of the antigen–antibody interaction with the exquisite sensitivity of fluorescence detection and quantitation yields one of the most widely applicable analytical tools in cell biology (*1*). Within the last decade, flow cytometry (FCM) has become an integral part of basic immunological research. Elaboration of this technology has been intensively stimulated by a rapidly growing sophistication in monoclonal antibody technology and vice versa (*2*).

The added specificity of monoclonal antibodies in immunocytochemical technology provides a consistent and reliable method for exploiting the range of pure antibodies and subclasses of antibodies. These antibodies provide a means of defining cell-surface, intracellular, and membrane epitopes for single cells as well as tissue sections. When these antibodies are "tagged" with a fluorescent reporter antibody, multiple markers are possible. In particular, methods using protein A or avidin–biotin complexed with alternative fluorescent tags as second steps have added significant latitude to the immunofluorescence technique. An increasing number of clinical laboratories are using flow cytometry to analyze cells stained with fluorescent antibodies, dyes, or receptors (*3–5*). Also, human genes coding for cell-surface molecules can be introduced into host cells using a variety of somatic cell genetic techniques (*6*), and flow cytometry can then be used to monitor the effectiveness of the genetic techniques. Today, choices have to be made for the most appropriate fluoro-

---

*The opinions or assertions contained herein are the private views of the author and are not to be construed as official or as reflecting the views of the Department of the Army or the Department of Defense.

From: *Methods in Molecular Biology, Vol. 115: Immunocytochemical Methods and Protocols*
Edited by: L. C. Javois © Humana Press Inc., Totowa, NJ

chromes, reagents, equipment, and preparative procedure. In the following example, cells will be labeled with a mouse antihuman monoclonal antibody followed by a fluoresceinated antimouse secondary antibody rendering a green fluorescently labeled product. This procedure can also be used for cell-enrichment techniques utilizing cell sorting.

2. Materials

1. Sample: cell suspension of at least 1 million cells.
2. Dulbecco's Phosphate-buffered saline (PBSG), pH 7.0–7.2 without Ca^{2+} and Mg^{2+}, but with 1.0 g glucose/L: 2.7 mM KCl, 1.2 mM KH_2PO_4, 138 mM NaCl, 8.1 mM Na_2HPO_4, and 5.6 mM D-glucose.
3. 10% Bovine serum albumin in PBSG.
4. Primary antibody: usually a monoclonal antibody diluted in PBSG with 10% bovine serum albumin (*see* **Note 1**).
5. Secondary antibody: usually fluorescent antimouse conjugate, stock solution diluted 1:20–1:500 in PBSG with 10% bovine serum albumin (*see* **Note 1**).
6. Test tubes: 100 × 13-mm polypropylene tubes.
7. Pipetors: 20–100 μL, 100–1000 μL.
8. Disposable pipet tips.
9. Ice bath.
10. Centrifuge, low-speed.
11. Vortex mixer.
12. Propidium iodide: stock solution 50 mg/mL in PBSG (*see* **Note 2**).

3. Method

1. Harvest and count the cells to be used for the experiment. For each antibody or experimental condition, label the tubes, one for the positive condition and one for the negative control (*see* **Note 3**).
2. Wash cells twice by resuspending the pellet in 2 mL of 4°C PBSG, vortexing the suspension, centrifuging at 300*g* for 5 min, decanting the supernatant, and vortexing the pellet to loosen cells (*see* **Note 4**).
3. Add 100 μL of 4°C PBSG/million cells.
4. Add the appropriate amount of primary monoclonal antibody to the positive tubes and PBSG to the negative tubes as controls (*see* **Note 1**).
5. Mix the tubes gently, and incubate the cell suspension for 30 min on ice.
6. Wash cells twice by resuspending the pellet in 2 mL of 4°C PBSG, vortexing the suspension, centrifuging at 300*g* for 5 min, decanting the supernatant, and vortexing the pellet to loosen cells (*see* **Note 4**).
7. Add 100 μL of 4°C PBSG.
8. Add appropriate amount of secondary fluorescent antibody (*see* **Note 1**).
9. Mix the tubes gently, and incubate the cell suspension for 30 min on ice.
10. Wash cells twice by resuspending the pellet in 2 mL of 4°C PBSG, vortexing the suspension, centrifuging at 300*g* for 5 min, decanting the supernatant, and vortexing the pellet to loosen cells (*see* **Note 4**).

11. Resuspend the pellet to a concentration of 1–2 million cells/mL of PBSG or other medium. Add propidium iodine (PI) to a final concentration of 500 ng/mL (*see* **Note 5**).

4. Notes

1. The working concentration of antibody must be determined empirically by serial dilution of the stock solution in PBSG with 10% bovine serum albumin. Usual concentrations are in the range of 10–20 µg/mL. Depending on the individual reagent, this could vary considerable. *See* Chapter 24 for additional instructions on performing titrations.
2. Propidium iodide is a possible carcinogen and should be handled appropriately. The stock solution should be stored refrigerated in the dark.
3. If the cells are tissue-culture cells intended for cell sorting, use tissue-culture medium in which the cells have been growing as the wash solution. The tissue-culture supernatant is withdrawn from the tissue-culture flask and filtered through a 0.22-µ filter to ensure sterility. The addition of this medium helps cells recover after sorting and increases the growth of cells when they are placed back into tissue culture.
4. One major pitfall is the centrifugation of the suspension. If the centrifugation is not long enough or if sufficient centrifugal force is not created, the cells can be "poured off." Conversely, if the centrifugation is too long or too much force is created, then the cells may clump. Therefore, it can be helpful to centrifuge a cell suspension at low speed and determine if a cell pellet results from this amount of centrifugal force. If a cell pellet does not form then more speed (centrifugal force) is needed.
5. It is useful to resuspend the stained cell pellet in 1 mL PBSG with 500 ng/mL propidium iodide (PI) to detect dead cells by the inclusion of PI and its resultant red fluorescence on the flow cytometer. Red fluorescence (PI) vs light scatter is used to exclude dead cells before collecting green (FITC) emitting cells of the viable cell population.

References

1. Bosman, F. T. (1983) Some recent developments in immunocytochemistry. *Histochem. J.* **15,** 189–200.
2. Kung, P. C., Talle M. A., DeMarie M. E., Butler M. S., Lifter J., and Goldstein, G. (1980) Strategies for generating monoclonal antibodies defining human T lymphocyte differentiation antigens. *Transplant. Proc. XIII* **3,** 141–146.
3. Othmer, M. and Zepp, F. (1992) Flow cytometric immunophenotyping: principles and pitfalls. *Eur. J. Pediatr.* **151,** 398–406.
4. Haaijman, J. J. (1988) Immunofluorescence: quantitative considerations. *Acta Histochem.* **35,** 77–83.
5. Zola, H., Flego, L., and Sheldon, A. (1992) Detection of cytokine receptors by high-sensitivity immunofluorescence/flow cytometry. *Immunobiology* **185,** 350–365.
6. Kamarck, M. E., Barbosa, J. A., Kuhn, L., Peters, P. G., Shulman, L., and Ruddle, F. H. (1983) Somatic cell genetics and flow cytometry. *Cytometry* **4,** 99–108.

33

Indirect Immunofluorescent Labeling
of Fixed Cells*

Robert E. Cunningham

1. Introduction

One of the major advantages of flow cytometry is the simultaneous evaluation of multiple markers, especially surface markers *(1)*. The detection of intracellular proteins is less well developed, in large part because antibodies can bind nonspecifically to dying cells and dead cell components, which leads to considerable biological noise in the fluorescence detectors. There is also noise caused by the intra- and/or intermolecular ionic interactions during the process of fixation, which reduces the ability to detect the desired protein(s) both intracellularly and extracellularly. This is a double edged sword for labeling cells. It is very important to start the staining procedure with a viable cell suspension. If the starting material is viable, at least one of the two problems associated with cell fixation and staining is remedied. The fixation protocols are varied not only for their uses of crosslinking agents, permeabilization agents, and/or precipitating agents, but also for time and temperature. This chapter includes protocols for surface staining followed by fixation/permeabilization for DNA staining, fixation/permeabilization followed by surface staining, and fixation/permeabilization followed by intracellular staining.

The fixation conditions used to prepare cells for antibody application are assumed to preserve the distributions of the protein(s) being examined *(2)*. Soluble proteins can be redistributed into inappropriate locations and

---

From: *Methods in Molecular Biology, Vol. 115: Immunocytochemical Methods and Protocols*
Edited by: L. C. Javois © Humana Press Inc., Totowa, NJ

can be differentially extracted from native locations during the permeabilization and fixation of the cells before antibody application *(3,4)*. Further, no cell aggregation or alteration of the intracellular antigenicity should occur in the permeabilization/fixation treatment. The fixation/stain methodology, with and without permeabilization, can be accomplished in various ways depending on the exact site of the organelle or cell constituent to be stained.

The stain/fixation method is usually used for surface markers that can withstand fixation and is followed by the application of a DNA-binding fluorochrome. The fixation/stain method is used not only for surface markers that can withstand fixation, but also for intracellular constituents, such as cytoplasmic proteins, nuclear membrane, and nuclear proteins. This is accomplished by using a crosslinking fixative (e.g., paraformaldehyde [PFA] or formalin) followed by a permeabilizing agent (e.g., Triton X-100, Tween-20, saponin, or lysolecithin). Some of the precipitating agents (e.g., ethanol, methanol, or acetone) can also be used for permeabilization after the initial fixation with PFA or formalin, or they can be used alone for both fixation and permeabilization (*see* Chapter 8).

Finally, the determination of methodology for cell staining must be evaluated based on the type of tissue or cells being examined. It is absolutely critical that the sample be a viable, single-cell suspension. Not only is this important during the staining and data collection, but it is also important in the analysis of the specimen as representative of the pathologic sample.

2. Materials

1. Cell suspension of $1-3 \times 10^6$ cells/sample.
2. Dulbecco's phosphate-buffered saline without calcium or magnesium, with 1.0g glucose/L (PBSG): 2.7 mM KCl, 1.2 mM KH$_2$PO$_4$, 138 mM NaCl, 8.1 mM Na$_2$HPO$_4$, and 5.6 mM D-glucose, pH 7.0–7.2.
3. Fixatives: 0.25% and 1.0% paraformaldehyde (PFA) in PBSG, 100% ethanol, 100% methanol, 0.1% Triton X-100, and 0.2% Tween-20 (*see* **Note 1**).
4. Primary antibody (e.g., mouse monoclonal antibody to a surface or intracellular antigen) (*see* **Note 2**).
5. Secondary antibody (e.g., fluorescently labeled antimouse conjugate) (*see* **Note 2**).
6. 10% Bovine serum albumin in PBSG.
7. Test tubes: 100 × 13-mm polypropylene tubes.
8. Pipetor(s): 100–1000 µL, with disposable tips.
9. Ice bath.
10. Low-speed centrifuge.
11. Vortex mixer.
12. Propidium iodide (PI): stock solution of 50 mg/mL in PBSG (*see* **Note 3**).
13. RNase A (DNase-free): stock solution of 1000 U diluted in 1 mL PBSG.

3. Methods

3.1. Staining Cell-Surface Antigens Prior to Fixation for Flow Cytometric Analysis

It is entirely possible that surface staining cannot be accomplished before fixation. Some antibody–antigen complexes cannot withstand chemical fixation and/or permeabilization. An empirical evaluation must be made. In this example, cells are first stained with a monoclonal antibody against a cell-surface receptor, fixed with ethanol, and then the DNA is stained with propidium iodide. The cells are analyzed for two-color fluorescence, the green of the fluorescein-labeled surface marker and the red of the labeled DNA intercalator. This approach works for antibody–antigens that are unaffected by fixation.

1. Harvest and count the cells; use 1×10^6 cells/sample.
2. Wash the cells twice by resuspending the pellet in 2 mL of 4°C PBSG, vortexing the suspension, centrifuging it at 300g for 5 min, decanting the supernatant, and vortexing the pellet to loosen the cells.
3. Resuspend the cells in 100 µL of PBSG.
4. Add the appropriate amount of primary monoclonal antibody diluted in PBSG with 10% bovine serum albumin (*see* **Note 2**).
5. Incubate on ice for 30 min.
6. Wash the cells twice by resuspending the pellet in 2 mL of 4°C PBSG, vortexing the suspension, centrifuging it at 300g for 5 min, decanting the supernatant, and vortexing the pellet to loosen the cells.
7. Add 100 µL PBSG.
8. Add the appropriate amount of fluorescent secondary antibody diluted in PBSG with 10% bovine serum albumin (*see* **Note 2**).
9. Incubate on ice for 30 min.
10. Wash the cells twice by resuspending the pellet in 2 mL of 4°C PBSG, vortexing the suspension, centrifuging it at 300g for 5 min, decanting the supernatant, and vortexing the pellet to loosen the cells.
11. Fix the cell pellet by adding 300 µL PBSG and dispersing the cell pellet; while vortexing the cell pellet, add 700 µL 100% –20°C ethanol dropwise.
12. Stain with propidium iodide by centrifuging the cells and decanting off the ethanol; for each million cells, dilute propidium iodide stock solution to 15 µg/mL in 1 mL of RNase stock solution; incubate the suspension at 4°C overnight for the best results (*see* **Notes 4** and **5**).

3.2. Fixation and Permeabilization Followed by Staining Cell-Surface Antigens for Flow Cytometric Analysis

1. Centrifuge the cell suspension to concentrate the cells; use $1–3 \times 10^6$ cells/sample; for each million cells, add 1–2 mL 0.25% (w/v) paraformaldehyde in PBSG at 4°C for 60 min (*see* **Note 6**).
2. Remove the fixative by centrifuging at 300g for 5 min and decanting.

3. Wash the cells twice by resuspending the pellet in 2 mL of 4°C PBSG, vortexing the suspension, centrifuging it at 300*g* for 5 min, decanting the supernatant, and vortexing the pellet to loosen the cells.
4. Add 1–2 mL 0.2% (v/v) Tween-20 in PBSG at 37°C for 15 min (*see* **Note 7**).
5. Wash the cells twice by resuspending the pellet in 2 mL of 4°C PBSG, vortexing the suspension, centrifuging it at 300*g* for 5 min, decanting the supernatant, and vortexing the pellet to loosen the cells.
6. Add the appropriate amount of primary monoclonal antibody diluted in PBSG with 10% bovine serum albumin (*see* **Note 2**).
7. Incubate on ice for 30 min.
8. Wash the cells twice by resuspending the pellet in 2 mL of 4°C PBSG, vortexing the suspension, centrifuging it at 300*g* for 5 min, decanting the supernatant, and vortexing the pellet to loosen the cells.
9. Add 100 μL PBSG.
10. Add the appropriate amount of secondary fluorescent antibody diluted in PBSG with 10% bovine serum albumin (*see* **Note 2**).
11. Incubate on ice for 30 min.
12. Wash the cells twice by resuspending the pellet in 2 mL of 4°C PBSG, vortexing the suspension, centrifuging it at 300*g* for 5 min, decanting the supernatant, and vortexing the pellet to loosen the cells.
13. Resuspend the cells in PBSG for analysis (*see* **Notes 5** and **8**).

3.3. Fixation and Permeabilization Followed by Staining of Intracellular Antigens for Flow Cytometry

1. Harvest and count the cells; use 1×10^6 cells per sample.
2. Wash the cells twice by resuspending the pellet in 2 mL of 4°C PBSG, vortexing the suspension, centrifuging it at 300*g* for 5 min, decanting the supernatant, and vortexing the pellet to loosen the cells.
3. Add 2 mL 1% (w/v) paraformaldehyde in PBSG at 20°C for 15 min.
4. Remove the fixative by centrifuging at 300*g* for 5 min and decanting.
5. Wash the cells twice by resuspending the pellet in 2 mL of 4°C PBSG, vortexing the suspension, centrifuging it at 300*g* for 5 min, decanting the supernatant, and vortexing the pellet to loosen the cells.
6. Add 2 mL 100% methanol at –20°C for 10 min.
7. Remove the fixative by centrifuging at 300*g* for 5 min and decanting.
8. Wash the cells twice by resuspending the pellet in 2 mL of 0.1% Triton X-100 (v/v) in PBSG, vortexing the suspension, centrifuging at 300*g* for 5 min, decanting the supernatant, and vortexing the pellet to loosen the cells.
9. Stain the cells by adding the appropriate amount of the primary monoclonal antibody diluted in PBSG with 10% bovine serum albumin (*see* **Note 2**).
10. Incubate on ice for 30 min.
11. Wash the cells twice by resuspending the pellet in 2 mL of 4°C PBSG, vortexing the suspension, centrifuging it at 300*g* for 5 min, decanting the supernatant, and vortexing the pellet to loosen the cells.

12. Add 100 μL PBSG.
13. Add the appropriate amount of secondary fluorescent antibody diluted in PBSG with 10% bovine serum albumin (*see* **Note 2**).
14. Incubate on ice for 30 min.
15. Wash the cells twice by resuspending the pellet in 2 mL of 4°C PBSG, vortexing the suspension, centrifuging it at 300*g* for 5 min, decanting the supernatant, and vortexing the pellet to loosen the cells.
16. Resuspend the cells in PBSG for analysis (*see* **Notes 5** and **8**).

4. Notes

1. Paraformaldehyde fixes by crosslinking surface proteins. Ethanol and methanol work through precipitation of membrane components. Triton X-100 and Tween-20 permeabilize membranes. Other permeabilizers are saponin (0.1–10% [v/v] in PBSG), L-lysophosphatidylcholine, and *n*-octyl-β-D-glucopyranoside (1–10 μg/mL in water).
2. The working concentration of antibody must be determined empirically by serial dilution of the stock solution in PBSG with 10% bovine serum albumin. Usual concentrations are in the range of 10–20 μg/mL. Depending on the individual reagent, this could vary considerably. *See* Chapter 24 for additional instructions on performing titrations.
3. Propidium iodide is a possible carcinogen and should be handled appropriately. The stock solution should be stored refrigerated in the dark.
4. If propidium iodide is used as the nuclear stain, then the incubation time at 4°C overnight yields the best results. Although the cell suspension may be examined after 30–45 min, the longer the propidium iodide is allowed to react with the DNA, the better the staining.
5. There is some evidence that increasing the saline concentration to 0.65–1.0 *M* NaCl in the staining solutions can enhance staining for the laser dyes and the fluorescently tagged antibodies. This can be accompanied by the use of 0.1% (v/v) Tween-20 in PBSG or 0.1% $NaBH_4$ in water to reduce autofluorescence, and thereby increase the signal-to-noise ratio of the fluorescent emission.
6. The fixation followed by permeabilization protocol can utilize several different methods of cytological fixation, including acetone (15 min), 95% ethanol (15 min), Bouin's fixative (5 h), 1.0% trypsin (15 min at 37°C), 10% buffered neutral formalin (24 h), methanol (3 min at –20°C), or 4% PFA. There is also a range of agents for permeabilization, such as 1% saponin at 0°C in PBSG containing 20% human serum or 0.1–1.0% (v/v) Triton X-100 in PBSG.
7. Permeabilization of cells with 0.1–0.2% detergent after paraformaldehyde fixation can leave an uneven cytoplasmic distribution of the labeled proteins, and some of the larger proteins are redistributed to the nuclei. Extraction with 1% detergent prior to fixation removes most, but not always all of the exogenous proteins from the cell remnants.
8. After intracellular and surface staining are complete, it is often possible to utilize a stain for nucleic acids, such as propidium iodide (PI), DAPI, mithramycin, or

7–aminoactinomycin D (7AAD) (*see* Chapter 30). The determination of which stain to use is dependent on the other fluorescent markers being used, spectral overlaps of the dyes, and the laser lines available for excitation (e.g., DAPI needs UV [345 nm] excitation light, whereas PI, 7AAD, and mithramycin use blue [488 nm] excitation light).

References

1. Vyth-Dreese, F. A., Kipp, J. B. A., and DeJohn, T. A. M. (1980) Simultaneous measurement of surface immunoglobulins and cell cycle phase of human lympho-cytes, in *Flow Cytometry IV* (Laerum, O. D., Lindo, T., and Thorn, E., eds.), Universitetsforlaget, Oslo, pp. 207–212.
2. Labalette-Houache, M., Torpier, G., Capron, A., and Dessaint, J. P. (1991) Improved permeabilization procedure for flow cytometric detection of internal antigens. Analysis of interleukin-2 production. *J. Immunol. Methods* **138,** 143–153.
3. Pollice, A. A., McCoy, J. P., Jr., Shackney, S. E., Smith, C. A., Agarwal, J., Burholt, D. R., Janocko, L. E., Hornicek, F. J., Singh, S. G., and Hartsock, R. J. (1992) Sequential paraformaldehyde and methanol fixation for simultaneous flow cytometric analysis of DNA, cell-surface proteins, and intracellular proteins. *Cytometry* **13,** 432–444.
4. Schmid, I., Uittengobaart, C. H., and Giogi, J. V. (1991) A gentle fixation and permeabilization method for combined cell-surface and intracellular staining with improved precision in DNA quantification. *Cytometry* **12,** 279–285.

34

Fluorescent Labeling of DNA*

Robert E. Cunningham

1. Introduction

Flow cytometry (FCM) is a high-precision technique for rapid analysis and sorting of cells and particles. In theory, it can be used to measure any cell component, provided that a fluorescent tracer is available that reacts specifically and stoichiometrically with that constituent. The technique provides statistical accuracy, reproducibility, and sensitivity.

The quantitative cytochemical determination of DNA has been carried out using cytofluorochemical stains and offers a direct measurement of the DNA content of individual cells in a population *(1,2)*. The fluorescence distribution produced by a cell suspension provides a representation of the cell-cycle distribution of this population. Based on histograms generated from FCM data of the DNA content of individual cells, three groups can be identified in an asynchronous and mitotically active cell population. Most cells are in a resting (G0/G1) phase, also known as Gap0 and Gap1. As cells enter the synthesis (S) phase, the amount of cellular DNA increases resulting in increased fluorescence. After S phase, cells enter the Gap2/Mitosis (G2/M) phase where very little additional DNA is synthesized and cell division occurs. Now the cells contain twice the amount of DNA with approximately twice the staining intensity of the G0/G1 phase.

This protocol uses propidium iodide (PI) as the fluorescent tracer for DNA content *(3–6)*. PI binds to both double-stranded DNA and double-stranded RNA. Therefore, RNase will be used to reduce the double-stranded RNA resulting in only DNA staining. For alternative DNA-specific ligands and dyes, *see* Chapter 30.

---

*The opinions or assertions contained herein are the private views of the author and are not to be construed as official or as reflecting the views of the Department of the Army or the Department of Defense.

From: *Methods in Molecular Biology, Vol. 115: Immunocytochemical Methods and Protocols*
Edited by: L. C. Javois © Humana Press Inc., Totowa, NJ

Finally, flow cytometric DNA staining has been regarded as an objective prognostic parameter in several types of human cancer, but it is important to remember that this is a "snapshot" of the cell cycle. From a DNA-content histogram, it cannot be determined if a cell is actively moving through the cell cycle, has slowed, or even stopped its traverse through the cell cycle.

2. Materials

1. Sample: cell suspension of at least 2×10^6 cells/tube.
2. 100% Ethanol stored at –20°C.
3. Dulbecco's phosphate-buffered saline (PBS), pH 7.0–7.2 without Ca^{2+} and Mg^{2+}, but with 1.0 g glucose/L: 2.7 mM KCl, 1.2 mM KH_2PO_4, 138 mM NaCl, 8.1 mM Na_2HPO_4, 5.6 mM D-glucose.
4. Propidium iodide: stock solution of 50 mg/mL in PBS (*see* **Note 1**).
5. RNase A (DNase-free): stock solution of 1000 U diluted in 1 mL PBS.
6. Low-speed centrifuge.
7. Test tubes: 100×13 mm polypropylene.
8. Ice bath.
9. Pipetors: 1–20, 20–100, 100–1000 μL with tips.
10. Vortex mixer.

3. Method

1. Harvest and count the cells; use 2×10^6 cells/sample.
2. Wash the cells twice by resuspending the pellet in 2 mL of 4°C PBS, vortexing the suspension, centrifuging it at 300g for 5 min, decanting the supernatant, and vortexing the pellet to loosen the cells.
3. Add 300 μL PBS, and vortex the suspension until no apparent clumps are visible; while vortexing the cells, add 700 μL 100% –20°C ethanol (*see* **Note 2**). At this time, the sample can be stored at 4°C.
4. Centrifuge the suspension at 300g for 5 min.
5. Decant the ethanol/PBS supernatant.
6. Stain with PI; for each million cells, dilute propidium iodide stock solution to 50 μg/mL in 1 mL of RNase A stock solution, add it to the cell pellet, and vortex (*see* **Note 1**).
7. Incubate for 20 min at room temperature before analysis (*see* **Notes 3** and **4**).

4. Notes

1. Propidium iodide is a possible carcinogen and should be handled appropriately. The stock solution should be stored refrigerated in the dark.
2. Other fixatives that may be considered are 4% paraformaldehyde (w/v) in PBS, 100% methanol, or acetone.
3. There are other fluorochromes that can be used for DNA analysis. The following three are the most straightforward and reproducible with regard to application: mithramycin, 4',6-diamidino-2-phenylindole hydrochloride (DAPI), and Hoechst 33258. Concentrations must be determined empirically (*see* Chapter 30).

4. The 20-min stain time at room temperature for PI is the minimum required for reproducible DNA staining. A preferable approach is to allow the cell suspension to incubate in the refrigerator overnight before analysis on the flow cytometer. The coefficient of variation is usually better, and the control size measurements more univariate. Concentration of PI can vary according to the material being evaluated. For dual staining of DNA and a surface marker, use 15 μg/mL concentration. For DNA staining alone, use up to 50 μg/mL. Doublets, higher aggregates, and cell debris can be excluded from analysis by using correlated area/peak measurements of DNA content histograms as described by Sharpless et al. *(7)*.

5. DNA content can be paired with cell size to define better the progression of cells through the cell cycle, and to distinguish between G0 and G1 cells and occasionally between G2 and M cells.

References

1. Joensuu, H. and Kallioniemi, O. P. (1989) Different opinions on classification of DNA histograms produced from paraffin-embedded tissue. *Cytometry* **10,** 711–717.
2. van Dam, P. A., Watson, J. V., Lowe, D. G., and Shepherd, J. H. (1992) Flow cytometric measurement of cell components other than DNA: virtues, limitations, and applications in gynecologic oncology. *Obstet. Gynecol.* **79,** 616–621.
3. Vindelov, L. L., Christensen, I. J., and Nissen, N. I. (1983) A detergent-trypsin method for the preparation of nuclei for flow cytometric DNA analysis. *Cytometry* **3,** 323–327.
4. Krishan, A. (1975) Rapid flow cytofluorometric analysis of mammalian cell cycle by propidium iodide staining. *J. Cell Biol.* **66,** 188–193.
5. Deitch, A. D., Law, H., and DeVere White, R. (1992) A stable propidium iodide staining procedure for flow cytometry. *J. Histochem. Cytochem.* **30,** 967–972.
6. Taylor, I. W. (1980) A rapid single step staining technique for DNA analysis by flow microfluorimetry. *J. Histochem. Cytochem.* **28,** 1021–1024.
7. Sharpless, T. F., Traganos, F., Darzynkiewicz, Z., and Melamed, M. R. (1975) Flow cytofluorimetry: discrimination between single cells and cell aggregates by direct size measurement. *Acta Cytol.* **19,** 577–581.

35

Deparaffinization and Processing of Pathologic Material*

Robert E. Cunningham

1. Introduction

DNA content has become an important diagnostic, as well as prognostic, method for clinical pathology and investigative oncology. Paraffin-embedded tissue can be examined by flow cytometric (FCM) methods for total DNA content and aneuploidy with respect to the classification of the original pathologic diagnosis. The relative significance of studies on archival material permits retrospective analysis on a great number of cases, studying different specimens of a tumor for intratumor heterogeneity while comparing results from previous pathologic evaluations (1). DNA content as measured in paraffin-embedded tissue is closely related to that obtained from fresh specimens. Still, a major drawback in the procedure is that only nuclei are recovered for analysis of DNA content.

A novel use of the nuclei isolated from paraffin-embedded tissue is to study a low number of nuclei by polymerase chain reaction (PCR). PCR has the end result of making DNA from a complementary DNA template (2,3). Nuclei can be sorted on the basis of cell cycle, digested to single-stranded DNA, and the desired segment of DNA amplified (4,5). Finally, the isolated nuclei can not only be stained for DNA content, but also for nuclear proteins, proliferation factors, and other nuclear proteins (6–8).

The method outlined here uses a modification of the Hedley technique (9,10) to prepare nuclear suspensions from the paraffin-embedded tissue samples. Microtome sections are dewaxed, hydrated, and incubated in pepsin with inter-

*The opinions or assertions contained herein are the private views of the author and are not to be construed as official or as reflecting the views of the Department of the Army or the Department of Defense.

From: *Methods in Molecular Biology, Vol. 115: Immunocytochemical Methods and Protocols*
Edited by: L. C. Javois © Humana Press Inc., Totowa, NJ

mittent vortexing and mechanical disruption to release the nuclei. After completion of the tissue digestion, the nuclei are either suspended in 70% ethanol for storage or stained with propidium iodide (PI) for FCM analysis. There are three alternative techniques for preparation of the nuclei: on microslides, in "tea bags," or in test tubes.

2. Materials

1. Paraffin-embedded tissue or cell block.
2. Xylene.
3. 100, 95, and 50% Ethanol.
4. Dulbecco's phosphate-buffered saline (PBS), pH 7.0–7.2, without Ca^{2+} and Mg^{2+}, with 1.0 g glucose/L: 2.7 mM KCl, 1.2 mM KH_2PO_4, 138 mM NaCl, 8.1 mM Na_2HPO_4, and 5.6 mM D-glucose.
5. Dulbecco's PBS, pH 1.5, adjusted with 1N HCl.
6. Pepsin: stock solution 10 mg/mL (10% w/v) in PBS at pH 1.5.
7. HEPES buffer: 10 mM in PBS, final pH 7.0.
8. Syringes: 2, 5, and 10 mL.
9. Needles: 16, 18, and 22 gage.
10. Disposable pipets.
11. Water bath at 37°C.
12. PI: stock solution of 50 mg/mL in PBS (*see* **Note 1**).
13. RNase A (DNase-free): stock solution of 1000 U diluted in 1.0 mL PBS, pH 7.0–7.2.
14. "Tea bags" (*see* **Note 2**).
15. Microslides, glass (*see* **Note 3**).
16. Surgical blades, no. 11.
17. Microfuge tubes.
18. 40-μm Nylon mesh.
19. 15-mL Conical centrifuge tubes, polycarbonate or polypropylene.
20. 1.5-mL Xylene-resistant tubes.
21. Vortex.
22. Low-speed centrifuge.

3. Methods

3.1. Slide Technique

1. Cut one thin (5-μm) section, and reserve it for hematoxylin and eosin staining (*see* **Notes 4** and **5**).
2. Cut 30- to 80-μm sections, and mount one per glass slide (*see* **Note 3**). Be careful when sections exceed 30 μm since they detach easily. Proceed as gently as possible.
3. Clear the sections in two changes of xylene for at least 10 min each.
4. Hydrate the sections through decreasing ethanol solutions (100, 95, 50%), two changes each for 5 min each; then place the sections in water.
5. Air-dry the sections.

6. If necessary, remove necrotic, inflammatory, and/or nontumorigenic portions of the tissue section by scraping with a sharp blade (*see* **Note 5**).
7. Dilute the stock pepsin solution 1:20 in PBS, pH 1.5, and add 200 µL to each slide covering the tissue section.
8. Use a no. 11 surgical blade to dislodge and disrupt each tissue section.
9. Aspirate the suspensions into a pipet, and place in a microfuge tube.
10. Incubate the tube for 30–60 min at 37°C, vortexing intermittently (*see* **Note 6**).
11. Homogenate the contents of the microfuge tube through sequentially smaller gage needles (16–22) until a smooth texture is obtained.
12. Draw all of the liquid up into the syringe, remove the needle, and place a small square of 40-µm mesh between the needle and syringe.
13. Push out the contents of the syringe (nuclei) through the mesh into a microfuge tube.
14. Add 1 mL 10 mM HEPES buffer in PBS, pH 7.0–7.2, to the tube.
15. Centrifuge at 300g for 5 min. Decant the supernatant, and then vortex the pellet.
16. Repeat **steps 14** and **15** (*see* **Note 7**).
17. From the stock solutions, prepare a mixture of 10 µg PI and 10 U RNase A in a total vol of 1 mL PBS. Add it to the isolated nuclear suspension.
18. Incubate for at least 20 min, and analyze with the flow cytometer for red fluorescence of PI (DNA) (*see* **Note 8**).

3.2. "Tea Bag" Technique

1. Cut two to three sections each 50 µm thick, and place in properly identified "tea bags" (*see* **Notes 2** and **4**).
2. Dewax the sections in two changes of xylene for 10 min each.
3. Hydrate through graded alcohols to water, two changes each for 10 min each: 100, 95, 50% ethanol, distilled water.
4. Open the "tea bags," remove tissue, and scrape it into a test tube.
5. Dilute the pepsin stock solution 1:20 in PBS, pH 1.5, and add 1.0 mL to each test tube.
6. Incubate 1–2 h at 37°C, vortexing intermittently (*see* **Note 6**).
7. Add 1 mL 10 mM HEPES in PBS, pH 7.0–7.2, to each tube.
8. Centrifuge the tubes at 300g for 5 min. Decant the supernatant, and vortex the pellet.
9. Repeat steps 7 and 8 (*see* **Note 7**).
10. From the stock solutions, prepare a mixture of 10 µg PI and 10 U RNaseA in a total volume of 1 mL PBS. Add it to the isolated nuclear suspension.
11. Incubate for at least 20 min, and analyze with the flow cytometer for red fluorescence of PI (DNA) (*see* **Note 8**).

3.3. Test-Tube Technique

1. Cut two to three sections each 50 µm thick, and place in properly identified 15-mL centrifuge tubes (*see* **Note 4**).
2. Dewax the sections in two changes of xylene for 10 min each.
3. Hydrate through graded alcohols to water, two changes each for 10 min each: 100, 95, 50% ethanol, distilled water.

4. Dilute the pepsin stock solution 1:20 in PBS, pH 1.5, and add 1.0 mL to each test tube.

5. Incubate 1–2 h at 37°C, and homogenize the sample with a 23-gage needle attached to a 2-mL syringe (*see* **Note 6**).

6. Add 1 mL 10 m*M* HEPES in PBS, pH 7.0–7.2 to each tube.

7. Centrifuge the tubes at 300*g* for 5 min. Decant the supernatant and vortex the pellet.

8. Repeat **steps 6** and **7** (*see* **Note 7**).

9. From the stock solutions, prepare a mixture of 10 µg PI and 10 U RNase A in a total volume of 1 mL PBS, and add it to the isolated nuclear suspension.

10. Incubate for at least 20 min, and analyze with the flow cytometer for red fluorescence of PI (DNA) (*see* **Note 8**).

4. Notes

1. PI is a possible carcinogen and should be handled appropriately. The stock solution should be stored refrigerated in the dark.

2. "Tea bags" are available from Shandon Southern Instruments, Inc., 515 Broad Street, Sewickly, PA 15143 (cat. no. 67740010).

3. Glass microslides should be precoated to facilitate adhesion. Use 0.1% poly-L-lysine (w/v) in distilled water, or purchase Superfrost Plus Treated Microscope Slides from Fisher Scientific (Pittsburgh, PA). Either works equally well.

4. Tissue thickness is dependent on the type of cell nuclei to be isolated. Endothelial nuclei tend to be long and narrow, whereas other cell types tend to have a more cuboidal shape, and thus, thinner sections can be utilized.

5. The presence of necrotic, inflammatory, and/or nontumorigenic tissue should be determined by a pathologist by reviewing the hematoxylin- and eosin-stained 5-µm section prepared in **step 1**.

6. The time necessary for nuclear isolation depends on the tissue type, extent of suboptimally fixed, partially autolyzed samples, and thickness of the tissue sections.

7. If the sample is to be stored before staining, while vortexing, add 750 µL cold (−20°C) 100% ethanol dropwise to each tube.

8. The longer the PI is allowed to react with the DNA, the better the staining. Overnight incubation at 4°C is best.

References

1. Coon, J. S., Landay, A. L., and Weinstein, R. S. (1986) Flow cytometric analysis of paraffin-embedded tumors: implications for diagnostic pathology. *Human Pathol.* **17,** 435–437.

2. Gibbs, R. A. (1990) DNA amplification by the polymerase chain reaction. *Anal. Chem.* **62,** 1202–1214.

3. Gyllensten, U. B. (1989) PCR and DNA sequencing. *Biotechniques* **7,** 700–708.

4. Rodu, B. (1990) The polymerase chain reaction: the revolution within. *Am. J. Med. Sci.* **299,** 210–216.

5. Erlich, H. A. (1989) Polymerase chain reaction. *J. Clin. Immunol.* **9,** 437–447.

6. Boschman, G. A., Buys, C. H., van der Veen, A. Y., Rens, W., Osinga, J., Slater, R. M., and Aten, J. A. (1993) Identification of a tumor marker chromosome by flow sorting, DNA amplification in vitro, and in situ hybridization of the amplified product. *Genes Chromosom. Cancer* **6,** 10–16.

7. Milan, D., Yerle, M., Schmitz, A., Chaput, B., Vaiman, M., Frelat, G., and Gellin, J. (1993) A PCR-based method to amplify DNA with random primers: determining the chromosomal content of porcine flow-karyotype peaks by chromosome painting. *Cytogenet. Cell Genet.* **62,** 139–141.

8. Maesawa, C., Tamura, G., Suzuki, Y., Ishida, K., Saito, K., and Satodate, R. (1992) Sensitive detection of p53 gene mutations in esophageal endoscopic biopsy specimens by cell sorting combined with polymerase chain reaction single-strand conformation polymorphism analysis. *Jpn. J. Cancer Res.* **83,** 1253–1256.

9. Hedley, D. W., Friedlander, M. L., Taylor, I. W., Rugg, C. A., and Musgrove, E. A. (1983) Method for analysis of cellular DNA content of paraffin-embedded pathological material using flow cytometry. *J. Histochem. Cytochem.* **31,** 1333–1335.

10. Hedley, D. W., Friedlander, M. L., and Taylor, I. W. (1985) Application of DNA flow cytometry to paraffin-embedded archival material for the study of aneuploidy and its clinical significance. *Cytometry* **6,** 327–333.

36

Assay for Phagocytosis

Liana Harvath and Douglas A. Terle

1. Introduction
1.1. Background

Phagocytic leukocytes (granulocytes, monocytes, and macrophages) provide a first line of host defense by their ingestion (phagocytosis) and killing of microorganisms. Leukocytes phagocytize particles through a complex series of events that include serum coating (opsonization), cellular recognition, particle adhesion to the cell surface, and internalization. Serum complement components are important for the particle adhesion to phagocytes, whereas IgG opsonization is necessary for particle internalization (reviewed in **ref. 1**).

A variety of assays have been developed to quantify phagocytic activity. These include: direct microscopic visualization *(2,3)*, spectrophotometric evaluation of phagocytized paraffin droplets containing dye *(4)*, scintillation counting of radiolabeled bacteria *(5)*, fluorometric *(6)*, and flow cytometric analysis of fluorescent particles *(7–13)*. The flow cytometric assay offers the advantage of rapid analysis of thousands of cells and quantification of the internalized particle density for each analyzed cell. The assay may be performed with purified leukocyte preparations *(7–13)* or anticoagulated whole blood *(14,15)*.

1.2. Basis of the Assay

Fluorescent particles that have been opsonized with serum or specific IgG are mixed with phagocytic leukocytes at 37°C and continuously mixed to optimize the cell–particle interaction. The reaction is stopped by the addition of ice-cold medium, and the free particles are washed away from the leukocytes by centrifugation. The cells are resuspended in cold medium and analyzed

From: *Methods in Molecular Biology, Vol. 115: Immunocytochemical Methods and Protocols*
Edited by: L. C. Javois © Humana Press Inc., Totowa, NJ

for fluorescent signal intensity, which directly corresponds to the number of particles associated with each cell. The extracellular fluorescence resulting from adherent, noninternalized particles can be quenched to determine the fluorescence of ingested particles.

2. Materials

1. Acid citrate dextrose solution-A (ACD-A): 22.0 g/L sodium citrate ($Na_3C_6H_5O_7 \cdot H_2O$), 8.0 g/L citric acid, and 24.5 g/L dextrose.
2. Normal human blood sample anticoagulated with ACD-A (15 mL ACD-A/ 100 mL blood).
3. Dextran, pyrogen-free: average mol wt 100,000–200,000 Dalton (United States Biochemical Corp., Cleveland, OH).
4. Ficoll-hypaque lymphocyte separation medium (Organon Teknika Corp., Durham, NC).
5. Hank's balanced salt solution (Gibco-BRL, Grand Island, NY).
6. Phosphate-buffered saline ([PBS]; Gibco-BRL).
7. Krebs' Ringer's PBS: PBS with 1.0 mM calcium, 1.5 mM magnesium, and 5.5 mM glucose, pH 7.3 ± 0.1.
8. 3.5% NaCl solution.
9. Sterile water for injection, USP (Abbott Laboratories, North Chicago, IL).
10. Polypropylene, 50-mL sterile, conical centrifuge tubes (Becton Dickinson Labware, Lincoln Park, NJ).
11. Polypropylene, 5.5-mL sterile tubes (Becton Dickinson Labware).
12. Fluorescent particles, such as one of the following: *Escherichia coli* Bioparticles, fluorescein-conjugated; *Staphylococcus aureus* Bioparticles, fluorescein-conjugated; Zymosan A Bioparticles, fluorescein-conjugated (Molecular Probes, Eugene, OR); or Fluoresbrite carboxy microspheres, 0.92 μm diameter (Polysciences, Warrington, PA).
13. Opsonizing reagent, such as normal human pooled serum or rabbit IgG, to *E. coli* or *S. aureus* Bioparticles (Molecular Probes).
14. Crystal violet solution: 2 mg/mL in 0.15 M NaCl.
15. Flow cytometer.
16. Ultrasonic bath (needed only if particles are clumped and cannot be uniformly suspended as single particles by vortex mixing).

3. Methods

3.1. Cell Isolation

A granulocyte (neutrophil) isolation procedure adapted from a method originally described by Boyum (*16*) with modifications (*17*) is outlined below.

1. Prepare a 5% dextran solution in PBS. This solution may be stored for 2 w at 4°C, provided it is kept free of contamination. If a cold solution is used, allow the solution to warm to room temperature before addition to blood samples.
2. Prepare a 3.5% solution of NaCl.

3. Obtain a fresh blood sample anticoagulated with ACD-A. Maintain the blood sample at room temperature; do not refrigerate. The cell isolation should be performed as soon as possible after the sample is collected.

4. Add 3 mL of 5% dextran/10 mL of blood, gently mix, and let stand at room temperature for 40–45 min to allow red blood cell sedimentation.

5. After the sedimentation step, carefully pipet off the straw-colored plasma layer to within 0.5 cm of the red blood cell interface.

6. Add ficoll-hypaque lymphocyte separation medium to tubes, and carefully layer the plasma fraction over the ficoll-hypaque medium at a final volume ratio of 2 parts ficoll-hypaque:3 parts plasma. If 50-mL conical polypropylene centrifuge tubes are used, add 20 mL of ficoll hypaque to the tube first; then layer 30 mL of plasma on top of the ficoll hypaque such that a sharp interface is visible between the two layers.

7. Centrifuge the tubes for 35 min at 500g with the centrifuge brake turned off.

8. After centrifugation, carefully pipet away the platelet- and mononuclear cell-containing layer, which forms at the plasma-ficoll-hypaque interface. Pipet away the remainder of the supernatant, and resuspend the pellet in 2–3 mL of PBS.

9. Add 24 mL of sterile water for injection, and gently mix (by inversion) the cell suspension for 20 s. Then add 8 mL of 3.5% NaCl to make the solution isotonic.

10. Add Hank's balanced salt solution (room temperature) to the mixture to bring the vol to 50 mL, and centrifuge the cells for 10 min at 500g.

11. Decant the supernatant, and wash the cells with Hank's balanced salt solution two more times. The cells should be relatively free of contaminating red blood cells and platelets after these three washing steps.

12. Perform a cell count. Final cell suspensions should contain 98% granulocytes.

3.2. Whole Blood Preparation

1. Obtain a fresh blood sample anticoagulated with ACD-A. Maintain the blood sample at room temperature; do not refrigerate the blood prior to study. The assay should be performed within 5 h after the blood is collected.

2. Pipet 100 µL of anticoagulated whole blood into sterile, 5.5-mL polypropylene tubes immediately before initiating the phagocytosis assay.

3.3. Particle Selection

A variety of particles have been utilized in flow cytometric phagocytosis assays, including latex microspheres *(7,8,12)*, bacteria *(10,13,16)*, zymosan *(9,13)*, and baker's yeast *(11)*. Most of these particles are commercially available as fluorescent reagents. However, microorganisms may also be directly fluorochrome-conjugated (*see* **Notes 1** and **2**).

When selecting a particle for analysis, it is important to consider the effects of particle size on fluorescence histogram coefficient of variation (CV). Small particles of uniform size, such as 1 µm diameter latex spheres, exhibit a sharp, narrow histogram with a relatively small CV (**Fig. 1A**). Spherical bacteria,

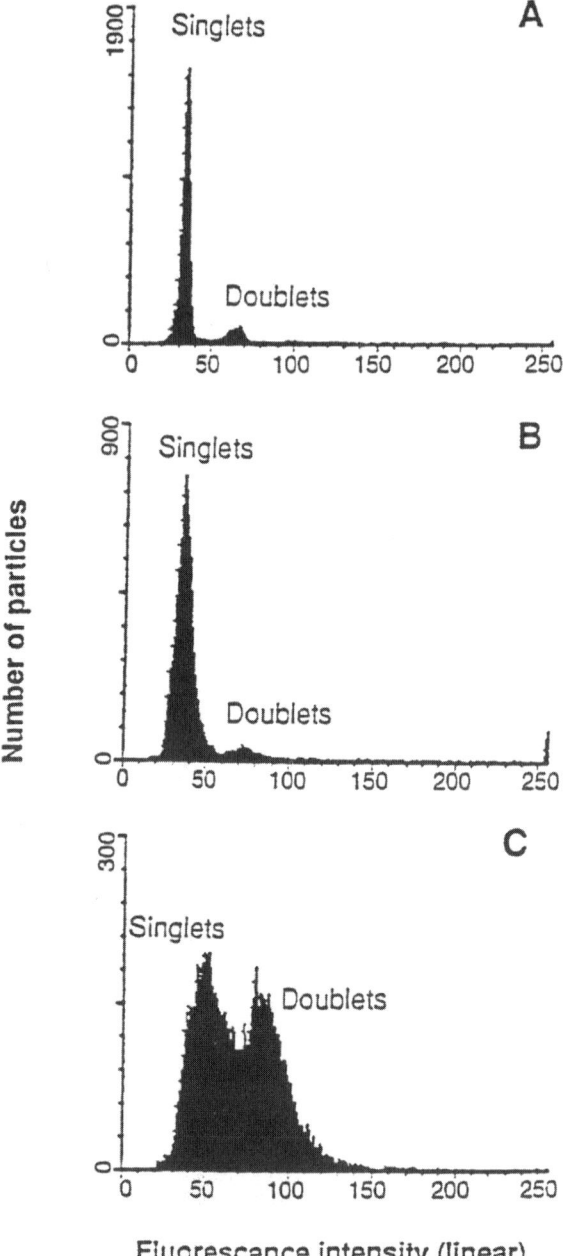

Fig. 1. Fluorescence histogram profiles of singlet and doublet 0.92-μm latex particles (**A**), *S. aureus* Bioparticles (**B**), and Zymosan Bioparticles (**C**).

such as *S. aureus*, exhibit a slightly wider histogram than latex spheres (**Fig. 1B**), whereas large zymosan particles have a broad histogram profile with a large CV (**Fig. 1C**). Single particles with a small CV are readily distinguished from pairs of particles with linear-scale fluorescence detection (**Fig. 1**).

When particles with a small CV are utilized in the phagocytosis assay, it is possible to observe as few as one particle per cell and characterize cell populations on the basis of the number of ingested particles per cell (*8,12*). If phagocytosis assays are performed with large fluorescent particles, quantification of internalized particles per cell is extremely difficult and often not possible. The investigator should determine which particle is most appropriate for the question under investigation. If one simply wishes to distinguish cells that have not phagocytized particles from cells that have internalized particles, then most fluorescent particles will be sufficient. However, if one needs to evaluate subsets of phagocytic cells on the basis of particles ingested per cell, it is important to utilize small particles of relatively uniform size, such as latex spheres or *S. aureus*.

3.4. Particle Opsonization

Particles are most efficiently phagocytized when they are opsonized with complement components or immunoglobulins. This is accomplished by incubating the particles with autologous or normal human pooled serum, or with specific antibodies that recognize antigens on the particle surface.

3.4.1. Opsonization with Normal Serum

1. Add fluorescent particles (10^8–10^9 particles/mL) to a 50% dilution of normal serum in Krebs' Ringer's PBS and incubate for 30 min at 37°C.
2. After opsonization, add particle suspensions to the phagocytosis assay mixture to a final concentration of 5% serum. Adjust particles to a density of 10^8/mL (*see* **Notes 3** and **4**).

3.4.2. Opsonization with Specific Immunoglobulin

1. Determine the titer of immunoglobulin solution for the microorganism to be utilized in the assay. Select a dilution of the immunoglobulin that does not cause aggregation of the particles. If *S. aureus* or *E. coli* Bioparticles from Molecular Probes are used, follow the directions provided by the manufacturer.
2. Incubate the particles with immunoglobulin for 60 min at 37°C, and wash the particles three times with PBS. Adjust particles to a density of 10^8/mL (*see* **Notes 3** and **4**).

3.5. Phagocytosis Assay
3.5.1. Assay Performed with Isolated Cells

1. Add 100 µL of granulocytes (10^7 cells/mL Krebs' Ringer's phosphate buffer) to a 6-mL polypropylene tube.

2. Add 10 µL of opsonized particles (10^8 particles/mL) to the tube containing neutrophils, and incubate in a shaking water bath at 37°C for 30 min (*see* **Notes 5** and **6**).
3. Prepare an identical sample as described in **steps 1** and **2**, but hold the sample at
   4°C as a control in which phagocytosis does not occur.
4. To stop the phagocytosis, add 2 mL of ice-cold PBS to the tubes, and wash the
   samples two times with cold buffer.
5. Resuspend the washed cells in 500 µL of cold PBS, maintain samples at 4°C, and
   analyze immediately with a flow cytometer. (*See* **Subheading 3.5.3.** for details
   on flow cytometer gates, determination of photodetector settings, and an example
   of results.)

3.5.2. Assay Performed with Whole Blood

1. Add 200 µL of anticoagulated whole blood to a 5.5-mL polypropylene sterile tube.
2. Add 10 µL of opsonized particles (10^8 particles/mL) to the tube containing neutrophils, and incubate in a shaking water bath at 37°C for 30 min (*see* **Notes 5** and **6**).
3. Prepare an identical sample as described in **steps 1** and **2**, but hold the sample at
   4°C as a control in which phagocytosis does not occur.
4. To stop the phagocytosis, add 2 mL of ice-cold PBS to the tubes, and wash the
   samples once with cold buffer.
5. Resuspend the pellet in 3 mL of sterile water for injection, and gently mix (by
   inversion) the cell suspension for 20 s. Then add 1 mL of 3.5% NaCl to make the
   solution isotonic. Centrifuge at 500*g* for 7 min to pellet the cells.
6. Resuspend the pellet in 500 µL of ice-cold PBS, and maintain the samples at 4°C
   until they are analyzed with a flow cytometer. (*See* **Subheading 3.5.3.** for details
   on flow cytometer gates, determination of photodetector settings, and an example
   of results.)

3.5.3. Flow Cytometric Analysis

1. Adjust the forward and right-angle light-scatter detectors so that the granulocyte population is clearly visible (**Fig. 2**). If whole blood is analyzed, three
   distinct clusters of cells should be observed (**Fig. 2A**), whereas isolated granulocytes should appear as a single cluster (**Fig. 2B**). Gate on the granulocyte
   population for analysis (indicated by the elliptical regions in **Fig. 2**).
2. Analyze the experimental sample (37°C reaction) by setting the fluorescence
   detector on linear amplification, and adjusting the amplifier gain such that distinct populations are clearly separable and on scale (**Fig. 3**). If the gain is set
   too low, the population peaks will be bunched together at the low end of the
   fluorescence histogram scale (**Fig. 3A**). In contrast, if the gain is set too high,
   the populations will be forced off scale (**Fig. 3B**). The appropriate gain setting
   is evident when several distinct population peaks are detected corresponding
   with cells that have not ingested particles and cells that have internalized 1, 2,
   3, or more particles (**Fig. 3C**) (*see* **Note 7**).
3. When the optimal fluorescence gain setting has been selected, analyze 10,000 cells,
   and record the histogram profile.

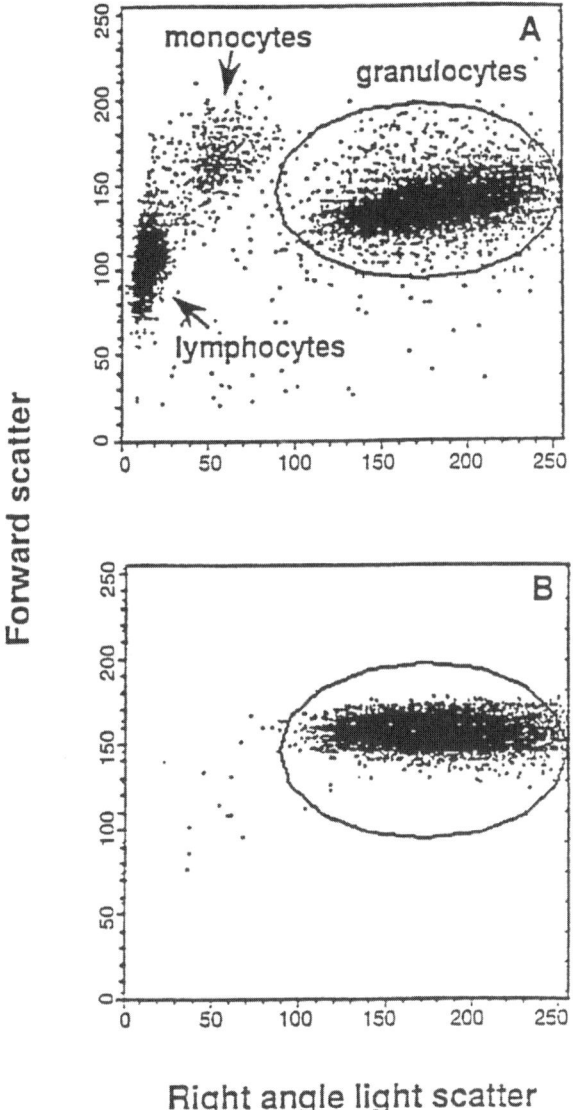

Fig. 2. Forward vs right-angle light-scatter profiles of granulocytes in an erythrocyte-lysed, whole-blood sample (**A**) and a purified granulocyte preparation (**B**).

4. Add 500 μL of crystal violet solution to the reaction mixture, and analyze another 10,000 cells. Crystal violet quenches extracellular fluorescein fluorescence of attached, uningested particles *(18)*. The histogram profile in the presence of crysta l violet represents the fluorescence of internalized particles (*see* **Note 8**).

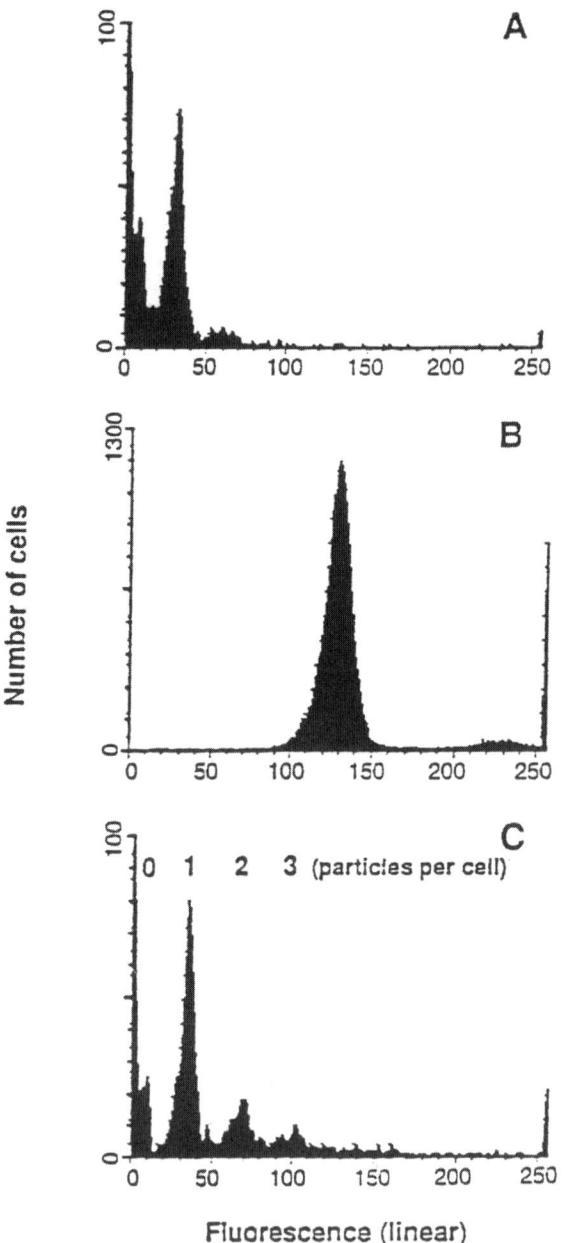

Fig. 3. Effects of photodetector settings on fluorescence histogram profiles of granulocytes exposed to opsonized 0.92-μm latex particles. Photodetector settings that are
too low (**A**) or too high (**B**) are compared with correctly adjusted photodetection (**C**).

4. Notes

1. Viable bacteria (10^9 organisms/mL) can be labeled in PBS, pH 7.4, containing 1 mg/mL of fluorescein isothiocyanate. The mixture is incubated for 30 min at room temperature and washed twice in PBS *(15)*.
2. Lyophilized bacteria or yeast (20 mg) can be labeled in 5 mL of 0.1 *M* carbonate-bicarbonate buffer, pH 9.0, containing 0.9% saline at 4°C. Fluorescein isothiocyanate (1 mg) is added to the mixture and incubated for 30 min at 4°C. The bacteria are washed several times with PBS *(6)*.
3. Particles with a diameter of 1 µm are difficult to count in a standard counting chamber and may require the aid of an electronic cell counter, such as the Coulter Counter (model ZBI).
4. If particles are aggregated after the opsonization procedure, vortex the particles vigorously. If vortexing does not disaggregate the particles, then place the tube containing the particles in an ultrasonic cleaning bath for 10 min.
5. The particle:cell ratio in this protocol is 10:1. Several studies have varied the particle:cell ratio from 5:1 to 100:1. If the particle density is large, phagocytes become loaded with large numbers of particles, and it is difficult to determine accurately the number of particles per phagocyte.
6. For best results, it is important to mix the samples during the incubation. If a shaking water bath is not available, a rocking platform or similar device may be placed in a 37°C incubator.
7. If distinct peaks cannot be detected with linear amplification gain, change the detector to log amplification. Larger particles, such as *E. coli* and zymosan, may require log fluorescence amplification gain.
8. Samples may be evaluated for the percentage of granulocytes with a specific phagocytic load (number of particles per cell) by setting regions for the fluorescence histogram of each population peak.

References

1. Brown, E. J. (1991) Complement receptors and phagocytosis. *Curr. Opin. Immunol.* **3,** 76–82.
2. Altman, A. J. and Stossel, T. P. (1974) Functional immaturity of bone marrow bands and polymorphonuclear leukocytes. *Br. J. Haematol.* **27,** 241–245.
3. Bjorksten, B., Petersen, P. K., Verhoef, J., and Quie, P. G. (1977) Limiting factors in bacterial phagocytosis by human polymorphonuclear leukocytes. *Acta Pathol. Microbiol. Scand. Sect. C.* **85,** 345–349.
4. Stossel, T. P. (1973) Evaluation of opsonic and leukocyte function with a spectrophotometric test in patients with infection and with phagocytic disorders. *Blood* **42,** 121–130.
5. Verhoef, J., Petersen, P. K., and Quie, P. G. (1977) Kinetics of staphylococcal opsonization, attachment, ingestion and killing by human polymorphonuclear leukocytes: a quantitative assay using [^{3}H]thymidine labeled bacteria. *J. Immunol. Methods* **14,** 303–311.
6. Oda, T. and Maeda, M. (1986) A new simple fluorometric assay for phagocytosis. *J. Immunol. Methods* **88,** 175–183.

7. Dunn, P. A. and Tyrer, H. W. (1981) Quantitation of neutrophil phagocytosis, using fluorescent latex beads. Correlation of microscopy and flow cytometry. *J. Lab. Clin. Med.* **98,** 374–381.

8. Steinkamp, J. A., Wilson, J. S., Saunders, G. C., and Stewart, C. C. (1981) Phagocytosis: flow cytometric quantitation with fluorescent microspheres. *Science* **215,** 64–66.

9. Bjerknes, R. and Bassoe, C.-F. (1983) Human leukocyte phagocytosis of zymosan particles measured by flow cytometry. *Acta Pathol. Microbiol. Immunol. Scand. Sect. C.* **91,** 341–348.

10. Bassoe, C.-F., Laerum, O. D., Solberg, C. O., and Haneberg, B. (1983) Phagocytosis of bacteria by human leukocytes measured by flow cytometry. *Proc. Soc. Exp. Biol. Med.* **174,** 182–186.

11. Derer, M., Walker, C., Kristensen, F., and Reinhardt, M. C. (1983) A simple and rapid flow cytometric method for routine assessment of baker's yeast uptake by human polymorphonuclear leukocytes. *J. Immunol. Methods* **61,** 359–365.

12. Parod, R. J. and Brain, J. D. (1983) Uptake of latex particles by macrophages: characterization using flow cytometry. *Am. J. Physiol.* **245,** (*Cell Physiol.* **14**), C227–C234.

13. Bassoe, C.-F. (1984) Processing *Staphylococcus aureus* and zymosan particles by human leukocytes measured by flow cytometry. *Cytometry* **5,** 86–91.

14. Trinkle, L. S., Wellhausen, S. R., and McLeish, K. R. (1987) A simultaneous flow cytometric measurement of neutrophil phagocytosis and oxidative burst in whole blood. *Diagnostic and Clin. Immunol.* **5,** 62–68.

15. Banfi, E., Cinco, M., Perticarari, S., and Presani, G. (1989) Rapid flow cytometric studies of *Borrelia bungdorferi* phagocytosis by human polymorphonuclear leukocytes. *J. Appl. Bacteriol.* **67,** 37–45.

16. Boyum, A. (1968) Isolation of mononuclear cells and granulocytes from human blood. *Scand. J. Clin. Lab. Invest.* **21(Suppl.),** 77–89.

17. Harvath, L., Balke, J. A., Christiansen, N. P., Russell, A. A., and Skubitz, K. M. (1991) Selected antibodies to leukocyte common antigen (CD45) inhibit human neutrophil chemotaxis. *J. Immunol.* **146,** 949–957.

18. Hed, J. (1977) The extinction of fluorescence by crystal violet and its use to differentiate between attached and ingested microorganisms in phagocytosis. *EMS Lett.* **1,** 357–361.

37

Assay for Filamentous Actin

Liana Harvath

1. Introduction
1.1. Background

Actin, a 43-kDa globular protein of the microfilamentous cytoskeleton, exists as globular actin monomers (G-actin) and polymerized filaments (F-actin) in most eukaryotic cells. Nucleated nonmuscle cells, including neutrophils, undergo dynamic changes in the amount of F-actin and its location during cell migration and plasma membrane ruffling *(1–3)*. The morphologic shape changes that accompany cell motility appear to involve the interaction of F-actin with force generating such proteins as myosin, and cytoskeletal proteins that bind to actin and control the reversible gelation and solation of the three-dimensional actin network *(4,5)*. Evidence that actin is an essential protein for cell motility has come from studies showing that: (1) agents that block F-actin formation, such as cytochalasins and botulinum C2 toxin, inhibit neutrophil migration in vitro *(6,7)*, and (2) neutrophils from patients who have actin dysfunction or abnormal concentrations of microfilamentous cytoskeletal proteins, which affect actin polymerization, have severe motility defects *(8–10)*. Current models of cell motility propose that actin is involved in the force generation for cell movement *(1,11)*.

Molecules (chemoattractants) that stimulate neutrophil-directed migration (chemotaxis) bind to distinct receptors on neutrophil plasma membranes (discussed in Chapter 38 of this text). Within seconds after chemoattractant binding, neutrophils exhibit rapid oscillations in actin polymerization and depolymerization *(12,13)*. The shape changes accompanying chemoattractant binding depend on the duration and extent of F-actin polymerization *(3)*. These quantitative studies of F-actin content were performed utilizing a flow cytometric assay that detects the fluorescence intensity of individual, fixed, permeabilized cells that have been stained with F-actin-specific, fluorescent phallotoxins *(14,15)*.

From: *Methods in Molecular Biology, Vol. 115: Immunocytochemical Methods and Protocols*
Edited by: L. C. Javois © Humana Press Inc., Totowa, NJ

1.2. Basis of the Assay

The relative F-actin changes in neutrophils are quantified by exposing the cells to a chemoattractant stimulus, and the reactions are stopped by formalin fixation of the cells. The cells are permeabilized by the addition of lysophosphatidyl choline to the samples, and fluorescently conjugated phallotoxin is added, which specifically labels F-actin, but not G-actin. After staining, the cells are washed to remove unbound phallotoxin, resuspended in phosphate-buffered saline (PBS), and analyzed for fluorescence intensity with a flow cytometer. All samples are compared with control, unstimulated (resting) neutrophils. The mean channel fluorescence (MCF) intensity of stimulated samples is divided by the MCF of resting cells and expressed as a relative F-actin ratio. Since phallotoxins bind well to both large and small F-actin polymers, but do not bind to monomeric G-actin, the fluorescence intensity of each cell is directly proportional to its F-actin content.

2. Materials

1. A flow cytometer is required for this assay.
2. Nitrobenzoxydiazole (NBD) phallacidin or rhodamine phalloidin; unconjugated phallacidin or phalloidin (Molecular Probes, Eugene, OR). Dissolve the phallotoxins in methanol to a final stock concentration of 3.3 μM, and store at $-20°C$ in the dark (*see* **Note 1**).
3. 37% Formaldehyde.
4. PBS (Gibco-BRL, Grand Island, NY).
5. Lysophosphatidyl choline (Sigma, St. Louis, MO). For assays in which fixation and permeabilization are performed in a single step, lysophosphatidyl choline is prepared as a stock solution of 1 mg/mL in 37% formalin. For assays in which fixation and permeabilization are performed as separate steps, the lysophosphatidyl choline is prepared as a stock solution of 1 mg/mL in PBS (*see* **Note 2**).
6. *N*-formyl-methionyl-leucyl-phenylalanine (FMLP) (Sigma): FMLP is prepared as a stock solution of 1 mM in ethanol and stored at $-20°C$.
7. Leukotriene B$_4$ (LTB$_4$) (BIOMOL Research Laboratories, Plymouth Meeting, PA): An aliquot is removed from the stock solution (stored at $-70°C$) and diluted in reaction buffer immediately before use in the assays.
8. Hank's balanced salt solution ([BSS]; GIBCO BRL): 50 mM phosphate, 150 mM NaCl, 4 mM KCl, 1.0 mM MgCl$_2$, and 1.2 mM CaCl$_2$. The Hank's BSS is buffered with 25 mM HEPES to pH 7.15 (as described in **ref. 2**).

3. Methods

3.1. Cell Preparation

1. Human neutrophils are isolated from acid citrate dextrose solution-A anticoagulated human blood by the method described by Boyum *(16)* with modifications *(17)* as described in detail in Chapter 36 of this text.

2. Each reaction contains 1×10^6 neutrophils/mL of HEPES-buffered Hank's BSS. In performing binding assays, it is important to work with a cell density where cell clumping is not occurring during the reaction. A cell density of 1×10^6 neutrophils/mL appears to be an optimal concentration for this assay.

3.2. Flow Cytometric Assay

Neutrophil F-actin content dynamically changes during chemoattractant stimulation in a dose-dependent manner *(14)*. The examples presented in this chapter are from experiments performed with human neutrophils stimulated at 37°C utilizing the technique described by Howard and Meyer *(14)*, briefly summarized below.

3.2.1. Protocol for Time-Course Evaluation of F-Actin Polymerization in Chemoattractant-Stimulated Human Neutrophils

1. Adjust the cells to a concentration of 1.1×10^6 cells/mL in Hank's BSS containing 25 m*M* HEPES. Prepare tubes containing 900 µL of cell suspension for each time-point (0–300 s). The tubes are warmed to 37°C in a water bath prior to and during chemoattractant exposure.
2. Add chemoattractant (10 µL of 10^{-5} *M* FMLP) to designated tubes, and mix the samples. Fix and permeabilize the samples in a single step with addition of 100 µL of 37% formalin containing 1 mg/mL of lysophosphatidyl choline at specified times after activation. Fix the cells for at least 5 min at 37°C (*see* **Note 3**).
3. Add to each reaction 50 µL of either rhodamine phalloidin or NBD-phallacidin stock solution (3.3 µ*M*), mix the samples and incubate for 10 min at 37°C (*see* **Note 4**).
4. Pellet the cells by centrifugation at 500*g* for 7 min. Decant the supernatants, and resuspend the cells in 1 mL of PBS.
5. The flow cytometer is set up to read linear-scale fluorescence intensities. For each sample, analyze 5000–10,000 cells. Analyze the resting, unstimulated control samples first to determine the appropriate gain/ voltage setting for the photomultiplier tube; then analyze the stimulated samples.
6. The MCF values are recorded for each sample, and the relative F-actin content is determined as: Stimulated Sample MCF/Resting (Control) MCF.

3.2.2. Examples of F-Actin Time-Dependent Changes in Chemoattractant-Stimulated Neutrophils

The data presented in **Fig. 1** illustrate the changes in F-actin polymerization and depolymerization during 5 min of FMLP (10^{-7} *M*) stimulation at 37°C. Human neutrophils often exhibit a biphasic F-actin response under these conditions *(18)*. All of the neutrophils exhibit changes in F-actin polymerization, as illustrated by the fluorescence histograms in **Fig. 2**, which

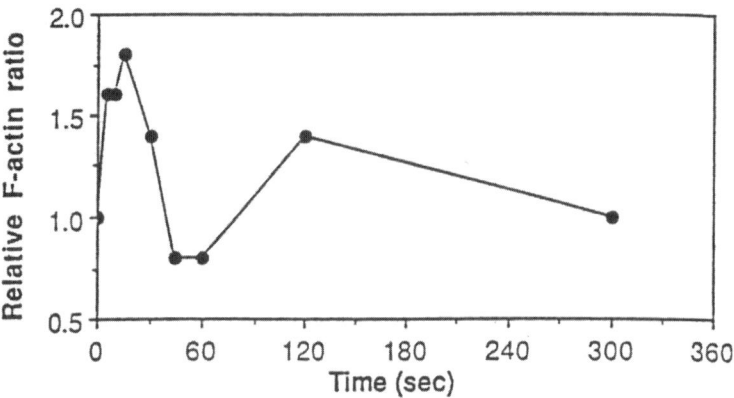

Fig. 1. Relative F-actin time-course response of human neutrophils exposed to $10^{-7} M$ FMLP at 37°C.

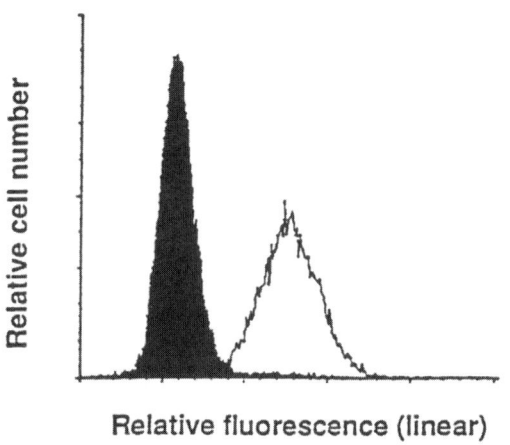

Relative fluorescence (linear)

Fig. 2. Relative F-actin fluorescence histograms of 10,000 neutrophils that are unstimulated (shaded curve) or stimulated with $10^{-7} M$ FMLP for 30 s at 37°C (open curve).

compares the fluorescence profile of 10,000 resting cells (shaded curve) with 10,000 FMLP-stimulated cells (open curve), 30 s after stimulation.

The effect of sequential chemoattractant exposure on neutrophil F-actin polymerization responses may also be evaluated with this technique. The data presented in **Fig. 3** were collected from an experiment in which neutrophils were exposed to $10^{-8} M$ LTB$_4$ for 5 min and then exposed to $10^{-7} M$ FMLP for 5 min (●), and compared with neutrophils exposed only to $10^{-7} M$ FMLP under identical incubation conditions (○).

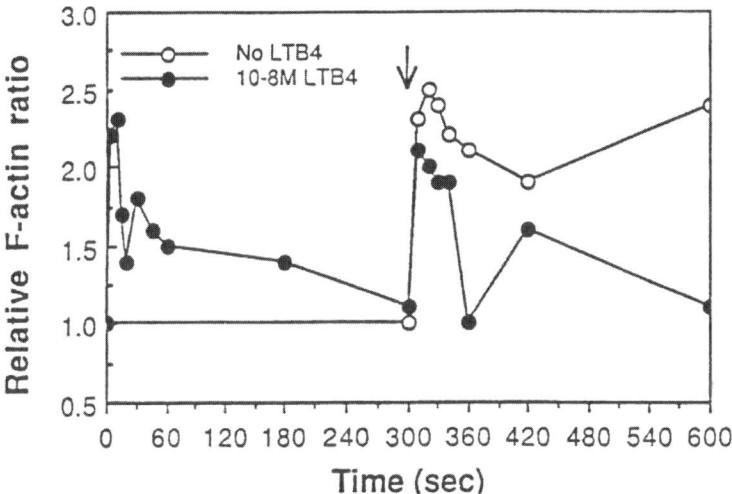

Fig. 3. Relative F-actin time-course responses of neutrophils sequentially exposed to 10^{-8} *M* LTB4 for 5 min and 10^{-7} *M* FMLP for 5 min (●) are compared with the response of neutrophils exposed to 10^{-7} *M* FMLP alone (○). The arrow indicates when 10^{-7} *M* FMLP was added to the cell suspensions.

3.2.3. Evaluation of Concentration-Dependent F-Actin Responses

To determine concentration-dependent F-actin responses, the protocol outlined in **Subheading 3.2.1.** is performed, except in **step 2**, the stimulus concentration is varied and all of the reactions are stopped at a fixed time.

3.2.4. Example of Concentration-Dependent F-Actin Responses in Chemoattractant-Stimulated Neutrophils

Data presented in **Fig. 4** illustrate the responses of neutrophils exposed to various FMLP concentrations for 30 s at 37°C. Human neutrophils exhibit a dose-dependent increase in F-actin polymerization, which is inhibited when cells are exposed to high concentrations of chemoattractant (*see* **Notes 5 and 6**).

4. Notes

1. The phallotoxins, phalloidin and phallacidin, are bicyclic peptides with mol wts of 789 and 847 Dalton, respectively. The NBD phallacidin and rhodamine phalloidin conjugates have been most frequently utilized in flow cytometric assays. The fluorescent phallotoxin conjugates have mol wts of 1000–1200 Dalton, are water-soluble, and stain actin at nanomolar concentrations (reviewed in **ref. *19***). Unconjugated phallotoxins should be obtained to verify the specificity of fluorescent phallotoxin staining to F-actin (*see* **Note 4**).

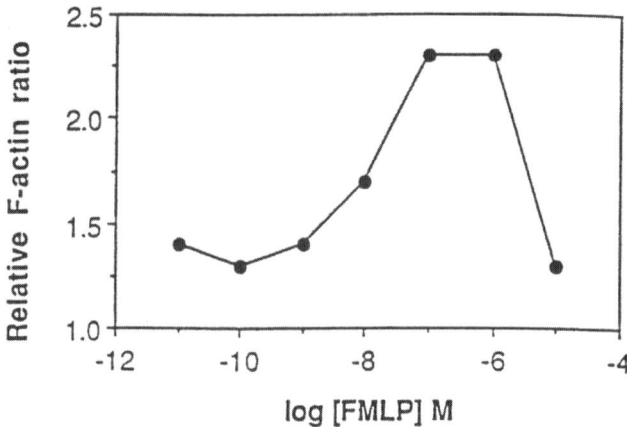

Fig. 4. Relative F-actin changes during a 30-s exposure to various concentrations of FMLP.

2. Both solutions are stored at 4°C and are stable for months. A precipitate may form during storage. It is readily dissolved by warming the solution at 37°C for a few minutes.

3. Samples may be fixed at **step 2** of **Subheading 3.2.1.** by the addition of 100 µL of 37% formalin, and stored at 4°C until it is convenient to continue with the staining and analysis. After refrigerated storage, the samples are permeabilized by the addition of 100 µL of 1 mg/mL lysophosphatidyl choline in PBS, and processed identically as described in **steps 3–6** of **Subheading 3.2.1.**

4. To verify the specificity of fluorescent phallotoxin binding, samples are prepared as described in **Subheading 3.2.1.**, except in **step 3**, 100-fold excess concentration of unlabeled phallotoxin is added simultaneously with the fluorescent phallotoxin. The specific binding of fluorescent phallotoxin binding to F-actin is determined as described in Chapter 38 of this text.

5. Although this chapter has focused on neutrophil activation events associated with F-actin changes, this assay has also been applied to studies of lymphocyte and platelet activation *(20–22)*.

6. A major limitation of flow cytometric analysis is that it provides data from individual cells at a single point in time and the same cells are not available for further analysis once they have passed through the flow cell of the instrument. Therefore, it is not possible to monitor a given cell over time for changes in fluorescence intensity or distribution of fluorescence signal. Such studies require microinjection of the fluorochrome into individual cells and fluorescence microscopy analysis.

References

1. Cassimeris, L. and Zigmond, S. H. (1990) Chemoattractant stimulation of polymorphonuclear leukocyte locomotion. *Sem. Cell Biol.* **1,** 125–134.

2. Howard, T. H. and Oresajo, C. O. (1985) The kinetics of chemotactic peptide-induced changes in F-actin content, F-actin distribution, and the shape of neutrophil. *J. Cell Biol.* **101,** 1078–1085.

3. Watts, R. G., Crispens, M. A., and Howard, T. H. (1991) A quantitative study of the role of F-actin in producing neutrophil shape. *Cell Motil. Cytoskel.* **19,** 159–168.

4. Omann, G. M., Allen, R. A., Bokach, G. M., Painter, R. G., Traynor, A. E., and Sklar, L. A. (1987) Signal transduction and cytoskeletal activation in the neutrophil. *Physiol. Rev.* **67,** 285–322.

5. Stossel, T. P., Chaponnier, C., Ezzell, R. M., Hartwig, J. H., Janmey, P. A., Kwiathkowski, D. J., Lind, S. E., Smith, D. B., Southwick, F. S., Yin, H. L., and Zaner, K. S. (1985) Nonmuscle actin-binding proteins. *Ann. Rev. Cell Biol.* **1,** 353–402.

6. Zigmond, S. H. and Hirsch, J. G. (1972) Effects of cytochalasin B on polymorphonuclear leukocyte locomotion, phagocytosis and glycolysis. *Exp. Cell Res.* **73,** 383–393.

7. Norgauer, J., Kownatzki, E., Seifert, R., and Aktories, K. (1988) Botulinum C2 toxin ADP-ribosylates actin and enhances O_2^- production and secretion but inhibits migration of activated human neutrophils. *J. Clin. Invest.* **82,** 1376–1382.

8. Boxer, L. A., Hedley-Whyte, E. T., and Stossel, T. P. (1974) Neutrophil actin dysfunction and abnormal neutrophil behavior. *N. Engl. J. Med.* **291,** 1093–1099.

9. Southwick, F. S., Dabiri, G. A., and Stossel, T. P. (1988) Neutrophil actin dysfunction is a genetic disorder associated with partial impairment of neutrophil actin assembly in three family members. *J. Clin. Invest.* **82,** 1525–1531.

10. Coates, T. D., Torkildson, J. C., Torres, M., Church, J. A., and Howard, T. H. (1991) An inherited defect of neutrophil motility and microfilamentous cytoskeleton associated with abnormalities of 47-Kd and 89-Kd proteins. *Blood* **78,** 1338–1346.

11. Singer, S. J. and Kupfer, A. (1986) The directed migration of eukaryotic cells. *Ann. Rev. Cell Biol.* **2,** 337–365.

12. Omann, G. M., Porasik, M. M., and Sklar, L. A. (1989) Oscillating actin polymerization/depolymerization responses in human polymorphonuclear leukocytes. *J. Biol. Chem.* **264,** 16,355–16,358.

13. Wymann, M. P., Kernen, P., Bengtsson, T., Andersson, T., Baggiolini, M., and Deranleau, D. A. (1990) Corresponding oscillations in neutrophil shape and filamentous actin content. *J. Biol. Chem.* **265,** 619–622.

14. Howard, T. H. and Meyer, W. H. (1984) Chemotactic peptide modulation of actin assembly and locomotion in neutrophils. *J. Cell Biol.* **98,** 1265–1271.

15. Howard, T. H. and Wang, D. (1987) Calcium ionophore, phorbol ester and chemotactic peptide-induced cytoskeleton reorganization in human neutrophils. *J. Clin. Invest.* **79,** 1359–1364.

16. Boyum, A. (1968) Isolation of mononuclear cells and granulocytes from human blood. *J. Clin. Lab. Invest.* **21(Suppl.),** 77–89.

17. Harvath, L., Balke, J. A., Christiansen, N. P., Russell, A. A., and Skubitz, K. M. (1991) Selected antibodies to leukocyte common antigen (CD45) inhibit human neutrophil chemotaxis. *J. Immunol.* **146,** 949–957.

18. Haugland, R. P. (1992) *Molecular Probes. Handbook of Fluorescent Probes and Research Chemicals,* 5th ed. (Larison, K. D., ed.), Molecular Probes, Inc., Eugene, OR, p. 205.
19. Harvath, L. (1990) Regulation of neutrophil chemotaxis: correlations with actin polymerization. *Cancer Invest.* **8(6),** 651–654.
20. DeBell, K. E., Conti, A., Alava, M. A., Hoffman, T., and Bonvini, E. (1992) Microfilament assembly modulates phospholipase C-mediated signal transduction by the TCR/CD3 in murine T helper lymphocytes. *J. Immunol.* **149,** 2271–2280.
21. Oda, A., Daley, J. F., Cabral, C., Kang, J., Smith, M., and Salzman, E. W. (1992) Heterogeneity in filamentous actin content among individual human blood platelets. *Blood* **79,** 920–927.
22. Kang, J., Cabral, C., Kushner, L., and Salzman, E. W. (1993) Membrane glycoproteins and platelet cytoskeleton in immune complex-induced platelet activation. *Blood* **81,** 1505–1512.

38

Assay for Chemoattractant Binding

Liana Harvath, Robert R. Aksamit, and Robert E. Cunningham

1. Introduction
1.1. Background

Phagocytic leukocytes play a major role in host defense because they rapidly migrate to sites of infection and destroy invading microorganisms. Specific signal molecules (chemoattractants), released by bacteria or endogenously generated by the host, can elicit directed leukocyte migration (chemotaxis) to the inflammatory site. Chemotaxis is initiated by the specific interaction (binding) of chemoattractants with leukocyte plasma membrane receptors. Polymorphonuclear leukocytes (neutrophils) migrate to a variety of chemoattractants, including N-formyl peptides, complement-derived C5a, leukotriene B_4, interleukin-8, and platelet-activating factor, each of which has a distinct receptor on the leukocyte plasma membrane.

The determination of chemoattractant binding can be performed with radiolabeled or fluorescently conjugated chemoattractants. Binding assays with radiolabeled chemoattractants require washing steps in which the unbound excess radioligand is removed from the leukocytes before the cells are placed in scintillation vials for analysis. In contrast, flow cytometric assays of fluorescent chemoattractant binding can be performed directly from the reaction mixture. This is possible because the sheath fluid dynamics around each leukocyte passing through the quartz flow cell of the flow cytometer serves as a complete washing step. When cells are analyzed directly from the reaction mixture, it is possible to obtain real-time binding kinetic data.

1.2. Basis of the Assay

The purpose of this assay is to detect saturable chemoattractant binding to specific chemoattractant receptors on leukocytes. A chemoattractant (fluorochrome

From: *Methods in Molecular Biology, Vol. 115: Immunocytochemical Methods and Protocols*
Edited by: L. C. Javois © Humana Press Inc., Totowa, NJ

conjugated by the investigator or obtained as a fluorescent conjugate from a commercial source) is evaluated in a functional assay to determine whether the conjugate retains biological activity. If the compound retains biological activity after conjugation, it is then evaluated for concentration-dependent binding. This is accomplished by adding various concentrations of the fluorescent chemoattractant to a fixed concentration of cells for a designated time at 0–4°C and evaluating the fluorescence signal intensity of several thousand cells for each chemoattractant concentration. Once a concentration range has been established, the binding specificity is then evaluated.

1.3. Determination of Specific Binding

Specific binding (total – nonspecific binding) is determined by quantifying the difference in fluorescence signal intensity between fluorescent chemoattractant bound in the absence of unlabeled chemoattractant (total binding) and the fluorescent chemoattractant bound in the presence of excess unlabeled chemoattractant (nonspecific binding). This is accomplished by comparing the cellular fluorescence of pairs of reactions at 0–4°C in which both samples contain identical concentrations of cells and fluorescent chemoattractant, but one of the samples contains an excess (100- to 1000-fold) of unlabeled chemoattractant as a competing ligand for the receptor.

1.4. Evaluation of Binding Kinetics

Since reaction mixtures may be directly sampled without a separate wash step, fluorescent chemoattractant binding to specific receptors on leukocytes over time is readily determined with the flow cytometer (1–4). Reaction mixtures may be sampled at selected time-points during the interaction of chemoattractant with cells (4), or flow cytometers may be equipped to monitor the binding association of ligand with cells continuously (3).

2. Materials

1. A flow cytometer is required for this assay.
2. Ficoll-hypaque, Lymphocyte Separation Medium (Organon Teknika, Durham, NC).
3. Dextran, pyrogen-free, 100,000–200,000 Dalton (U.S. Biochemical, Cleveland, OH).
4. Sterile water for injection, USP (Abbott Laboratories, North Chicago, IL).
5. N-Formyl-Nle-Leu-Phe-Nle-Tyr-Lys (Sigma, St. Louis, MO).
6. N-Formyl-Nle-Leu-Phe-Nle-Tyr-Lys fluorescein conjugate or rhodamine conjugate (Molecular Probes, Eugene, OR).
7. N-Formyl-Met-Leu-Phe-Lys (FMLPK) (Peninsula Labs, Belmont, CA) (see Note 1).
8. Fluorescein isothiocyanate (Molecular Probes): 10 mM in 0.2 M sodium borate; adjust to pH 9.5 by adding sodium hydroxide (see Note 1).
9. Glacial acetic acid (see Note 1).

10. Acetonitrile, HPLC-grade (*see* **Note 1**).
11. HPLC system and a C_{18} reverse-phase HPLC column (*see* **Note 1**).
12. 48-Well microchemotaxis chamber (NeuroProbe, Cabin John, MD) (*see* **Note 2**).
13. 5 µm-Pore, polyvinylpyrrolidone-free polycarbonate membranes (NeuroProbe) (*see* **Note 2**).
14. Hank's balanced salt solution (BSS) (Gibco-BRL, Grand Island, NY) (*see* **Note 2**).
15. Diff-Quik (Dade Diagnostics, Aguada, Puerto Rico) (*see* **Note 2**).
16. Phosphate-buffered saline (PBS) (Gibco-BRL).
17. Ice-water bath.
18. Vortex mixer.

3. Methods

3.1. N-Formyl Peptide Fluorochrome Labeling

Some *N*-formyl peptide (FP) chemoattractants that are directly conjugated with fluorochrome (FP-Fl) are commercially available (Molecular Probes) and may be utilized in the assay. Alternatively, the peptide may be fluorochrome-conjugated and purified by the investigator.

The data presented in this chapter are from experiments with the fluorescein-conjugated *N*-formyl peptide (*N*-formyl-Met-Leu-Phe-Lys; FMLPK-Fl), which we prepared with the reagents and reactions protected from the light as much as possible as described below.

1. Dissolve each milligram of FMLPK in 0.21 mL of 10 mM FITC solution. Check the pH, and adjust to pH 9.5 by adding a few µL of 1 M NaOH.
2. Incubate the mixture at room temperature for 1 h.
3. Add glacial acetic acid dropwise until the pH is 2.5, and an orange pre cipitate forms.
4. Add acetonitrile to give a final acetonitrile concentration of 30%. The orange precipitate easily dissolves under these conditions.
5. Purify the FMLPK-Fl on a C_{18} reverse-phase HPLC column by isocratic elution with a solution of 30% acetonitrile and 0.2 M acetic acid. Monitor the effluent at 254 nm. The peak corresponding to FMLPK-Fl should be clearly resolved from unreacted FITC and its breakdown products. Confirm the identity of FMLPK-Fl by performing an amino acid analysis of an acid-hydrolyzed sample.

3.2. Cell Preparation

1. Human neutrophils are isolated from acid citrate dextrose solution-A anti-coagulated human blood by the method described by Boyum *(5)* with modifications *(6)* as described in detail in Chapter 36 of this text.
2. Each reaction contains 1×10^6 neutrophils/mL of PBS. It is important to work with a cell density where cell clumping is not occurring during the reaction. A density of 1×10^6 neutrophils/mL appears to be an optimal concentration for this assay. Cells should be kept at 0–4°C after isolation.

3.3. Bioactivity Evaluation

It is important to examine the fluorescent conjugate for biological activity and confirm that the coupling procedure does not destroy the chemoattractant stimulatory activity. We evaluate neutrophil chemotaxis in a 48-well micro-chemotaxis chamber (7) as previously described in detail (8) and briefly outlined below.

1. Prepare 10-fold dilutions of FP and FP-Fl chemoattractant in Hank's BSS (e.g., $10^{-12} M$–$10^{-5} M$ concentration range).
2. Pipet 25 µL of chemoattractant in each lower well of the chemotaxis chamber.
3. Position a polyvinylpyrrolidone-free polycarbonate membrane containing 5-µm pores over the lower wells, position the silicone gasket, and assemble the top plate to the chamber.
4. Add 45 µL of cell suspension that is at a density of 1×10^6 cells/mL of Hank's BSS to each upper well of the chamber.
5. Incubate the chamber for 35 min at 37°C in humid air.
6. After the incubation, wipe off the nonmigrating cells, fix the membrane in methanol, and stain the with Diff-Quik stain.
7. Air-dry the filter, apply immersion oil, and count the stained nuclei of migrated cells.
8. The cell migration responses of neutrophils to various concentrations of FP (●) and FP-Fl (○) should be comparable, as shown in **Fig. 1**.

3.4. Flow Cytometric Binding Assay

3.4.1. Concentration-Dependent Specific Binding

3.4.1.1. PROTOCOL

Outlined below is a brief protocol for evaluating concentration-dependent specific binding of FMLPK-Fl.

1. Set up a pair of tubes for each concentration of FMLPK-Fl to be tested. If a concentration range of $4 \times 10^{-11} M$ to $4 \times 10^{-8} M$ is evaluated, label each set of tubes for concentration (i.e., 4×10^{-11}, 1×10^{-11}... $4 \times 10^{-8} M$).
2. Add 1 mL of 1×10^6 neutrophils/mL to each tube, and place in an ice-water bath.
3. To one tube of each set, add 1000-fold excess of unlabeled FMLPK. For example, one tube of the $4 \times 10^{-11} M$ FMLPK-Fl set should have $4 \times 10^{-8} M$ FMLPK, one tube of the $1 \times 10^{-11} M$ FMLPK-Fl set should have $1 \times 10^{-8} M$ FMLPK, and so on. Mix the samples well, and place in the ice-water bath.
4. To each set of tubes, add the designated concentration of fluorescent FMLPK-Fl, mix the tubes, and place in the ice-water bath for 15 min. Samples should be mixed by gentle vortexing occasionally during the incubation.
5. Before analyzing the samples, set the fluorescence amplifier gain on linear-scale detection and test an unstained sample of the cell preparation to determine the

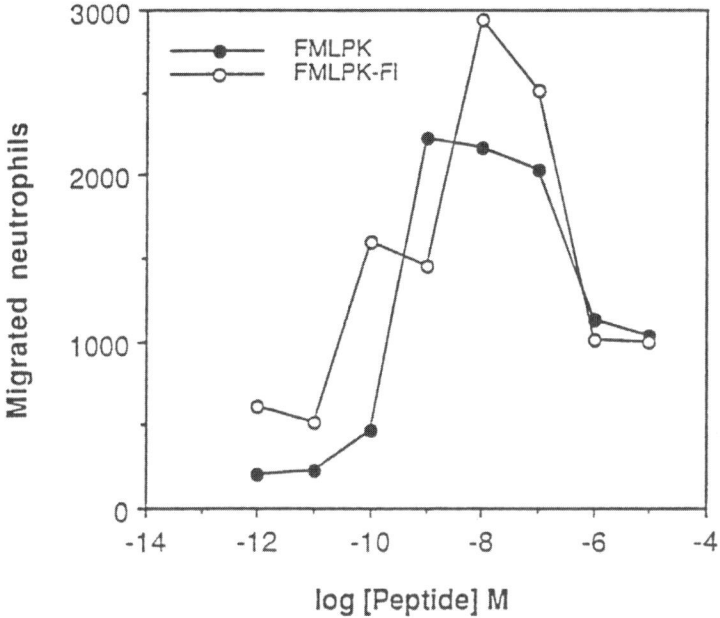

Fig. 1. Neutrophil chemotactic responses to FMLPK (●) and fluorescent FMLPK-Fl (○). Data are presented as the number of migrated neutrophils/mm² polycarbonate membrane filter surface.

autofluorescence of the neutrophils. Record the autofluorescence histogram profile, and proceed with analysis of the labeled samples.

3.4.1.2. EXAMPLES OF BINDING RESULTS

Since all peripheral blood neutrophils have chemoattractant receptors for *N*-formyl peptide, the entire population of neutrophils should exhibit a concentration-dependent shift in fluorescence intensity when exposed to FP-Fl. The mean channel values of the normal distribution fluorescence histograms are recorded and plotted vs the concentration of fluorescent FMLPK-Fl (**Fig. 2**). The data presented in **Fig. 2** are from an experiment in which paired samples of neutrophils were incubated for 15 min at 0–4°C in the presence (-x-) or absence (○) of 1000-fold excess unlabeled FMLPK. The excess unlabeled FMLPK is added to the reaction mixtures before fluorescent FMLPK-Fl is mixed with the cells to inhibit the specific binding of FMLPK-Fl to the neutrophil receptors competitively. The specific binding of FMLPK-Fl (●) is calculated by subtracting the nonspecific binding (-x-) from the total binding (○). As illustrated in **Fig. 2**, the specific binding should be saturable (fluorescence reaches a plateau at the highest concentrations of FMLPK-Fl tested) (*see* **Notes 3** and **4**).

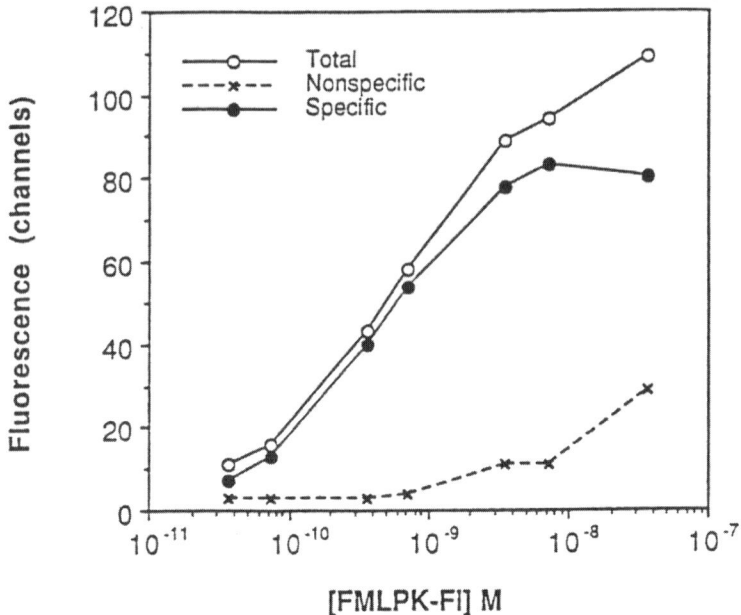

Fig. 2. Concentration-dependent binding of fluorescent FMLPK-Fl to neutrophils. The specific binding of FMLPK-Fl (●) is determined by subtracting the nonspecific binding (-x-) from the total binding (○). Nonspecific binding was evaluated in the presence of 1000-fold excess unlabeled FMLPK.

3.4.2. Analysis of Binding Kinetics

3.4.2.1. PROTOCOL

N-formyl peptides rapidly bind to neutrophils *(1–4)*. The association rate may be monitored by sampling a reaction mixture at selected times *(4)* or by continuous analysis of the reaction *(3)*.

1. Determine the half-maximal and maximal binding concentrations of FP-Fl from a concentration dependent specific binding experiment as described in **Subheading 3.4.1.1.**
2. Select a concentration in the half-maximal to maximal binding concentration range (e.g., 1×10^{-9} *M* FP-Fl) and set up a reaction mixture containing 1×10^6 neutrophils/mL with and without 1000-fold excess unlabeled FP-Fl.
3. Adjust the flow cytometer as described in **step 5** of **Subheading 3.4.1.1.**, and record the autofluorescence as time 0. Add the FP-Fl, mix, and analyze the reaction mixture at designated time intervals or continuously over a specified time period.

3.4.2.2. EXAMPLES OF TIME-DEPENDENT BINDING DATA

Data collected from a 4°C reaction containing 1×10^6 neutrophils exposed to 1×10^{-9} *M* FMLPK-Fl and sampled at 1-min intervals are summarized in

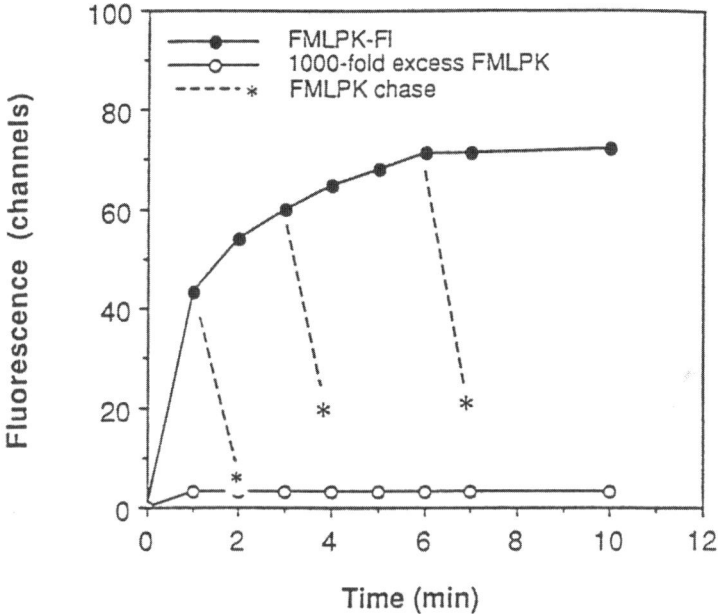

Fig. 3. Time-course of FMLPK-Fl binding to neutrophils determined by sampling a reaction containing neutrophils exposed to 1×10^{-9} *M* FMLPK-Fl at 1-min intervals. The bound FMLPK-Fl is competitively displaced by a 1000-fold excess chase of FMLPK (- - $_*$) if added within the first minute of the reaction. The continuous presence of 1000-fold excess FMLPK (○) completely inhibits FMLPK-Fl binding.

Fig. 3. If 1000-fold excess (10^{-6} *M* FMLPK) is added within 1 min after FMLPK-Fl exposure, the fluorescent ligand can be completely displaced from the receptors (- - $_*$). Displacement of the FMLPK-Fl with 1000-fold FMLPK is less complete if neutrophils have been exposed to the fluorescent ligand for 2 min.

An example of a continuously monitored reaction of 1×10^6 neutrophils before and after exposure to 1×10^{-9} *M* is presented in **Fig. 4**. Most binding occurs within the first minute of fluorescent ligand exposure. A small, but detectable displacement of FMLPK-Fl is observed when 10-fold excess FMLPK is added to the reaction.

The data presented in **Figs. 3** and **4** are examples of the types of kinetic binding data that are readily acquired with commercially available flow cytometers. Quantitative, real-time analysis of fluorescent *N*-formyl peptide association with neutrophil receptors has been described by Fay et al. *(4)*, and this publication should be consulted for detailed protocols required for quantitative kinetic assays (*see* **Notes 5** and **6**).

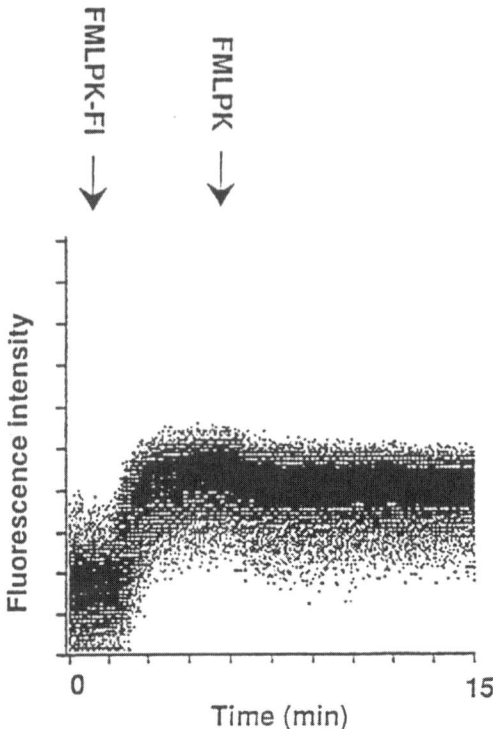

Fig. 4. Time-course of FMLPK-Fl binding to neutrophils determined by continuous monitoring of the reaction mixture. Arrows indicate the addition of $1 \times 10^{-9} M$ fluorescent FMLPK-Fl to the reaction mixture and displacement with a 10-fold excess of unlabeled FMLPK.

4. Notes

1. If *N*-formyl peptide chemoattractants conjugated with fluorochrome are purchased from a commercial source, the reagents listed in **Subheading 2.** referencing this note are not needed. Although this chapter has focused on *N*-formyl peptides, other chemoattractants, such as complement-derived C5a, have also been fluorescently labeled and evaluated in flow cytometric assays *(9–11)*.

2. The reagents listed in **Subheading 2.** referencing this note are not needed if a chemotaxis assay is not performed.

3. If the specific binding is not saturable or the nonspecific binding is >20%, it is important to determine whether unconjugated fluorochrome is present in the FP-Fl preparation. HPLC-purified FP-Fl should be free of unconjugated fluorochrome. Unconjugated fluorochrome binds to cells and causes nonspecific fluorescence. Repeated freezing and thawing of FP-FL can cause dissociation of the fluorochrome and should be avoided.

4. Although neutrophils exhibit homogeneous binding of *N*-formyl peptide, it should be noted that a subpopulation of human monocytes does not appear to express receptors for *N*-formyl peptide *(12)*. When flow cytometric binding studies are performed on a heterogeneous population of cells, it is important to evaluate the subpopulations separately. This can be accomplished by gating on the positive subpopulation (excluding the receptor-negative subpopulation) to determine the changes in mean channel fluorescence as a function of fluorescent chemoattractant concentration.

5. Flow cytometric evaluation of fluorescent chemoattractant binding to leukocytes offers several advantages over conventional radioligand assays:
   a. Separate washing steps to remove unbound ligand are not necessary.
   b. Individual cells, rather than a bulk population, are evaluated, which permits the identification of receptor-negative cell subpopulations.
   c. Reaction mixtures may be directly monitored before, during, and after the addition of fluorescent-conjugated ligand and unlabeled competitive ligand so that real-time analysis of binding kinetics is possible.
   d. Handling of radioisotopes is not required.

6. A limitation of the flow cytometric binding assay has been the precise determination of the receptor affinity and calculation of the receptors per cell. This limitation appears to have been overcome by the development of fluorescein and phycoerythrin compensation-calibration standards (Flow Cytometry Standards Corp., Research Triangle Park, NC). These standards have made it possible to quantify the fluorescence intensity of samples labeled with fluorescein or phycoerythrin, and relate the intensity to molecules of equivalent soluble fluorochrome. These standards have been utilized in quantitative studies of neutrophil chemoattractant–ligand interaction *(4)*.

References

1. Sklar, L. A. and Finney, D. A. (1982) Analysis of ligand-receptor interactions with the fluorescence activated cell sorter. *Cytometry* **3,** 161–165.
2. Finney, D. A. and Sklar, L. A. (1983) Ligand/receptor internalization: a kinetic, flow cytometric analysis of the internalization of *N*-formyl peptides by human neutrophils. *Cytometry* **4,** 54–60.
3. Omann, G. M., Coppersmith, W., Finney, D. A., and Sklar, L. A. (1985) A convenient on-line device for reagent addition, sample mixing, and temperature control of cell suspensions in flow cytometry. *Cytometry* **6,** 69–73.
4. Fay, S. P., Posner, R. G., Swann, W. N., and Sklar, L. A. (1991) Real-time analysis of the assembly of ligand, receptor, and G protein by quantitative fluorescence flow cytometry. *Biochemistry* **30,** 5066–5075.
5. Boyum, A. (1968) Isolation of mononuclear cells and granulocytes from human blood. *Scand. J. Clin. Lab. Invest.* **21(Suppl.),** 77–89.
6. Harvath, L., Balke, J. A., Christiansen, N. P., Russell, A. A., and Skubitz, K. M. (1991) Selected antibodies to leukocyte common antigen (CD45) inhibit human neutrophil chemotaxis. *J. Immunol.* **146,** 949–957.

7. Falk, W., Goodwin, R. H., and Leonard, E. J. (1979) A 48-well microchemotaxis assembly for rapid and accurate measurement of leukocyte migration. *J. Immunol. Methods* **33,** 239–247.

8. Harvath, L., Falk, W., and Leonard, E. J. (1980) Rapid quantitation of neutrophil chemotaxis: use of a polyvinylpyrrolidone-free polycarbonate membrane in a multiwell assembly. *J. Immunol. Methods* **37,** 39–45.

9. Van Epps, D. E., and Chennoweth, D. E. (1984) Analysis of the binding of fluorescent C5a and C3a to human peripheral blood leukocytes. *J. Immunol.* **132,** 2862–2867.

10. Yancey, K. B., Lawley, T. J., Dersookian, M., and Harvath, L. (1989) Analysis of the interaction of human C5a and C5a des Arg with human monocytes and neutrophils: flow cytometric and chemotaxis studies. *J. Invest. Dermatol.* **92,** 184–189.

11. Van Epps, D. E., Simpson, S., Bender, J. G., and Chenoweth, D. E. (1990) Regulation of C5a and formyl peptide receptor expression on human polymorphonuclear leukocytes. *J. Immunol.* **144,** 1062–1068.

12. Leonard, E. J., Noer, K., and Skeel, A. (1985) Analysis of human monocyte chemoattractant binding by flow cytometry. *J. Leukocyte Biol.* **38,** 403–413.

39

Assay for Oxidative Metabolism

Liana Harvath and Douglas A. Terle

1. Introduction
1.1. Overview

Phagocytic leukocytes that are exposed to opsonized particles, chemoattractants, or selected cytokines undergo a rapid burst in oxygen consumption and activation of the enzyme responsible for the oxidative metabolic burst, NADPH oxidase (reviewed in **ref. *1***). Active NADPH oxidase catalyzes the reaction:

$$NADPH + 2O_2 \rightarrow NADP^+ + 2O_2^- + H^+$$

The O_2^- formed by NADPH oxidase activity can rapidly be converted into H_2O_2 and other toxic species that destroy microorganisms and impart injury to surrounding host tissue. The most direct evidence for the role of NADPH oxidase in host defense has come from studies of patients who have genetic defects in NADPH oxidase activity (chronic granulomatous disease). Chronic granulomatous disease patients suffer from recurrent, severe bacterial infections, which are often fatal in early childhood.

Assays that evaluate phagocyte oxidative metabolic activity include light microscopic visualization of nitroblue tetrazolium dye reduction in cells *(2)*, and spectrophotometric quantification of O_2^- *(3)* and H_2O_2 *(4)* production. The spectrophotometric-based assays are frequently utilized in a microtiter plate format for rapid screening of O_2^- and H_2O_2 *(5)* production, and provide quantitative information on the bulk population of phagocytes under investigation. A flow cytometric assay for quantitative evaluation of oxidative metabolism in individual cells has been developed by Bass et al. *(6)* that has been adapted for studies of phagocytes in anticoagulated whole blood *(7,8)* as well as purified cells. The major advantage of the flow cytometric assay is the opportunity it

From: *Methods in Molecular Biology, Vol. 115: Immunocytochemical Methods and Protocols*
Edited by: L. C. Javois © Humana Press Inc., Totowa, NJ

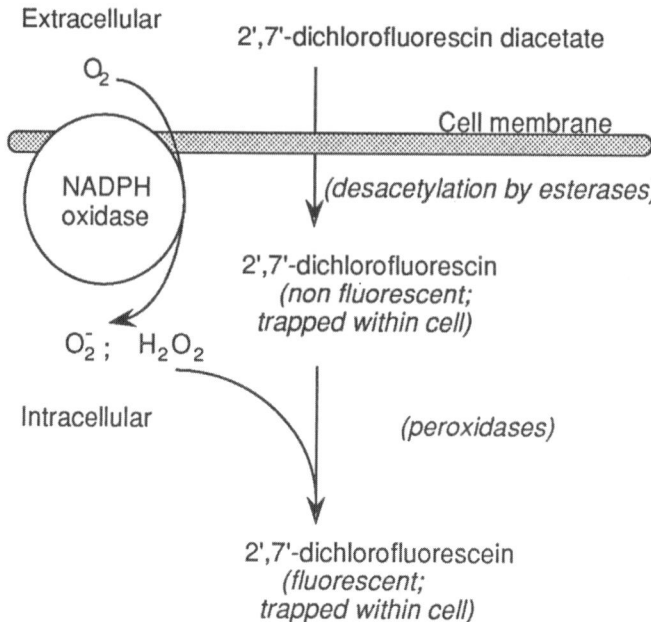

Fig. 1. Schematic representation of cell loading with 2',7'-dichlorofluorescin diacetate and the oxidation events that occur when the phagocyte NADPH oxidase is activated.

provides to quantify rapidly the individual oxidative metabolic activity of thousands of phagocytes.

1.2. Basis of the Assay

The nonfluorescent probe 2',7'-dichlorofluorescin diacetate is mixed with neutrophils in suspension to load the cells with the probe. Once 2',7'-dichlorofluorescin diacetate is inside the cell, it is desacetylated by intracellular esterases forming the product 2',7'-dichlorofluorescin, which is nonfluorescent and remains trapped within the cells (**Fig. 1**). When neutrophils are exposed to a stimulus that activates the oxidative burst, hydrogen peroxide (generated during neutrophil activation) reacts with the intracellular probe to form the fluorescent compound 2',7'-dichlorofluorescein. The fluorescent signal intensity, which is evaluated with a flow cytometer, is directly related to the oxidative metabolic activity of the cells. Phagocyte responses to soluble stimuli, such as phorbol myristate acetate *(6)*, as well as phagocytic stimuli *(7–9)* have been studied with this assay. The concepts of this assay have also been applied to a technique in which immune complexes are labeled with the nonfluorescent, reduced form of 1',7'-dichlorofluorescin and utilized as the stimulus of phagocyte oxidative metabolism via Fc receptors *(10)*.

2. Materials

1. A flow cytometer is required for this assay.
2. Acid citrate dextrose solution-A (ACD-A): 22.0 g/L sodium citrate ($Na_3C_6H_5O_7 \cdot H_2O$), 8.0 g/L citric acid, and 24.5 g/L dextrose.
3. Normal human blood sample anticoagulated with ACD-A (15 mL ACD-A/ 100 mL blood).
4. Dextran, pyrogen-free: average mol wt 100,000–200,000 Dalton (United States Biochemical Corp., Cleveland, OH).
5. Ficoll-hypaque lymphocyte separation medium (Organon Teknika Corp., Durham, NC).
6. Hank's balanced salt solution (BSS) (Gibco-BRL, Grand Island, NY).
7. Phosphate-buffered saline (PBS), without calcium and magnesium (Gibco-BRL).
8. Krebs' Ringer's PBS: PBS with 1.0 mM calcium, 1.5 mM magnesium, and 5.5 mM glucose, pH 7.3 ± 0.1.
9. 3.5% NaCl solution.
10. Sterile water for injection, USP (Abbott Laboratories, North Chicago, IL).
11. Polypropylene, 50-mL sterile, conical centrifuge tubes (Becton Dickinson Labware, Lincoln Park, NJ).
12. Polypropylene, 6-mL sterile tubes (Becton Dickinson Labware).
13. 2',7'-Dichlorofluorescin diacetate (DCFH-DA; Molecular Probes, Eugene, OR).
14. Phorbol myristate acetate (PMA; Sigma, St. Louis, MO).
15. Ethylenediaminetetraacetic acid (EDTA; Sigma).
16. *S. aureus* Bioparticles, labeled with Texas Red (Molecular Probes).
17. Opsonizing reagent, such as normal human pooled serum or rabbit IgG, to *S. aureus* Bioparticles (Molecular Probes).
18. 37°C Shaking water bath.

3. Methods

3.1. Cell Isolation

Human neutrophils are isolated from acid citrate dextrose solution-A anti-coagulated human blood by the method described by Boyum *(11)* with modifications *(12)* as described in detail in Chapter 36, **Subheading 3.1.** of this text.

3.2. Assay for Isolated Cells

1. Adjust cells to a density of 1×10^6 cells/mL in PBS without calcium and without magnesium. Prepare approx 10 mL of cell suspension.
2. Add 2',7' DCFH-DA to the cell suspension to a final concentration of 5 µM. Incubate the cells with DCFH-DA for 15 min at 37°C, with gentle agitation during the incubation.
3. After the 15-min incubation, remove a 1-mL aliquot, and add EDTA to the aliquot to a final concentration of 5 mM. This sample serves as the basal, unstimulated control value.
4. Remove 3 mL of dye-loaded cell suspension from the mixture in **step 2**, and place 1 mL each in three separate tubes. Add the soluble stimulus, PMA, to each

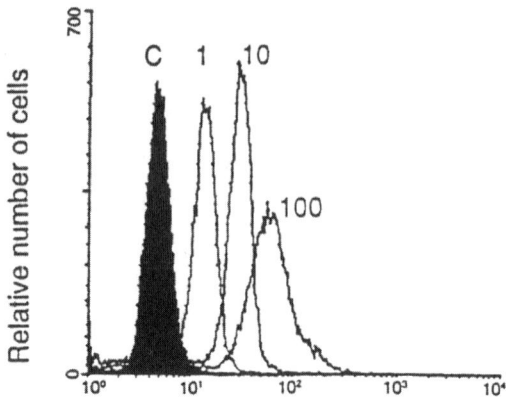

Fluorescence intensity (log)

Fig. 2. Fluorescence histogram profiles from the dose-dependent oxidative metabolic responses of granulocytes stimulated with 1, 10, or 100 ng/mL of PMA and unstimulated, 2',7'-dichlorofluorescin-loaded control cells (C). Each histogram represents the analysis of 10,000 cells for each reaction condition.

      tube at final concentrations of 1, 10, and 100 ng/mL. Incubate these reactions for 15 min at 37°C (*see* **Note 1**).

5. After the 15-min incubation, add EDTA to a final concentration of 5 mM to each of the tubes. Analyze the samples as soon as possible or store them at 4°C until analysis.

6. Analyze the samples on the flow cytometer. (*See* Chapter 36, **Fig. 2**, for normal light-scatter profiles.) When the fluorescence gain is properly adjusted, the unstimulated control cells (shaded curve, **Fig. 2**) should be visible on scale with the PMA-stimulated cells (open curves, **Fig. 2**). PMA (1–100 ng/mL) should stimulate a dose-dependent (graded) response. All normal granulocytes should exhibit an oxidative metabolic response to PMA (*see* **Note 2**).

3.3. Assay for Whole Blood

1. Incubate 1 mL of anticoagulated whole blood with 2',7' DCFH-DA (final concentration of 50 μM) for 10 min at 37°C in a shaking water bath.

2. Remove a 100-μL aliquot as an unstimulated, control sample.

3. Add 10 μL of opsonized particles (10^8 particles/mL; *see* Chapter 36, **Subheading 3.4.2.** for opsonization protocol) to a tube containing 100 μL of whole blood containing cells loaded 2',7' DCFH-DA, and incubate in a shaking water bath at 37°C for 30 min.

4. Hyptonically lyse the red blood cells as described in Chapter 36, **Subheading 3.5.2.** Resuspend the leukocyte pellet in 1 mL of PBS.

5. Adjust the forward and right-angle light-scatter detectors so that the granulocyte population is clearly visible (*see* Chapter 32, **Subheading 3.5.3., Fig. 2**).

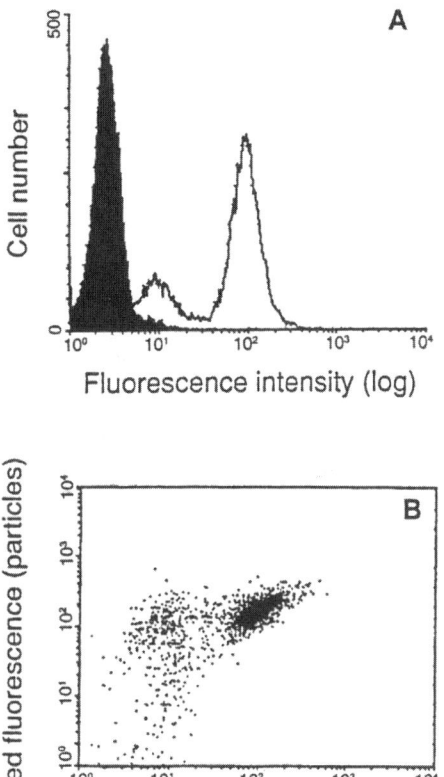

Fig. 3. Fluorescence profiles of 2',7'-dichlorofluorescin-loaded cells assayed in whole blood. (**A**) Compares the fluorescence histograms of unstimulated, control cells (shaded curve) with granulocytes exposed to opsonized *S. aureus* (open curve). (**B**) illustrates the two-color analysis profile of the granulocytes that were exposed to Texas Red-labeled *S. aureus*. Red fluorescence is the result of particle association with each granulocyte, whereas green fluorescence is the result of the oxidation of 2',7'-dichlorofluorescin to 2',7'-dichlorofluorescein (DCF). The red and green fluorescence analyses were performed with log-scale detection amplification for each fluorochrome.

Gate on the granulocyte population, and exclude the lymphocytes and monocytes from analysis.

6. When the photodetector for fluorescein fluorescence is properly set on the flow cytometer, the dye-loaded, unstimulated cells (**Fig. 3A**, shaded curve) should be visible on scale with the dye-loaded, particle-stimulated cells (**Fig. 3A**, open curve) (*see* **Note 2**).

7. If fluorescent particles, such as Texas Red-labeled *S. aureus*, are the particulate stimuli, a two-color analysis of the granulocytes can be performed in which the red fluorescence signal intensity (representing the number of particles associated with each granulocyte) is compared with the green fluorescence signal intensity (representing the oxidative metabolic activity of each cell). Cells that have internalized particles and generated hydrogen peroxide exhibit both bright red and green fluorescence intensities (**Fig. 3B**, upper right cell cluster). Cells that have fluorescent particles adherent to their surface, but have less oxidative metabolic activity, have bright red fluorescence and a lower level green fluorescence than actively phagocytizing cells (**Fig. 3B**, upper left cell cluster). Cells that have very few particles associated with their cell surface and little oxidative metabolic activity have low levels of red and green fluorescence (**Fig. 3B**, bottom left cells) (*see* **Notes 3** and **4**).

4. Notes

1. PMA is one of the most potent stimuli of phagocyte oxidative metabolism and should be used as a positive control for isolated cell preparations in this assay. Other soluble stimuli or opsonized particles may also be evaluated with this assay.
2. To obtain the control and stimulated fluorescence histograms on the same scale, it is usually necessary to analyze the samples with fluorescence log amplification setting on the flow cytometer.
3. Several protocols have been published for measurement of oxidative metabolism during phagocytosis and should be consulted for detailed information *(7–9)*.
4. Immune complexes have been conjugated to 2',7'-dichlorofluorescin and utilized as the phagocyte stimulus for a flow cytometric assay of oxidative metabolism *(10)*. The reagents and a detailed protocol for this assay are available as the Fc-OxyBURST™ Kit from Molecular Probes.

References

1. Morel, F., Doussiere, J., and Vignais, P. V. (1991) The superoxide generating oxidase of phagocytic cells. Physiologic, molecular and pathologic aspects. *Eur. J. Biochem.* **201,** 523–546.
2. Park, B. H., Fikrig, S. M., and Smithwick, E. M. (1968) Infection and nitrobluetetrazolium reduction by neutrophils. *Lancet* **7,** 532–534.
3. Babior, B. M., Kipnis, R. S., and Curnutte, J. T. (1973) Biological defense mechanisms. The production by leukocytes of superoxide, a potent bactericidal agent. *J. Clin. Invest.* **52,** 741–744.
4. Root, R. K., Metcalf, J., Oshino, N., and Chance, B. (1975) H_2O_2 release from human granulocytes during phagocytosis. Documentation, quantitation and some regulating factors. *J. Clin. Invest.* **55,** 945–955.
5. Pick, E. and Mizel, D. (1981) Rapid microassays for the measurement of superoxide and hydrogen peroxide production by macrophages in culture using an automatic enzyme immunoassay reader. *J. Immunol. Methods* **46,** 211–226.

6. Bass, D. A., Parce, J. W., DeChatelet, L. R., Szejda, P., Seeds, M. C., and Thomas, M. (1983) Flow cytometric studies of oxidative product formation by neutrophils: a graded response to membrane stimulation. *J. Immunol.* **130,** 1910–1917.

7. Trinkle, L. S., Wellhausen, S. R., and McLeish, K. R. (1987) A simultaneous flow cytometric measurement of neutrophil phagocytosis and oxidative burst in whole blood. *Diagnostic Clin. Immunol.* **5,** 62–68.

8. Hasui, M., Hirabayashi, Y., and Kobayashi, Y. (1989) Simultaneous measurement by flow cytometry of phagocytosis and hydrogen peroxide production of neutrophils in whole blood. *J. Immunol. Methods* **117,** 53–58.

9. Szejda, P., Parce, J. W., Seeds, M. C., and Bass, D. A. (1984) Flow cytometric quantitation of oxygen product formation by polymorphonuclear leukocytes during phagocytosis. *J. Immunol.* **133,** 3303–3307.

10. Ryan, T. C., Weil, G. J., Newberger, P. E., Haugland, R., and Simons, E. R. (1990) Measurement of superoxide release in the phagovacuoles of immune complexes-stimulated human neutrophils. *J. Immunol. Methods* **130,** 223–233.

11. Boyum, A. (1968) Isolation of mononuclear cells and granulocytes from human blood. *Scand. J. Clin. Lab. Invest.* **21(Suppl.),** 77–89.

12. Harvath, L., Balke, J. A., Christiansen, N. P., Russell, A. A., and Skubitz, K. M. (1991) Selected antibodies to leukocyte common antigen (CD45) inhibit human neutrophil chemotaxis. *J. Immunol.* **146,** 949–957.

V

Colloidal Gold Detection Systems for Electron Microscopic Analyses

40

Fixation and Embedding

Constance Oliver and Maria Celia Jamur

1. Introduction

For electron microscopic immunocytochemistry, the fixation procedure is always a compromise between good morphological preservation and retention of antigenicity *(1,2)*. It is preferable to fix tissues by perfusion, but if that is not possible, the time between removal of the tissue and fixation should be kept as short as possible. The fixative used will depend on whether the immunogold labeling will be done before or after the samples are embedded and on how resistant the antigen is to the fixative. In order to preserve the ultrastructural detail, it is desirable to include glutaraldehyde in the fixative. However, some antigens are extremely sensitive to crosslinking by glutaraldehyde, and the concentration of glutaraldehyde will have to be reduced or, in the most extreme cases, omitted entirely. For pre-embedding staining, it is possible to immunolabel the samples prior to fixation or after only a very mild fixation. Following incubation in the primary antibody or at subsequent steps, the samples can then be refixed in a stronger fixative, such as 2% glutaraldehyde, to give good morphological preservation. For postembedding labeling, it is not possible to refix the tissue after immunolabeling. Therefore, the composition of the initial fixative must be such that morphological detail and antigenicity are both preserved. The composition of the fixative will have to be determined empirically for each antigen. However, a fixative containing 0.5% glutaraldehyde and 2% formaldehyde generally is suitable for a wide variety of antigens. Adding 0.1% picric acid to the fixative can also help preserve morphological detail while retaining antigenicity.

In addition to composition, time and temperature of fixation can also affect the ability to immunolabel a particular antigen. It may be possible to preserve antigenicity by reducing the time of fixation, by altering the temperature of

From: *Methods in Molecular Biology, Vol. 115: Immunocytochemical Methods and Protocols*
Edited by: L. C. Javois © Humana Press Inc., Totowa, NJ

the fixative, or by doing both simultaneously. Many antigens can be preserved quite successfully by fixing at room temperature, but in some instances, fixation and processing at 4°C will help preserve antigenicity. Alternatively, warming (37–50°C) can increase penetration of the fixative into the tissue, and thus, fixation times can be reduced. Recently, microwave fixation has been introduced as an alternative to standard immersion fixation for electron microscopy *(3,4)*. Samples are immersed in fixative and microwaved for a few seconds. The combination of the chemical fixative and the heating action of the microwave gives good morphological preservation while retaining antigenicity.

Virtually any embedding resin may be used for immunogold staining. However, for postembedding methods, the resin can affect the immunostaining *(2,5)*. The hydrophilic resins, such as LR Gold, LR White, and Lowicryl, generally give better results than the epoxy-based hydrophobic resins, such as the Epon substitutes or Spurr.

Fixation by rapid freezing followed by either freeze substitution or cryosectioning can also overcome some of the problems of standard immersion fixation and resin embedding. These are more specialized techniques and will not be dealt with here. Discussion of the methods can be found in Polak and Varndell *(6)*, Hayat *(7)*, and Verkleij and Leunissen *(8)*.

2. Materials

1. 70% EM-grade glutaraldehyde.
2. 20% EM-grade formaldehyde (*see* **Note 1**).
3. Dulbecco's phosphate-buffered saline (PBS): 100 mg anhydrous calcium chloride, 200 mg potassium chloride, 200 mg monobasic potassium phosphate, 100 mg magnesium chloride · 6 H_2O, 8 g sodium chloride, and 2.16 g dibasic sodium phosphate · 7 H_2O; bring volume to 1 L with deionized glass-distilled water, pH 7.4.
4. Dulbecco's PBS without calcium and magnesium: 200 mg potassium chloride, 200 mg monobasic potassium phosphate, 8 g sodium chloride, and 2.16 g dibasic sodium phosphate ·7 H_2O; bring volume to 1 L with deionized distilled water, pH 7.4.
5. 0.1M Cacodylate buffer: 21.4 g cacodylic acid · 3 H_2O, sodium salt; bring volume to 1 L with deionized glass-distilled water; adjust pH to 7.4 with HCl.
6. 0.5% Glutaraldehyde-2% formaldehyde fixative: 0.5 mL 70% glutaraldehyde, 7 mL 20% formaldehyde, 3 mg calcium chloride, and 62.5 mL PBS or 0.1M cacodylate buffer, pH 7.4.
7. Saturated picric acid solution: Fill 500-g bottle of picric acid with deionized glass-distilled water (*see* **Note 2**).
8. 0.1M glycine: 750 mg glycine; bring volume to 100 mL with PBS, pH 7.4.
9. Osmium tetroxide (4% stock): 1 g osmium tetroxide and 25 mL deionized glass-distilled water (*see* **Note 3**).

10. Microwave oven (*see* **Note 4**).
11. Giemsa solution (Sigma Chemicals).
12. Electrophoresis-grade agar.
13. 35-mm plastic tissue culture dishes.
14. Epon 812 substitute; Epon A: 62 mL Epon 812 substitute and 100 mL dodecenyl succinic anhydride (DDSA); Epon B: 100 mL Epon 812 substitute and 89 mL nadic methyl anhydride (NMA); mix together 10 mL Epon A and 15 mL Epon B; add 0.5 mL 2,4,6-tri (dimethylaminomethyl) phenol (DMP-30), and mix well (*see* **Note 5**).
15. Araldite 502: 100 g araldite 502 resin and 75 g DDSA; add 2.5–3.5 g 2,4,6-tri (dimethylaminomethyl) phenol (DMP–30) just prior to use (*see* **Note 6**).
16. Spurr resin: 26 g nonenyl succinic anhydride (NSA), 10 g vinyl–4-cyclohexene dioxide (ERL 4206), 4 g DER 736, and 0.4 g 2-Dimethylaminothanol (DMAE) (*see* **Note 7**).
17. Lowicryl K4M: 4.0 g K4M crosslinker, 26.0 g K4M monomer, and 150 mg initiator; mix just prior to use, and degas under vacuum for 15–30 min (*see* **Note 8**).
18. LR White resin: Store at 4°C; let the bottle come to room temperature before opening; stable for 1 yr.
19. LR Gold resin: 25 mL LR Gold resin and 125 mg benzoin methyl ether; mix just prior to use.

3. Methods
3.1. Fixation

1. Rinse samples twice in PBS, and fix for 30 min to 1 h at room temperature in 0.5% glutaraldehyde and 2% formaldehyde in PBS, or fix by perfusion with 0.5% glutaraldehyde and 2% formaldehyde (*see* **Note 9**).
2. Following fixation, rinse the samples three to five times over a period of 30 min in PBS (*see* **Note 10**).
3. Quench free aldehyde groups by rinsing the samples for 5 min in $0.1M$ glycine in PBS (*see* **Note 11**).
4. Samples may be fixed in 2% OsO_4 in $0.1M$ cacodylate buffer for 1 h at room temperature and then rinsed in $0.1M$ cacodylate buffer prior to embedding.

3.2. Microwave Fixation

1. Place a glass beaker containing 100 mL of distilled water in the right rear corner of the microwave oven.
2. Preheat the water load for 2 min to warm up the magnetron.
3. Replace the water load with 100 mL distilled water at 25°C.
4. Using Giemsa/agar blocks (*see* **Note 12**), map the microwave oven (*see* **Note 13**).
5. Place the samples in fixative, one at a time, at the location determined during calibration, and irradiate at full power for the appropriate length of time (*see* **Note 14**).
6. Immediately after irradiation quench the samples by rinsing them with PBS or $0.1M$ cacodylate buffer at 4°C.

7. Rinse the sample in 0.1*M* cacodylate buffer.
8. Post-fix for 1 h in 2% OsO_4 in 0.1*M* cacodylate buffer and then rinse in 0.1*M* cacodylate buffer prior to embedding.

3.3. Embedding

3.3.1. Epon 812 Substitutes

1. Dehydrate the tissue, and infiltrate according to the following schedule:
   50% Ethanol, 15 min
   75% Ethanol, 15 min
   95% Ethanol, 15 min
   100% Ethanol, 15 min
   100% Ethanol, 15 min
   100% Propylene oxide, 15 min
   100% Propylene oxide, 30 min
   1:1 Complete resin:propylene oxide, 1–2 h
   2:1 Complete resin:propylene oxide, 1–2 h
   100% Complete resin, 3–6 h (*see* **Notes 15** and **16**).
2. Transfer to polyethylene or gelatin-embedding capsules containing complete resin. Polymerize overnight at 45°C and then for 24 h at 60°C.

3.3.2. Araldite

1. Dehydrate the tissue, and infiltrate according to the following schedule:
   50% Ethanol, 15 min
   75% Ethanol, 15 min
   95% Ethanol, 15 min
   100% Ethanol, 15 min
   100% Ethanol, 15 min
   100% Propylene oxide, 15 min
   100% Propylene oxide, 30 min
   1:1 Complete resin:propylene oxide, 1–2 h
   2:1 Complete resin:propylene oxide, 1–2 h
   100% Complete resin, 3–6 h.
2. Transfer to polyethylene or gelatin-embedding capsules containing complete resin. Polymerize overnight at 35°C, 24 h at 45°C, and 24 h at 60°C. Alternatively, araldite may be polymerized overnight at 60°C.

3.3.3. Spurr

1. Dehydrate the tissue, and infiltrate according to the following schedule:
   50% Ethanol, 15 min
   75% Ethanol, 15 min
   95% Ethanol, 15 min
   100% Ethanol, 15 min
   100% Ethanol, 15 min

100% Propylene oxide, 15 min
100% Propylene oxide, 30 min
1:1 Complete resin:propylene oxide, 1–2 h
2:1 Complete resin:propylene oxide, 3–6 h
100% Complete resin, overnight.
2. Transfer to polyethylene or gelatin-embedding capsules containing complete resin. Polymerize overnight at 70°C.

3.3.4. Lowicryl (9)

1. Dehydrate the tissue, and infiltrate with resin according to the following schedule:
30% Methanol, 5 min, 4°C
50% Methanol, 5 min, 4°C
70% Methanol, 5 min, –10°C
90% Methanol, 30 min, –20°C
1:1 Complete resin:90% methanol, 60 min, –20°C
2:1 Complete resin:90% methanol, 60 min, –20°C
100% Complete resin, 60 min, –20°C
100% Complete resin, overnight, –20°C.
2. Transfer tissue to polyethylene-embedding capsules or gelatin capsules, fill the capsules with complete resin, and cap. Oxygen will inhibit polymerization.
3. Polymerize the plastic with UV light (366 nm, two 15-W Sylvania F15T8/BLB bulbs) placed 10 cm from the capsules. Expose the capsules to the UV light for 24 h at –20°C (*see* **Notes 17** and **18**).

3.3.5. LR White

1. Dehydrate the tissue, and infiltrate with resin according to the following schedule:
50% Ethanol, 30 min
75% Ethanol, 30 min
1:1 LR White:75% ethanol, 1–2 h
2:1 LR White:75% ethanol, 1–2 h
100% LR White, 1–2 h
100% LR White, 3 h to overnight.
2. Transfer tissue to polyethylene-embedding capsules or gelatin capsules, fill the capsules with 100% LR White, and cap. Oxygen will inhibit polymerization.
3. Polymerize 24 h at 50–55°C (*see* **Note 19**).

3.3.6. LR Gold (10)

1. Dehydrate the tissue, and infiltrate with resin according to the following schedule; samples should be in capped glass vials during dehydration and infiltration:
50% Acetone, 5 min, 0°C
50% Acetone, 45 min, 0°C
70% Acetone, 45 min, 0°C
90% Acetone, 45 min, 0°C
1:1 LR Gold:acetone, 60 min, –20°C

7:3 LR Gold:acetone, 60 min, –20°C
100% LR Gold, 60 min, –20°C
100% LR Gold, 5 h to overnight, –20°C
100% LR Gold + initiator (0.5% benzoin methyl ether), 60 min, –20°C
100% LR Gold + initiator, 60 min, –20°C
100% LR Gold + initiator, 5 h to overnight, –20°C.

2. Transfer tissue to polyethylene-embedding capsules or gelatin capsules, fill the capsules with 100% LR Gold + initiator, and cap. Oxygen will inhibit polymerization.
3. Polymerize the plastic with UV light (366 nm, two 15-W Sylvania F15T8/BLB bulbs) placed 10 cm from the capsules. Expose the capsules to the UV light for 24 h at –20°C.

4. Notes

1. The use of distilled formaldehyde, not formalin, which contains alcohol, is recommended. Freshly prepared paraformaldehyde can also be used, especially if large volumes of fixative are needed for perfusion fixation. To prepare an 8% solution of paraformaldehyde, **in a fume hood** add 2 g of paraformaldehyde (trioxymethylene) powder to 25 mL of deionized glass-distilled water. With constant stirring, heat solution to 60–70°C. Once the solution has reached the proper temperature, continue to stir for 15 min. The solution will be milky. Add one to two drops of 1 N NaOH, with stirring, until the solution clears. A slight milkiness may persist. Cool and filter through Whatman No. 1 filter paper. This solution should be used the same day that it is prepared.
2. The picric acid will dissolve, resulting in a saturated solution. It will keep indefinitely at room temperature. Do not let picric acid dry, since the powder is explosive.
3. Clean the vial and score it with a diamond scribe. Place it in a clean glass bottle, and add the water. With a glass rod, break the vial. The solution is stable if kept refrigerated and protected from light. The stock solution should be stored away from other chemicals, since osmium vapors may escape from the bottle. Alternatively, aqueous solutions of osmium tetroxide in sealed glass ampules are available commercially. Dilute them to 1–2% with 0.1M cacodylate buffer for use.
4. A conventional microwave oven, with a maximal power output of 550 W and an operating frequency of 2450 MHz, or an oven designed specifically for laboratory use can be used. The microwave should have a fixed plate on the bottom, not a turntable.
5. Epon A and B solutions may be stored at 4°C for at least 6 mo. Let them come to room temperature before opening.
6. Mix components together in a glass screw-cap bottle. The resin solution may be stored at 4°C for at least 6 mo. Let it come to room temperature before opening.
7. Mix components together just before use. Store unused resin under vacuum; it is stable for 2–3 d. Store opened bottles of NSA under vacuum.
8. Avoid contact with the skin, and avoid breathing the vapors. Wear chemically resistant gloves, and use the resin in a fume hood.
9. The composition of the fixative may vary depending on the sensitivity of the antigen. For some antigens, it may be necessary to omit the glutaraldehyde

entirely and fix in 4% formaldehyde in PBS for 1–4 h. Picric acid (0.1%) may also be added to the fixative. For very sensitive antigens, it may be desirable to fix and process tissue at 4°C.

10. Phosphate buffer (0.1 M), pH 7.4, may be substituted for PBS.

11. Following fixation in formaldehyde or glutaraldehyde, tissue is generally quenched. The purpose of this step is to block any free aldehyde groups that remain after fixation and washing. This step is especially critical after glutaraldehyde fixation. Glutaraldehyde is a bifunctional aldehyde. During fixation, one end may bind to cellular constituents, leaving the other end free to react. If this end is not blocked, it can bind to the protein in the blocking solution or to the primary antibody. Although any small molecular-weight compound containing an amino group may be used, the most commonly used quenching agents are glycine, ammonium chloride, and sodium borohydride.

12. Giemsa agar blocks are made by preparing a solution of 2% agar in 0.9% saline and then adding Giemsa solution to the liquid agar to a final concentration of 0.5%. The Giemsa agar solution is then poured into flat embedding molds and allowed to solidify. The Giemsa/agar blocks can be trimmed into cubes (0.5 cm^3) or used as they come from the molds.

13. Microwaves are not dispersed evenly over the oven, leaving "hot" and "cold" spots on the oven floor. The radiation pattern inside the oven may be mapped using either an array of neon bulbs or with Giemsa/agar blocks. To map the microwave oven with Giemsa/agar blocks, immerse the blocks in 5 mL of fixative solution in a 35-mm tissue culture dish. Use only one block per dish. Place the dish (one at time, use a fresh dish for each run) at various locations on the floor of the microwave unit and expose to microwave radiation at 100% power. The amount of microwave radiation that an area is receiving can be determined by the changes in the Giemsa/agar blocks. The correct amount of irradiation is achieved if the Giemsa/agar block turns from blue to violet without showing any signs of melting. If level of radiation is too low, there will be no change in the block, and if it is too high the block will show signs of melting. Once the ideal position for the sample has been determined, subsequent samples should be placed in the same location. A gridded acetate sheet may be placed in the bottom of the oven to act as a reference guide.

14. To ensure that the sample receives the correct amount of irradiation, place a Giemsa/agar block in the same type of container with the same amount of fixative to be used for the sample at the predetermined location on the floor of the oven and irradiate. 0.05% Glutaraldehyde, 2% formaldehyde, 0.025% CaCl$_2$ in 0.1M cacodylate buffer, pH 7.4, is a reliable fixative for microwave fixation. The sample received the correct amount of irradiation if the Giemsa/agar block turns from blue to violet without showing any signs of melting. The irradiation time can be adjusted to achieve optimum sample fixation.

15. Other solvents, such as methanol or acetone, can be used during dehydration, depending on the tissue. However, acetone should not be used with LR White.

16. During infiltration of samples with embedding resin, samples should be placed on a rotator to provide adequate mixing of the sample with the resin. Infiltration

schedules for embedding resins will vary with the size of the sample, the viscosity of the resin, and the density of the tissue. Larger samples, more viscous resins, and dense tissue will all require longer times to ensure adequate infiltration of the sample. Certain samples, such as pellets of cultured cells and lung, may trap air or solvents in the samples. More complete infiltration and better polymerization of the resin will be obtained if following infiltration, specimens are transferred to embedding capsules containing just enough resin to cover the samples, and the samples are placed under vacuum overnight. The capsules can then be filled with resin, and samples can be polymerized.

17. For complete polymerization, it may be necessary to continue curing the blocks by UV for up to 2 wk at either –20°C or room temperature. Temperature for infiltration and polymerization of Lowicryl K4M can be as low as –35°C.

18. Samples that are embedded in resins that are polymerized by UV light (Lowicryl, LR White) should not be osmicated. The osmium may interfere with the polymerization by preventing the light from penetrating the samples.

19. LR White is very hygroscopic and will readily adsorb water. If the 1:1 LR White:75% ethanol is cloudy, increase the concentration of ethanol up to 95%.

References

1. Brandtzaeg, P. (1982) Tissue preparation methods for immunocytochemistry, in *Techniques in Immunocytochemistry,* vol. 1 (Bullock, G. R. and Petrusz, P., eds.), Academic, New York, pp. 1–76.
2. Larsson, L. I. (1988) Fixation and tissue pretreatment, in *Immunocytochemistry: Theory and Practice* (Larsson, L.-I., ed.), CRC, Boca Raton, FL, pp. 41–76.
3. Login, G. R. and Dvorak, A. M. (1988) Microwave fixation provides excellent preservation of tissue, cells and antigens for light and electron microscopy. *Histochem. J.* **20,** 373–387.
4. Jamur, M. C., Faraco, C. D., Lunardi, L. O., Siraganian, R. P., and Oliver, C. (1995) Microwave fixation improves antigenicity of glutaraldehyde-sensitive antigens while preserving ultrastructural detail. *J. Histochem. Cytochem.* **43,** 307–311.
5. Causton, B. E. (1984) The choice of resins for electron immunocytochemistry, in *Immunolabelling for Electron Microscopy* (Polak, J. M. and Varndell, I. M., eds.), Elsevier, New York, pp. 29–70.
6. Polak, J. M. and Varndell, I. M. (1984) *Immunolabelling for Electron Microscopy.* Elsevier, New York.
7. Hayat, M. A. (1989) *Colloidal Gold: Principles, Methods, and Applications,* vol. 1. Academic, New York.
8. Verkleij, A. J. and Leunissen, J. L. M. (1989) *Immuno-gold Labeling in Cell Biology.* CRC, Boca Raton, FL.
9. Bendayan, M. (1984) Protein A-gold electron microscopic immunocytochemistry: methods, applications, and limitations. *J. Electron Micros. Tech.* **1,** 243–270.
10. Berryman, M. A. and Rodewald, R. D. (1990) An enhanced method for postembedding immunocytochemical staining which preserves cell membranes. *J. Histochem. Cytochem.* **38,** 159–170.

41

Preparation of Colloidal Gold

Constance Oliver

1. Introduction

Since their introduction by Faulk and Taylor (*1*), colloidal gold probes have become widely used for immunocytochemical staining at the electron microscopic level. Many different methods of producing colloidal gold sols have been published (*2–10*). Gold sols are producing by boiling a solution of tetrachloroauric acid with a reducing agent. At the beginning of the reduction process, gold atoms are liberated from the chloroauric acid. The gold atoms aggregate forming microcrystals. As more chloroauric acid is reduced, the microcrystals grow in size until all of the chloroauric acid is reduced. The type of reducing agent and the concentration of components determine the ratio between nucleation and growth, and thus determine the final particle size. Gold sols can be made in the laboratory that have a particle size ranging from 2 to 40 nm, depending on the type and concentration of the reducing agent. The methods given below for various sizes of gold are relatively simple and very reproducible from batch to batch.

2. Materials

1. Detergent, such as Haemasol, Micro, or 7X.
2. Deionized, glass-distilled water.
3. 10% Chloroauric acid: 1-g glass vial chloroauric acid and 10 mL deionized glass-distilled water (*see* **Note 1**).
4. 1% Trisodium citrate: 100 mg trisodium citrate and 10 mL deionized glass-distilled water; dissolve sodium citrate in water; prepare immediately prior to use.
5. 1% Tannic acid: 100 mg low-mol-wt tannic acid (Mallinckrodt 8835) and 10 mL deionized glass-distilled water; dissolve tannic acid in water; prepare immediately prior to use (*see* **Note 2**).

From: *Methods in Molecular Biology, Vol. 115: Immunocytochemical Methods and Protocols*
Edited by: L. C. Javois © Humana Press Inc., Totowa, NJ

6. 0.025 *M* Potassium carbonate: 34 mg potassium carbonate; bring the volume to 10 mL with deionized glass-distilled water.
7. 1 *M* Sodium thiocyanate: 81 mg sodium thiocyanate and 10 mL deionized glass-distilled water; dissolve sodium thiocyanate in water; prepare immediately before use.
8. 0.2 *M* Potassium carbonate: 276 mg potassium carbonate and 10 mL deionized glass-distilled water.
9. Stirring hot plate and Teflon™ stir bars.
10. Glassware: 250-mL volumetric flask, two 100-mL screw-cap glass bottles, and 50-mL beaker.

3. Methods
3.1. Preparation of Glassware

1. Clean glassware by boiling in detergent, such as Haemasol, Micro, or 7X (*see* **Note 3**).
2. Rinse free of detergent, and rinse an additional 10–15 times in deionized glass-distilled water (*see* **Note 4**).

3.2. Sodium Citrate Gold, 15 nm (2,5)

1. Add 10 µL of 10% chloroauric acid to 100 mL of deionized glass-distilled water (*see* **Note 1**).
2. Bring the solution to a boil (*see* **Note 5**).
3. While vigorously stirring the solution using a Teflon™ stir bar and a stirring hot plate, quickly add 4 mL of freshly prepared 1% aqueous trisodium citrate (*see* **Note 6**).
4. Continue to boil the solution, with stirring. The solution will first turn blue and then red in about 5–7 min. After 2–3 additional min, the solution will reach its end point, a red-orange color.
5. Let it boil an additional 5 min, then cool, and transfer it to a clean screw-cap glass bottle. The gold sol is stable if kept refrigerated and protected from light.

3.3. Tannic Acid–Sodium Citrate Gold, 6 nm (9)

1. Place 79 mL deionized glass-distilled water and 100 µL of 10% chloroauric acid in a clean 250-mL Erlenmyer flask (*see* **Note 1**).
2. In a 50-mL beaker, mix 4 mL of 1% freshly prepared aqueous trisodium citrate, 0.5 mL 1% tannic acid, and 0.5 mL 0.025 *M* potassium carbonate (to adjust pH). Add 15 mL of deionized glass-distilled water (*see* **Note 7**).
3. Heat the gold solution and the reducing solution separately on a stirring hot plate.
4. When the gold solution reaches 60°C, stir vigorously and quickly add the reducing solution.
5. Continue stirring and maintain temperature at approx 60°C until solution turns red.
6. Heat to boiling after the sol is formed. Solution evaporation can be avoided by using a relux column or covering the top of the flask with a glass slide. Alternatively, a siliconized 250-mL volumetic flask can be used.

7. Cool the solution, and transfer it to a clean screw-cap bottle. Gold sol is stable if kept refrigerated and protected from light.

3.4. Thiocyanate Gold, 2.8 nm (10)

1. Add 0.3 mL of 1 M sodium thiocyanate, with stirring, to 50 mL of deionized glass-distilled water containing 0.5 mL 10% chloroauric acid and 0.75 mL of 0.2 M potassium carbonate.
2. Continue to stir for 15–30 min. A straw yellow color will develop.
3. Let it stand overnight in the dark at room temperature to complete the reaction (*see* **Notes 8** and **9**).

4. Notes

1. To prepare 10% chloroauric acid solution, remove the label from the glass vial containing 1 g of chloroauric acid, and thoroughly clean the vial. Rinse in deionized glass-distilled water. Score the vial with a diamond scribe, but do not break it. Add 10 mL of deionized glass-distilled water to clean glass bottle. Add the vial of chloroauric acid to the bottle with the water. Break the vial into the water using a clean glass rod. The resulting solution is stable for years if kept refrigerated and protected from light. Do not use chloroauric acid that was not sealed in a glass vial. Powdered chloroauric acid is very hygroscopic and adsorbs water readily.
2. The tannic acid should be low mol wt. Mallinckrodt 8835 tannic acid is predominately a mixture of tetra- and pentagalloyl glucose and should be used for making gold sols. This tannic acid is available from suppliers of electron microscopic supplies.
3. All glassware, stir bars, and so forth, used in preparing gold sols must be free of any contamination or the sol will not form properly. Glassware may be siliconized if desired.
4. The water used in all solutions and to wash and rinse the glassware should be of the highest purity possible. Sterile deionized glass-distilled water is recommended for all procedures.
5. Care should be taken not to allow the solution to evaporate, thereby changing the concentration of the components. A siliconized 250-mL volumetric flask works well. Alternatively, a reflux apparatus can be used.
6. For 19 nm gold, add 3 mL of 1% trisodium citrate; for 12.5 nm gold, add 6 mL of 1% trisodium citrate.
7. For 15 nm gold, add 20 μL 1% tannic acid and 19.98 mL deionized glass-distilled water; for 10 nm gold, add 100 μL 1% tannic acid and 19.9 mL deionized glass-distilled water; for 3.5 nm gold, add 3 mL 1% tannic acid, 3 mL 0.025 M potassium carbonate, and 14 mL deionized glass-distilled water. Sodium citrate remains constant at 4 mL.
8. Gold colloid will aggregate with age. Use within 3 d of preparation.
9. The size of the gold colloid may be determined in the electron microscope by first wetting a formvar-carbon-coated grid with 0.01% bacitracin. Using a tri-

angle of filter paper, moistened at one end, wick off the bacitracin, and apply a drop of gold colloid to the grid. After 1–2 min, wick off the excess gold colloid. Air-dry and examine it in the electron microscope.

References

1. Faulk, W. and Taylor, G. (1971) An immunocolloid method for the electron microscope. *Immunochemistry* **8,** 1081–1083.
2. Bendayan, M. (1984) Protein-A gold electron microscopic immunocytochemistry: methods, applications, and limitations. *J. Electron Microsc. Tech.* **1,** 243–270.
3. Leunissen, J. L. M. and DeMey, J. R. (1989) Preparation of gold probes, in *Immuno-gold Labeling in Cell Biology* (Verkleij, A. J. and Leunissen, J. L. M., eds.), CRC, Boca Raton, FL, pp. 3–16.
4. Handley, D. A. (1989) Methods for synthesis of colloidal gold, in *Colloidal Gold,* vol. 1 (Hayat, M. A., ed.), Academic, New York, pp. 13–33.
5. Frens, G. (1973) Controlled nucleation for the regulation of particle size in monodisperse gold suspensions. *Nature Phys. Sci.* **241,** 20–22.
6. Horisberger, M. and Rosset, J. (1977) Colloidal gold, a useful marker for transmission and scanning electron microscopy. *J. Histochem. Cytochem.* **25,** 295–305.
7. Roth, J. (1982) The preparation of protein A-gold complexes with 3 nm and 15 nm gold particles and their use in labeling multiple antigens on ultra-thin sections. *Histochem. J.* **14,** 791–801.
8. Slot, J. W. and Geuze, H. J. (1985) A new method of preparing gold probes for multiple-labeling cytochemistry. *Eur. J. Cell Biol.* **38,** 87–93.
9. Muhlpfordt, J. A. (1982) The preparation of colloidal gold particles using tannic acid as an additional reducing agent. *Experientia* **38,** 1127–1128.
10. Bashong, W., Lucocq, J. M., and Roth, J. (1985) "Thiocyanate-Gold:" small (2–3 nm) colloidal gold for affinity cytochemical labeling in electron microscopy. *Histochem.* **83,** 409–411.

42

Conjugation of Colloidal Gold to Proteins

Constance Oliver

1. Introduction

The ability to conjugate proteins to colloidal gold sols provides a wide variety of probes for electron microscopy. In addition to antibodies, protein A, lectins, enzymes, toxins, and other proteins have all been conjugated to colloidal gold *(1–7)*. The nature of the interaction between the colloidal gold and the protein is poorly understood. Colloidal gold is a negatively charged lyophobic sol. The surface of the particle displays not only electrostatic characteristics, but also hydrophobic properties. In conjugating proteins to a gold sol, the electrostatic interactions must be reduced so that the hydrophobic interactions can prevail. This is accomplished by adjusting the pH of the gold sol to approx 0.5 pH unit higher than the pI of the protein being conjugated. Roth *(7)* gives a table of optimum pH for a number of commonly used proteins. Once the pH is properly adjusted, the net charge of the protein is zero or slightly negative. This prevents the aggregation of the protein owing to electrostatic attraction while maintaining the hydrophobic interactions and facilitates the conjugation of the protein to the gold.

2. Materials

1. Cold water fish gelatin (Sigma, St. Louis, MO): Warm in 37°C water bath to liquefy; prepare dilute solutions just prior to use; keep stock bottle at 4°C.
2. 0.1 *N* Hydrochloric acid (HCl): 8.15 mL concentrated HCl and 1 L deionized glass-distilled water.
3. 0.2 *M* Potassium carbonate (K_2CO_3): 276 mg K_2CO_3; bring volume to 10 mL with deionized glass-distilled water.
4. Dulbecco's phosphate-buffered saline (PBS): 100 mg anhydrous calcium chloride, 200 mg potassium chloride, 200 mg monobasic potassium phosphate,

From: *Methods in Molecular Biology, Vol. 115: Immunocytochemical Methods and Protocols*
Edited by: L. C. Javois © Humana Press Inc., Totowa, NJ

100 mg magnesium chloride · 6 H_2O, 8 g sodium chloride, and 2.16 g dibasic sodium phosphate · 7 H_2O; bring volume to 1 L with deionized glass-distilled water, pH 7.4.

5. 10% Polyethylene glycol (PEG): 100 mg polyethylene glycol (20,000 mol wt), and 10 mL deionized glass-distilled water; dissolve PEG in deionized glass-distilled water; prepare just before use.

6. 10% Sodium chloride: 100 mg NaCl, and 10 mL deionized glass-distilled water; dissolve NaCl in deionized glass-distilled water; prepare shortly before using.

7. 0.2 *M* Borate–NaCl buffer: 76.3 g sodium borate · 10 H_2O and 9 g sodium chloride; bring volume to 1 L with deionized glass-distilled water, pH 9.0.

8. Tris-buffered saline (TBS): 2.4 g Tris-HCl and 8.76 g sodium chloride; bring vol to 1 L with deionized glass-distilled water, pH 7.4. 10 m*M* sodium azide (650 mg) may be added as a preservative.

3. Methods

3.1. General Method for Conjugating Proteins to Colloidal Gold

1. Adjust pH to at least 0.5 pH point on the basic side of the pI of the protein to be adsorbed, using either 0.2 *M* K_2CO_3 or 0.1 *N* HCl. The pIs for some commonly used proteins can be found in Roth *(7)*.

2. Add two drops of 1% PEG (20,000 mol wt) to 5-mL aliquots of the colloidal gold to avoid clogging the pH meter when adjusting the pH. **Do not add the PEG to the colloidal gold stock**. Discard aliquots after reading pH. Alternatively, narrow-range pH paper can be used.

3. Perform a saturation isotherm to determine the protein:gold ratio for each protein and each size colloidal gold (*see* **Note 1**).

4. Add 10 mL of colloidal gold to 0.2 mL of protein solution containing the minimal amount of protein needed to stabilize the gold plus 10% (as determined in **step 3**) (*see* **Note 2**).

5. Let the solution stand 10 min, and add 1% PEG to a final concentration of 0.04%.

6. Let the solution stand another 30 min, and then centrifuge it for 45 min at 60,000*g*.

7. Remove and discard the clear supernatant.

8. Resuspend the **soft gold pellet** in 1.5 mL PBS with 0.04% PEG. The hard gold pellet is uncoated gold.

9. For use: dilute the stock solution 1:10–1:20 with PBS containing 0.02% PEG.

3.2. Preparation of Protein A Gold (2)

1. Adjust pH of colloidal gold to 6.0 with either 0.1 *N* HCl or 0.2 *M* K_2CO_3 (*see* **Note 3**).

2. Add 10 mL colloidal gold to 0.3 mg Protein A dissolved in 0.2 mL water for 15 nm gold.

3. Let the solution stand 10 min; add 1% PEG to a final concentration of 0.04%.

4. Let the solution stand 30 min, and then centrifuge for 45 min at 60,000*g*.

5. Remove the supernatant, and resuspend the soft pellet in 1.5 mL PBS containing 0.04% PEG. Store at 4°C.

6. For use: dilute 1:10–1:20 in PBS containing 0.02% PEG.

3.3. Preparation of Antibodies Conjugated to Colloidal Gold (8)

1. Use affinity-purified antibodies.
2. Dialyze the antibody: For monoclonal antibodies, etc., already diluted with 0.2 M borate–NaCl buffer, dialyze against 2 mM borate–NaCl buffer, pH 9.0 for 2 h; for all other antibodies, first dialyze overnight against 0.2 M borate buffer–NaCl, pH 9.0, and then for 2 h against 2 mM borate buffer, pH 9.0. Do not leave the antibodies in the 2 mM borate buffer for extended periods or the antibodies may aggregate.
3. Determine the optimal amount of antibody needed by running a saturation isotherm (*see* **Notes 1** and **4**).
4. Add the colloidal gold to the antibody with stirring, and allow to react for 10–30 min. Use 10 mL of colloidal gold and the amount of antibody as determined in **step 3**.
5. In order to stabilize the probes, add warmed (37°C) cold-water fish gelatin to give a final concentration of 1%.
6. Remove excess antibody by centrifuging the colloidal gold preparation twice on a glycerol step gradient (*see* **Note 5**).
7. Remove the supernatant and discard it. Collect the soft pellet (it may be suspended in 1–2 mL of TBS to reduce viscosity), and dialyze for 1 h at room temperature against TBS to remove the glycerol.
8. Repeat the centrifugation. Remove and discard the supernatant, and resuspend the soft pellet in 1–2 mL TBS containing 1% cold-water fish gelatin.
9. Dialyze for 1 h at room temperature against TBS.
10. Dilute the resulting preparation with TBS containing 1% cold-water fish gelatin, and store in the refrigerator.

4. Notes

1. The minimal amount of protein necessary to stabilize the gold is determined by adding 1 mL of the colloidal gold to 0.1 mL of serial aqueous dilutions of the protein. The order of addition of gold to the protein is critical. After 10 min, add 0.1 mL of 10% NaCl to each tube. If there is not enough protein to stabilize the gold, the solution will change from red to blue. For gold sols prepared with tannic acid, it may be necessary to add 0.1% H_2O_2 to the preparations in order to visualize the color change. If the color change cannot be assessed visually, it can be assessed spectrophotometrically (maximum absorbance of 510–550 nm).
2. If you wish to make larger quantities of gold, prepare multiple 10-mL aliquots rather then one large batch.
3. To measure the pH, remove 5-mL aliquots from the stock of colloidal gold, and add two drops of 1% polyethylene glycol. **Do not add the PEG to the colloidal gold stock**. Discard aliquots after reading pH.
4. If the amount of antibody is limited, the isotherm may be run using 100 µL of colloidal gold and 10 µL each of antibody and NaCl.
5. Prepare the glycerol step gradients by layering 1 mL each 60, 40, and 20% glycerol in TBS containing 1% cold-water fish gelatin over a cushion of 0.5 mL

100% glycerol. Keep the gradients on ice once they are made. Layer the colloidal gold over the gradients, and centrifuge at 4°C for 1 h in a swinging bucket rotor at $r_{av} = 200,000g$.

References

1. Leunissen, J. L. M., and DeMey, J. R. (1989) Preparation of gold probes, in *Immuno-gold Labeling in Cell Biology* (Verkleij, A. J. and Leunissen, J. L. M., eds.), CRC, Boca Raton, FL, pp. 3–16.
2. Bendayan, M. (1984) Protein A-gold electron microscopic immunocytochemistry: methods, applications, and limitations. *J. Electron Micros. Technol.* **1**, 243–270.
3. Bendayan, M. (1989) Protein A-gold and protein G-gold postembedding immuno-electron microscopy, in *Colloidal Gold*, vol. 1 (Hayat, M. A., ed.), Academic, New York, pp. 34–96.
4. Horisberger, M. (1981) Colloidal gold: a cytochemical marker for light and fluorescent microscopy and for transmission and scanning electron microscopy, in *Scanning Electron Microscopy II* (Johari, O., ed.), SEM, Inc., AMF O'Hare, Chicago, pp. 9–31.
5. Horisberger, M. (1989) Quantitative aspects of labeling colloidal gold with proteins, in *Immuno-gold Labeling in Cell Biology* (Verkleij, A. J. and Leunissen, J. L. M., eds.), CRC, Boca Raton, FL, pp. 49–60.
6. Geoghegan, W. D. and Ackerman, G. A. (1977) Adsorption of horseradish peroxidase, ovomucoid and anti-immunoglobulin to colloidal gold for the indirect detection of concanavalin A, wheat germ agglutinin and goat anti-human immunoglobulin G on cell surfaces at the electron microscopic level: a new method, theory and application. *J. Histochem. Cytochem.* **25**, 1187–1200.
7. Roth, J. (1983) The colloidal gold marker system for light and electron microscopic cytochemistry, in *Techniques in Immunocytochemistry*, vol. 2 (Bullock, G. R. and Petrusz, P., eds.), Academic, New York, pp. 217–284.
8. Birrell, G. B., Hedberg, K. K., and Griffith, P. H. (1987) Pitfalls of immunogold labeling: analysis by light microscopy, transmission electron microscopy, and photoelectron microscopy. *J. Histochem. Cytochem.* **35**, 843–853.

43

Colloidal Gold/Streptavidin Methods

Constance Oliver

1. Introduction

Biotin–avidin detection systems are widely used in both immunocytochemistry and molecular biology *(1,2)* (*see* Chapter 25). They take advantage of the high affinity of biotin, a low-molecular-weight vitamin, for avidin, an egg-white protein. The avidin–biotin complex has one of the highest dissociation constants known, $10^{-15} M$. This high dissociation constant has made it a convenient system for linking indicators to antibodies. However, egg-white avidin binds nonspecifically to many tissue sites. This nonspecific binding has been attributed both to its high isoelectric point (pI = 10) and to the fact that the protein is glycosylated *(3,4)*. Because of the problem of nonspecific binding of avidin, streptavidin has largely replaced avidin for immunocytochemical procedures. Streptavidin, produced by *Streptomyces avidinni*, has properties that are very similar to avidin, but it is not glycosylated *(5)*. Streptavidin generally gives little or no background staining, and is superior to avidin for immunocytochemical uses.

Streptavidin/colloidal gold-biotin detection systems for electron microscopy are most commonly used in postembedding immunocytochemistry. A bridging technique is generally used. In this method, the primary antibody is unlabeled, the secondary antibody is biotinylated, and the colloidal gold is conjugated to streptavidin. In certain applications, the primary antibody may be biotinylated and no bridging antibody is needed. It is also possible to use streptavidin–biotin complexes during detection. These complexes contain multiple gold particles and can enhance a weak signal.

High-quality reagents needed for biotin–avidin immunostaining are all available commercially. Kits are also available commercially for biotinylating antibodies (*see* Chapter 7). The major consideration in biotinylating antibodies is the use of biotin with a carbon spacer arm at least 1 nm long, since the binding site of biotin on avidin and probably streptavidin is in a deep depression *(4)*.

From: *Methods in Molecular Biology, Vol. 115: Immunocytochemical Methods and Protocols*
Edited by: L. C. Javois © Humana Press Inc., Totowa, NJ

2. Materials

1. Dulbecco's phosphate-buffered saline (PBS): 100 mg anhydrous calcium chloride, 200 mg potassium chloride, 200 mg monobasic potassium phosphate, 100 mg magnesium chloride · 6 H_2O; 8 g sodium chloride, and 2.16 g dibasic sodium phosphate · 7 H_2O; bring vol to 1 L with deionized glass-distilled water, pH 7.4.
2. Fixative: 2% EM-grade formaldehyde in PBS and 0.5% EM-grade glutaraldehyde in PBS.
3. 0.1 *M* Cacodylate buffer: 21.4 g cacodylic acid, sodium salt · 3 H_2O; bring vol to 1 L with deionized glass-distilled water; adjust pH to 7.4 with HCl.
4. 0.1 *M* Glycine: 750 mg glycine; bring vol to 100 mL with PBS, pH 7.4.
5. Osmium tetroxide (4% stock): 1-g glass vial osmium tetroxide, and 25 mL deionized glass-distilled water (*see* **Note 1**).
6. Nickel grids: Clean in acetone or alcohol before use.
7. Saturated solution of sodium *meta*-periodate prepared daily in deionized glass-distilled water.
8. Tris-buffered saline (TBS): 2.4 g Tris-HCl and 8.76 g NaCl; bring vol to 1 L with deionized glass-distilled water, pH 7.4.
9. High-salt Tween-20 buffer: 2.4 g Tris-HCl, 29.2 g sodium chloride, and 1 mL Tween-20; bring vol to 1 L with deionized glass-distilled water, pH 7.4.
10. 1% Bovine serum albumin (BSA): 1 g BSA and 100 mL TBS or high-salt Tween–20 buffer; add BSA to buffer with stirring.
11. Dulbecco's PBS without calcium and magnesium: 200 mg potassium chloride, 200 mg monobasic potassium phosphate, 8 g sodium chloride, and 2.16 g dibasic sodium phosphate · 7 H_2O; bring vol to 1 L with deionized glass-distilled water, pH 7.4.
12. Primary antibody diluted in PBS without calcium and magnesium, and 1% BSA or TBS with 1% BSA (*see* **Note 2**).
13. Biotinylated secondary antibody (*see* Chapter 7) diluted in PBS without calcium or magnesium with 1% BSA or TBS with 1% BSA.
14. Colloidal gold conjugated with streptavidin (*see* **Note 3**).
15. 0.1 *M* Maleate buffer: 1.16 g maleic acid and 3.5 g sucrose; add deionized glass-distilled water to make 100 mL, and adjust to pH 6.5.
16. 2% Uranyl acetate: 2 g uranyl acetate and 100 mL maleate buffer; add uranyl acetate to the buffer, and adjust to pH 6.0.
17. Reynolds' lead citrate *(6)*: 1.33 g lead nitrate, 1.76 g trisodium citrate · 2H_2O; 30 mL deionized glass-distilled water, and 8.0 mL 1 *N* NaOH (*see* **Note 3**).
18. 0.05 Borate buffer: 1.9 g sodium borate · 10 H_2O; bring vol to 100 mL with deionized glass-distilled water, pH 8.6.

3. Method

3.1. Postembedding Labeling with Streptavidin-Gold

1. Rinse the samples twice in PBS, and fix for 30 min to 1 h at room temperature in 0.5% glutaraldehyde and 2% formaldehyde in PBS (*see* **Note 5**).

2. Following fixation, rinse the samples three to five times over a period of 30 min in PBS (*see* **Note 6**).
3. Rinse the samples for 5 min in 0.1 *M* glycine in PBS to quench the free aldehyde groups (*see* **Note 7**).
4. Then rinse the samples in buffer, and postfix in 1–2% osmium tetroxide for 1 h at room temperature (*see* **Note 8**).
5. Dehydrate the samples, and embed (*see* **Note 9**).
6. Cut the resulting blocks with a diamond knife, and mount sections on nickel grids (*see* **Note 10**).
7. Etch the plastic in the sections slightly in order to allow for penetration of reagents into the samples. Float the grids section side down on drops of a freshly prepared saturated solution of sodium *meta*-periodate (*see* **Note 11**).
8. Rinse the grids three to five times in deionized glass-distilled water.
9. Block nonspecific binding by incubating the sections in 1% BSA or normal serum in PBS without calcium and magnesium, or in TBS for 15 min.
10. Rinse the grids five times in PBS without calcium and magnesium or in TBS.
11. Incubate for 1–2 h at room temperature in primary antibody diluted in PBS without calcium and magnesium containing 1% BSA or in TBS containing 1% BSA.
12. Rinse as in **step 10**.
13. Incubate the grids in biotinylated secondary antibody diluted in PBS without calcium and magnesium containing 1% BSA or in TBS plus 1% BSA for 30 min at room temperature.
14. Rinse as in **step 10**.
15. Incubate grids for 30 min at room temperature in colloidal gold conjugated with streptavidin.
16. Rinse as in **step 10**.
17. Rinse the grids in a stream of deionized glass-distilled water from a squirt bottle and dry.
18. At this point, the grids may be examined in the electron microscope to evaluate the staining.
19. Stain the grids with uranyl acetate for 5–10 min, rinse in water, and stain with lead citrate for 1–2 min. They are now ready for final examination in the electron microscope (*see* **Note 12**).

4. Notes

1. Clean the vial, and score it with a diamond scribe. Place it in a clean glass bottle, and add the water. With a glass rod, break vial. The solution is stable if kept refrigerated and protected from light. The stock solution should be stored away from other chemicals since osmium vapors may escape from the bottle. Alternatively, aqueous solutions of osmium tetroxide in sealed glass ampules are available commercially. Dilute to 1–2% with cacodylate buffer for use.
2. The concentration of the primary antibody can range from 1–20 µg/mL with 5 µg/mL being average. If background staining is high, the antibody may be diluted in high-salt Tween-20 buffer.

3. The colloidal gold should be diluted in PBS without calcium and magnesium containing 1% BSA or 0.01% polyethylene glycol (mol wt 20,000), or in TBS plus 1% BSA or 0.01% polyethylene glycol (mol wt 20,000).

4. Add the lead nitrate and sodium citrate to the water in a 50-mL volumetric flask. Shake vigorously for 1 min and allow to stand with intermittent shaking for 30 min. Add 8.0 mL 1 *N* NaOH. Dilute the suspension to 50 mL with deionized glass-distilled water, and mix by inversion. The solution is stable up to 6 m. Turbidity can be removed by centrifugation.

5. The composition of the fixative may vary depending on the sensitivity of the antigen. For some antigens, it may be necessary to reduce the glutaraldehyde concentration or omit it entirely, and fix in 4% formaldehyde in PBS for 1–4 h.

6. 0.1 *M* Phosphate or cacodylate buffer, pH 7.4, may be substituted for PBS.

7. The amino group on the glycine will bind to any free aldehyde groups left on the cells after fixation and prevent them from binding to the antibody or blocking solution, thus reducing the nonspecific background.

8. If embedding in a resin that is polymerized by UV light, such as Lowicryl and LR Gold, do not osmicate the samples. Any effect of the osmium on antigenicity can usually be eliminated by using sodium *meta*-periodate during etching to oxidize the unbound osmium (*see* **step 7**).

9. The choice of embedding resin can affect the degree of immunostaining. Although all embedding resins may be used, the epoxy resins, such as Epon substitutes and Spurr, may reduce the intensity of staining. The acrylic resins (LR White, LR Gold, and Lowicryl) are more hydrophilic and usually result in better immuno-labeling (*see* Chapter 40).

10. It is often advisable to cut the sections slightly thicker than normal (light gold). The sections will adhere to the grids better during the staining process if, after the sections are picked up, the grids are placed in a 50°C oven for an hour. The nickel grids can become magnetized and should be handled with nonmagnetic forceps. They may also need to be degaussed before viewing in the electron microscope, and the astigmatism may need to be adjusted for each grid.

11. If the primary antibody is directed against a carbohydrate-containing epitope, keep the exposure to the *meta*-periodate as brief as possible, since the *meta*-periodate may remove the carbohydrate. If this is a problem, it may be necessary to etch in hydrogen peroxide or sodium ethoxide. The incubation can be done in Petri dishes lined with parafilm or in spot plates. The time required for etching depends on the embedding resin, and can vary from 30 min for Epon substitutes to <5 min for LR White and LR Gold. From this step until the staining is completed, the grids should not be allowed to dry.

12. If the nonspecific background is high, it may be because of the streptavidin binding to endogenous biotin. In that instance, it is necessary to block the endogenous biotin with streptavidin. After etching, grids are floated on drops of unconjugated streptavidin (5 g/mL) in TBS for 15 min. Rinse five times in TBS. The streptavidin must then be reacted with unconjugated biotin or it will react with the biotinylated antibody. Float grids on biotin (5 g/mL) in TBS for 15 min.

Rinse five times in TBS. At this point, the sections may be blocked again with BSA or incubated with the primary antibody.

References

1. Bonnard, C., Papermaster, D. S., and Kraehenbuhl, J.-P. (1984) The streptavidin–biotin bridge technique: application in light and electron microscope immunocytochemistry, in *Immunolabelling for Electron Microscopy* (Polak, J. M. and Varndell, I. M., eds.), Elsevier, New York, pp. 95–111.
2. Larsson, L.-I. (1988) Immunocytochemical detection systems, in *Immunocytochemistry: Theory and Practice* (Larsson, L.-I., ed.), CRC, Boca Raton, FL, pp. 77–146.
3. Wooley, D. W. and Longsworth, L. G. (1942) Isolation of an antibiotin factor from egg white. *J. Biol. Chem.* **142,** 285–290.
4. Green, N. M. (1975) Avidin. *Adv. Prot. Res.* **29,** 85–133.
5. Chaiet, L. and Wolf, F. J. (1964) The properties of streptavidin, a biotin-binding protein produced by *Streptomycetes. Arch. Biochem. Biophys.* **106,** 1–5.
6. Reynolds, E. S. (1963) The use of lead citrate at high pH as an electron opaque stain in electron microscopy. *J. Cell Biol.* **17,** 208–212.

44

Pre-Embedding Labeling Methods

Constance Oliver

1. Introduction

Colloidal gold conjugates generally do not readily penetrate cells, even after permeabilization. Therefore, their use in pre-embedding immunostaining has been restricted to labeling cell-surface antigens for scanning (*1*) or transmission electron (**Fig. 1A**) microscopy or for tracing endocytic pathways in living cells (**Fig. 1B**). Recently, 1-nm gold conjugates that do penetrate cells and tissues much more readily have been used successfully to immunolabel intracellular structures (*2,3*). For pre-embedding labeling, all of the immunostaining is done prior to embedding the tissue in resin or preparing the samples for scanning electron microscopy. This method is especially useful if the antigen to be detected is sensitive to fixation. The immunostaining may be done on unfixed or lightly (4% formaldehyde) fixed samples. Following the immunolabeling, the samples may then be refixed in 2% glutaraldehyde–2% formaldehyde to give good ultrastructural preservation.

When immunolabeling cell-surface components, it is important to be aware that redistribution of membrane components may be induced by the immunolabeling. Crosslinking of membrane components in unfixed or lightly fixed cells by antibodies may result in aggregation, capping, or internalization of the molecule of interest. Usually, brief fixation in a fixative containing low concentrations of glutaraldehyde is sufficient to prevent redistribution. Fixation in formaldehyde alone or performing the immunostaining at 4°C may not be sufficient to prevent lateral diffusion of molecules in the plasma membrane.

From: *Methods in Molecular Biology, Vol. 115: Immunocytochemical Methods and Protocols*
Edited by: L. C. Javois © Humana Press Inc., Totowa, NJ

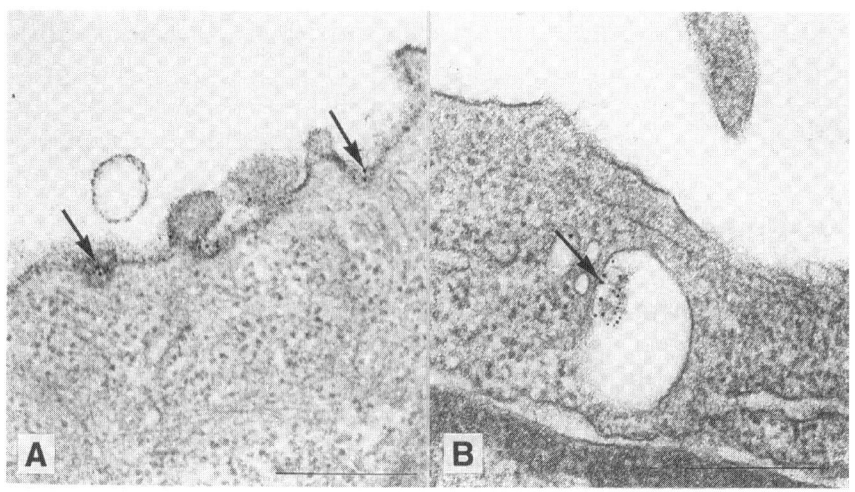

Fig. 1. Transmission electron micrograph of RBL–2H3 cells with colloidal gold conjugated to a monoclonal antibody against the IgE receptor. (**A**) Gold conjugate is localized primarily in coated pits. (**B**) Five minutes after exposing cells to antibody-coated gold, gold particles are localized in early endosomes. Bar = 0.5 μm.

2. Materials

1. Dulbecco's phosphate-buffered saline (PBS): 100 mg anhydrous calcium chloride, 200 mg potassium chloride, 200 mg monobasic potassium phosphate, 100 mg magnesium chloride · 6 H_2O; 8 g sodium chloride, and 2.16 g dibasic sodium phosphate · 7 H_2O; bring vol to 1 L with deionized glass-distilled water, pH 7.4.
2. 0.5% Glutaraldehyde–2% formaldehyde fixative: 0.5 mL 70% EM-grade glutaraldehyde, 7 mL 20% EM-grade formaldehyde, 3 mg calcium chloride, and 62.5 mL 0.1 *M* cacodylate buffer or PBS, pH 7.4 (*see* **Note 1**).
3. 0.1 *M* Glycine: 75 mg glycine, bring the vol to 100 mL with PBS.
4. Dulbecco's PBS without calcium and magnesium: 200 mg potassium chloride, 200 mg monobasic potassium phosphate, 8 g sodium chloride, and 2.16 g dibasic sodium phosphate · 7 H_2O; bring vol to 1 L with deionized glass-distilled water, pH 7.4.
5. 1% Bovine serum albumin (BSA): 1 g BSA and 100 mL PBS without calcium and magnesium; add BSA to PBS with stirring. Make fresh daily.
6. Primary antibody (*see* **Note 2**).
7. Colloidal gold conjugate (*see* **Note 3**).
8. 0.1 *M* Cacodylate buffer: 21.4 g cacodylic acid · 3 H_2O, sodium salt; bring vol to 1 L with deionized glass-distilled water; adjust pH to 7.4 with HCl.

3. Methods

3.1. Cell-Surface Labeling

3.1.1. Unlabeled Primary Antibody

1. Rinse cells twice in PBS, and fix for 30 min to 1 h at room temperature in 0.5% glutaraldehyde and 2% formaldehyde in PBS (*see* **Note 4**).
2. Following fixation, rinse the cells three to five times over a period of 30 min in PBS (*see* **Note 5**).
3. Rinse the cells for 5 min in 0.1 *M* glycine in PBS to quench free aldehyde groups (*see* **Note 6**).
4. Rinse the cells for 15 min in a solution of 1% BSA in PBS without calcium and magnesium to block nonspecific binding of the primary antibody (*see* **Note 7**).
5. Expose the cells to the primary antibody. Incubation times usually range from 1–2 h at room temperature.
6. Rinse cells five times in PBS without calcium and magnesium over a period of 30 min to remove unbound primary antibody (*see* **Note 8**).
7. Incubate the cells in colloidal gold conjugate for 30 min at room temperature (*see* **Note 3**).
8. Rinse cells in PBS without calcium and magnesium five times over a period of 30 min.
9. For better morphological preservation, refix the cells in 2% glutaraldehyde–2% formaldehyde in 0.1 *M* cacodylate buffer, pH 7.4 for 1 h at room temperature.
10. Rinse in cacodylate buffer, postfix in 1–2% osmium tetroxide, and embed as usual (*see* Chapter 40; **Note 9**).

3.1.2. Primary Antibody Conjugated to Colloidal Gold

1. Perform **steps 1–4** as outlined in **Subheading 3.1.1.**
2. Incubate the cells in primary antibody conjugated to colloidal gold diluted in PBS without calcium and magnesium containing 1% BSA for 1 h at room temperature (*see* **Note 2**).
3. Continue with **step 8** in **Subheading 3.1.1.**

3.1.3. Tracing Endocytic Pathways

1. Rinse cells in sterile medium containing 1–2% BSA.
2. Add ligand conjugated to the colloidal gold, i.e., colloidal gold conjugated with antibody to a specific receptor.
3. Allow to bind at 4°C for 30–60 min, or return to incubator immediately.
4. At various time intervals, stop endocytosis with cold (4°C) PBS or balanced salt solution.
5. Rinse twice in cold PBS.
6. Fix and embed as usual (*see* Chapter 40; **Note 9**).

4. Notes

1. The purity of the reagents is critical. Always use the highest quality available, i.e., affinity-purified antibodies, EM-grade glutaraldehyde, distilled

formaldehyde, or freshly prepared paraformaldehyde. The use of distilled **form-aldehyde**, not formalin, which contains alcohol, is recommended. Freshly prepared paraformaldehyde can also be used, especially if large volumes of fixative are needed. To prepare an 8% solution of paraformaldehyde, **in a fume hood** add 2 g of paraformaldehyde (trioxymethylene) powder to 25 mL of double-distilled water. With constant stirring, heat solution to 60–70°C. Once the solution has reached the proper temperature, continue to stir for 15 min. The solution will be milky. Add one to two drops of 1 *N* NaOH, with stirring, until the solution clears. A slight milkiness may persist. Cool and filter through Whatman no. 1 filter paper. This solution should be used the same day that it is prepared.

2. Primary antibody is usually at a concentration of 1–5 µg/mL, in PBS without calcium and magnesium, containing 1% BSA.

3. The colloidal gold conjugate should be directed against the primary antibody and diluted to 1–5 µg/mL in PBS without calcium and magnesium plus 1% BSA.

4. The composition of the fixative may vary depending on the sensitivity of the antigen. For some antigens, it may be necessary to omit the glutaraldehyde entirely and fix in 4% formaldehyde in PBS for 1–4 h.

5. 0.1 *M* Phosphate buffer, pH 7.4, may be substituted for PBS.

6. Following fixation in formaldehyde or glutaraldehyde, tissue is generally quenched. The purpose of this step is to block any free aldehyde groups that remain after fixation and washing. This step is especially critical after glutaraldehyde fixation. Glutaraldehyde is a bifunctional aldehyde. During fixation, one end may bind to cellular constituents, leaving the other end free to react. If this end is not blocked, it can bind to the protein in the blocking solution or to the primary antibody, increasing the nonspecific background. Although any small molecular-weight compound containing an amino group may be used, the most commonly used quenching agents are glycine, ammonium chloride, and sodium borohydride.

7. Following queching, the samples are blocked to reduce nonspecific binding of the primary antibody and to reduce the background. The proteins in the blocking solution are chosen so as not to react with the secondary antibody or detection system. They bind to sites in the samples that bind proteins through nonspecific interactions, such as charge. Although 2–10% BSA is most commonly used, other proteins, such as ova albumin, normal serum, or IgG fractions, can also be used. The exact composition of the blocking solution may have to be determined empirically if background staining is high.

8. If the nonspecific staining is high, 0.1 *M* EDTA can also be added to the rinse solutions.

9. Appropriate controls should always be run with any immunocytochemical procedure. Controls may include omitting the primary antibody, substituting pre-immune serum, normal serum, or normal IgG for the primary antibody, adsorbing the primary antibody against the antigen, or immunostaining with an unrelated antibody.

References

1. Becker, R. P. and Johari, O. (eds.) (1979) *Cell Surface Labeling*, SEM, Inc., AMF O'Hare, Chicago.
2. Vandre, D. D. and Burry, R. W. (1992) Immunoelectron microscopic localization of phosphoproteins associated with the mitotic spindle. *J. Histochem. Cytochem.* **40,** 1837–1847.
3. Burry, R. W., Vandre, D. D., and Hayes, D. M. (1992) Silver enhancement of gold antibody probes in pre-embedding electron microscopic immunocytochemistry. *J. Histochem. Cytochem.* **40,** 1849–1856.

45

Postembedding Labeling Methods

Constance Oliver

1. Introduction

Since it was first introduced *(1)*, postembedding immunogold labeling has become the most widely used method of immunolabeling for electron microscopy *(2–8)*. For postembedding labeling, samples are first fixed, embedded, and sectioned. All immunostaining is performed on sections mounted on grids. The immunostaining may be done with a direct labeling technique with the primary antibody conjugated to colloidal gold, or by an indirect method where the primary antibody is unlabeled and the gold is conjugated to a secondary or tertiary antibody, protein A, protein G, and so forth. Colloidal gold has several advantages as a marker for electron microscopy. The gold particles are distinct and readily visualized, they are easily quantifiable, and they do not diffuse on the sections.

The most popular reagents for postembedding colloidal gold immunocytochemistry are the protein A and protein G conjugates. Protein A, a cell-wall component of *Staphylococcus aureus*, has a high affinity toward the F_c region of IgG molecules from a number of species, including rabbit and human *(9)*. However, protein A has a low affinity for most classes of mouse IgG molecules. Protein G, an immunoglobulin-binding molecule present in the cell wall of the group G *Streptococcal* strain (G 148), has a high affinity for immunoglobulins from a wider range of species, including mouse *(10)*. Thus, protein G has been used with mouse monoclonal antibodies.

Colloidal gold–antibody conjugates are also widely used, especially in indirect immunocytochemical methods. However, the stability of these gold conjugates is less than that for protein A, and depending on the antibody, the affinity may be less. Colloidal gold conjugated to avidin (*see*

From: *Methods in Molecular Biology, Vol. 115: Immunocytochemical Methods and Protocols*
Edited by: L. C. Javois © Humana Press Inc., Totowa, NJ

Chapter 43), lectins *(11)*, and enzymes *(12)*, has also been used for post-embedding staining.

2. Materials

1. Dulbecco's phosphate-buffered saline (PBS): 100 mg anhydrous calcium chloride, 200 mg potassium chloride, 200 mg monobasic potassium phosphate, 100 mg magnesium chloride · 6 H_2O, 8 g sodium chloride, and 2.16 g dibasic sodium phosphate · 7 H_2O; bring vol to 1 L with deionized glass-distilled water, pH 7.4.
2. 0.5% Glutaraldehyde–2% formaldehyde fixative: 0.5 mL 70% EM-grade glutaraldehyde, 7 mL 20% EM-grade formaldehyde, 3 mg calcium chloride, and 61 mL 0.1 *M* cacodylate buffer or PBS, pH 7.4 (*see* **Note 1**).
3. 0.1 *M* Glycine: 750 mg glycine; bring vol to 100 mL with PBS, pH 7.4.
4. Osmium tetroxide (4% stock): 1 g glass vial osmium tetroxide and 25 mL deionized glass-distilled water (*see* **Note 2**).
5. Embedding medium, e.g., LR Gold and initiator (*see* Chapter 40).
6. Nickel grids: Clean in acetone or alcohol just prior to use.
7. Saturated aqueous solution of sodium *meta*-periodate; prepare daily.
8. Tris-buffered saline (TBS): 2.4 g Tris-HCl and 8.76 g NaCl; bring vol to 1 L with deionized glass-distilled water, pH 7.4.
9. 1% Bovine serum albumin (BSA): 1 g BSA and 100 mL TBS (*see* **Note 3**).
10. High-salt Tween-20 buffer: 2.4 g Tris-HCl, 29.2 g sodium chloride and 1 mL Tween-20; bring vol to 1 L with deionized glass-distilled water, pH 7.4.
11. Dulbecco's PBS without calcium and magnesium: 200 mg potassium chloride, 200 mg monobasic potassium phosphate, 8 g sodium chloride, and 2.16 g dibasic sodium phosphate · 7 H_2O; bring vol to 1 L with deionized glass-distilled water, pH 7.4.
12. Primary antibody (*see* **Note 4**).
13. Secondary antibody (*see* **Note 5**).
14. Colloidal gold conjugate (*see* **Note 6**).
15. 0.1 *M* Cacodylate buffer: 21.4 g cacodylic acid · 3 H_2O, sodium salt; bring vol to 1 L with deionized glass-distilled water; adjust pH to 7.4 with HCl.
16. Saturated picric acid solution: Fill 500 g bottle of picric acid with deionized glass-distilled water (*see* **Note 7**).
17. Membrane enhancement fixative: 1 mL 70% glutaraldehyde, 14 mL 20% formaldehyde, 3 mg calcium chloride, 150 µL saturated picric acid, and 55 mL 0.1 *M* cacodylate buffer, pH 7.4.
18. Acetone.
19. Polyethylene embedding capsules or gelatin capsules.
20. UV light source, e.g., 366 nm, two 15-W Sylvania F15t8/BLB bulbs.
21. 0.1 *M* Maleate buffer: 1.16 g maleic acid and 3.5 g sucrose; add deionized glass-distilled water to make 100 mL; adjust pH to 6.5.
22. 2% Uranyl acetate: 2 g uranyl acetate and 100 mL maleate buffer; add uranyl acetate to buffer; adjust pH to 6.0.
23. Reynolds' lead citrate *(13)*: 1.33 g lead nitrate, 1.76 g trisodium citrate · 2H_2O, 30 mL deionized glass-distilled water, and 8.0 mL 1 *N* NaOH (*see* **Note 8**).

3. Methods

3.1. Standard Method

1. Rinse samples twice in PBS, and fix for 30 min to 1 h at room temperature in 0.5% glutaraldehyde and 2% formaldehyde in PBS (*see* **Note 9**).
2. Following fixation, rinse the samples three to five times over a period of 30 min in PBS (*see* **Note 10**).
3. Quench free aldehyde groups by rinsing the samples for 5 min in 0.1 *M* glycine in PBS (*see* **Note 11**).
4. Rinse the samples in cacodylate buffer or PBS, and postfix in 1–2% osmium tetroxide for 1 h at room temperature (*see* **Note 12**).
5. Following postfixation, dehydrate the samples and embed (*see* **Note 13**).
6. Cut the resulting blocks with a diamond knife, and mount sections on nickel grids. It is often advisable to cut the sections slightly thicker than normal such that the interference color of the sections is light gold (*see* **Note 14**).
7. Etch the plastic in the sections slightly in order to allow for penetration of reagents into the samples (*see* **Note 15**).
8. Rinse three to five times in deionized glass-distilled water.
9. Block nonspecific binding by incubating sections in 1% BSA or 5% normal serum in PBS without calcium and magnesium, or in TBS for 15 min (*see* **Note 16**).
10. Rinse five times in PBS without calcium and magnesium, or in TBS.
11. Incubate the sections for 1–2 h at room temperature in primary antibody diluted in PBS without calcium and magnesium containing 1% BSA or in TBS containing 1% BSA (*see* **Note 4**).
12. Rinse as in **step 10**.
13. If an unlabeled secondary antibody is necessary as a bridging antibody, incubate the grids in the secondary antibody diluted in PBS without calcium and magnesium containing 1% BSA or in TBS plus 1% BSA for 30 min at room temperature, and then rinse as in step 10 (*see* **Note 5**).
14. Incubate grids for 30 min at room temperature in colloidal gold conjugated with protein A, protein G, immunoglobulin, or if a biotinylated secondary antibody was used, streptavidin (*see* **Note 6**).
15. Rinse as in **step 10**.
16. Rinse the grids in a stream of deionized glass-distilled water from a squirt bottle and dry.
17. At this point, the grids may be examined in the electron microscope to evaluate the staining.
18. Stain the grids with uranyl acetate and lead citrate for final examination in the electron microscope (*see* **Notes 16** and **17**).

3.2. Membrane Enhancement Method (14)

1. Fix samples in membrane enhancement fixative for 2–3 h at room temperature.
2. Rinse tissue five times with cold (4°C) PBS or cacodylate buffer over a period of 1–2 h (*see* **Note 18**).

3. Rinse samples with 0.1 *M* glycine in phosphate or cacodylate buffer for 1 h at 4°C to quench free aldehyde groups.

4. If phosphate buffer has been used, rinse tissue five times over a period of 1 h in 0.1 *M* maleate buffer containing 3.5% sucrose, pH 6.5, to remove phosphate ions.

5. Stain the tissue *en bloc* with 2% uranyl acetate in maleate–sucrose buffer, pH 6.0, for 2 h at 4°C in the dark.

6. Dehydrate the tissue, and infiltrate with resin according to the following schedule:
   50% Acetone, 5 min, 0°C
   50% Acetone, 45 min, 0°C
   70% Acetone, 45 min, 0°C
   90% Acetone, 45 min, 0°C
   1:1 LR Gold:acetone, 60 min, –20°C
   7:3 LR Gold:acetone, 60 min, –20°C
   100% LR Gold, 60 min, –20°C
   100% LR Gold, 5 h to overnight, –20°C
   100% LR Gold + initiator (0.5% benzoin methyl ether), 60 min, –20°C
   100% LR Gold + initiator, 60 min, –20°C
   100% LR Gold + initiator, 5 h to overnight, –20°C (*see* **Note 19**).

7. Transfer tissue to polyethylene embedding capsules or gelatin capsules, fill the capsules 100% LR Gold + initiator and cap (*see* **Note 20**).

8. Polymerize the plastic with UV light placed 10 cm from the capsules. Expose the capsules to the UV light for 24 h at –20°C (*see* **Note 21**).

9. Cut gold sections with a diamond knife, and mount on nickel grids. To increase the adherence of the sections to the grids, after sectioning, the grids may be placed in a 60°C oven for 1 h.

10. Hydrate sections for 5 min in TBS (*see* **Note 22**).

11. Rinse in distilled water.

12. Block sections with 1% BSA or 5% normal goat serum in TBS.

13. Incubate grids for 60 min with primary antibody diluted with TBS containing 1% BSA or 5% normal goat serum (*see* **Note 4**).

14. Rinse grids five times in TBS.

15. If a secondary antibody is necessary as a bridging antibody, incubate the grids in the secondary antibody diluted in TBS plus 1% BSA for 30 min at room temperature and then rinse five times in TBS.

16. Incubate grids for 30 min at room temperature in colloidal gold conjugated with protein A, protein G, immunoglobulin, or if a biotinylated secondary antibody was used, streptavidin (*see* **Note 6**).

17. Rinse grids five times in TBS.

18. Rinse in deionized glass-distilled water.

19. Fix with 2% aqueous glutaraldehyde for 5 min.

20. Rinse grids three times in deionized glass-distilled water.

21. In order to enhance membrane contrast, stain for 15 min in 2% aqueous osmium tetroxide.

22. Rinse five times in deionized glass-distilled water.

23. Stain in Reynolds' *(13)* lead citrate for 1–2 min.
24. Rinse grids five times in deionized glass-distilled water and in a stream of deionized glass-distilled water from a squirt bottle.
25. Dry grids by placing on fine filter paper, i.e., Whatman no. 50.
26. Examine with the electron microscope (*see* **Notes 17** and **18**).

4. Notes

1. The use of distilled **formaldehyde**, not formalin, which contains alcohol, is recommended. Freshly prepared paraformaldehyde can also be used, especially if large volumes of fixative are needed. To prepare an 8% solution of paraformaldehyde, **in a fume hood** add 2 g of paraformaldehyde (trioxymethylene) powder to 25 mL of double-distilled water. With constant stirring, heat solution to 60–70°C. Once the solution has reached the proper temperature, continue to stir for 15 min. The solution will be milky. Add one to two drops of 1 *N* NaOH, with stirring, until the solution clears. A slight milkiness may persist. Cool and filter through Whatman no. 1 filter paper. This solution should be used the same day that it is prepared.

2. Clean the vial, and score with a diamond scribe. Place in a clean glass bottle, and add water. With a glass rod, break the vial. The solution is stable if kept refrigerated and protected from light. **Stock solution should be stored away from other chemicals since osmium vapors may escape from bottle.** Alternatively, aqueous solutions of osmium tetroxide in sealed glass ampules are available commercially. Dilute to 1–2% with cacodylate buffer for use.

3. Add the BSA to the buffer with stirring. High-salt Tween-20 buffer or PBS without calcium and magnesium may be substituted for TBS.

4. The concentration of the primary antibody can range from 1–20 µg/mL with 5 µg/mL being average. If nonspecific staining is a problem, the antibody may be diluted in high-salt Tween-20 buffer.

5. The bridging antibody is directed against IgG of the same species as the primary antibody. For example, if the primary antibody is a mouse monoclonal, then the bridging antibody would be an antimouse IgG.

6. The colloidal gold should be diluted in PBS without calcium and magnesium containing 1% BSA or 0.01% polyethylene glycol (mol wt 20,000), or in TBS plus 1% BSA or 0.01% polyethylene glycol (mol wt 20,000).

7. Let the picric acid/water stand; it will keep indefinitely at room temperature. Do not let the picric acid dry since the powder is explosive.

8. Add 1.33 g lead nitrate and 1.76 g sodium citrate to 30 mL glass-distilled water in a 50-mL volumetric flask. Shake vigorously for 1 min and allow to stand with intermittent shaking for 30 min. Add 8.0 mL 1 *N* NaOH. Dilute the suspension to 50 mL with deionized glass-distilled water and mix by inversion. The solution is stable up to 6 mo. Turbidity can be removed by centrifugation.

9. The composition of the fixative may vary depending on the sensitivity of the antigen. For some antigens, it may be necessary to reduce the glutaraldehyde concentration, or omit it entirely and fix in 4% formaldehyde in PBS for 1–4 h.

10. 0.1 *M* Phosphate buffer, pH 7.4, may be substituted for PBS.

11. The amino group on the glycine will bind to any free aldehyde groups left on the cells after fixation, and prevent them from binding to the antibody or blocking solution and increasing the nonspecific background. Following fixation in form-aldehyde or glutaraldehyde, tissue is generally quenched. The purpose of this step is to block any free aldehyde groups that remain after fixation and washing. This step is especially critical after glutaraldehyde fixation. Glutaraldehyde is a bifunctional aldehyde. During fixation, one end may bind to cellular constitu-ents, leaving the other end free to react. If this end is not blocked, it can bind to the protein in the blocking solution or to the primary antibody. Although any small molecular-weight compound containing an amino group may be used, the most commonly used quenching agents are glycine, ammonium chloride, and sodium borohydride.

12. If the samples are to be embedded in a resin that is polymerized by UV, such as Lowicryl or LR Gold, do not postfix in osmium tetroxide. Any effect of the osmium on antigenicity can usually be eliminated by using sodium *meta*-periodate during etching (*see* **step 7**).

13. The choice of embedding resin can affect the degree of immunostaining. Although all embedding resins may be used, the epoxy resins, such as Epon substitutes and Spurr, can reduce the intensity of staining. The acrylic resins (LR White, LR Gold, and Lowicryl) are more hydrophilic and usually result in better immunolabeling (*see* Chapter 40).

14. The sections will adhere to the grids better during the staining process if after the sections are picked up, the grids are placed in a 60°C oven for an hour. The nickel grids can become magnetized and should be handled with nonmagnetic forceps. They may also need to be degaussed before viewing in the electron microscope, and the astigmatism may need to be adjusted for each grid.

15. Grids are floated, section side down, on drops of a freshly prepared saturated solution of sodium *meta*-periodate. If the primary antibody is directed against a carbohydrate-containing epitope, keep the exposure to the *meta*-periodate as brief as possible, because the *meta*-periodate may remove the carbohydrate. If this is a problem, it may be necessary to etch in hydrogen peroxide or sodium ethoxide. The incubation can be done in Petri dishes lined with parafilm or in spot plates. The time required for etching depends on the embedding resin and can vary from 30 min for Epon substitutes to <5 min for LR White and LR Gold. From this step until the staining is completed, the grids should not be allowed to dry.

16. Following quenching, the samples are blocked to reduce nonspecific binding of the primary antibody and to reduce the background. The proteins in the blocking solu-tion are chosen so as not to react with the secondary antibody or detection system. They bind to sites in the samples that bind proteins through nonspecific interac-tions such as charge. Although 2–10% BSA is most commonly used, other proteins (e.g., ova albumin, normal serum, or IgG fractions) can also be used. The exact composition of the blocking solution may have to be determined empirically if background staining is high.

17. If the nonspecific staining is high, 0.1 *M* EDTA can also be added to the rinse solution.
18. Appropriate controls should always be run with any immunocytochemical procedure. Controls may include omitting the primary antibody, substituting preimmune serum, normal serum, or normal IgG for the primary antibody, adsorbing the primary antibody against the antigen, or immunostaining with an unrelated antibody.
19. To enhance preservation, 3–5% purified sucrose and 0.5 m*M* calcium chloride may be added to the buffer.
20. Samples should be in capped glass vials during dehydration and infiltration.
21. Oxygen will inhibit polymerization.
22. Instead of hydrating the sections, they may be etched for 1 or 2 min with a freshly prepared saturated solution of sodium *meta*-periodate.

References

1. Romano, E. L. and Romano, M. (1984) Historical aspects, in *Immunolabelling for Electron Microscopy* (Polak, J. M. and Varndell, I. M., eds.), Elsevier, New York, pp. 3–16.
2. Roth, J. (1982) The protein A-gold (pAg) technique—a qualitative and quantitative approach for antigen localization on thin sections, in *Techniques in Immunocytochemistry*, vol. 1 (Bullock, G. R. and Petrusz, P., eds.), Academic, New York, pp. 107–154.
3. Roth, J. (1984) The protein A-gold technique for antigen localization in tissue sections by light and electron microscopy, in *Immunolabelling for Electron Microscopy* (Polak, J. M. and Varndell, I. M., eds.), Elsevier, New York, pp. 113–122.
4. Slot, J. W. and Geuze, H. J. (1984) Gold markers for single and double immunolabeling of ultrathin cryosections, in *Immunolabelling for Electron Microscopy* (Polak, J. M. and Varndell, I. M., eds.), Elsevier, New York, pp. 129–142.
5. Bendayan, M. (1989) Protein A-gold electron microscopic immunocytochemistry: methods, applications, and limitations. *J. Electron Microsc. Tech.* **1,** 243–270.
6. Bendayan, M. (1989) Protein A-gold and protein G-gold postembedding immuno-electron microscopy, in *Colloidal Gold*, vol. 1 (Hayat, M. A., ed.), Academic, New York, pp. 34–95.
7. Bendayan, M. and Stephens, H. (1984) Double immunostaining procedures: techniques and applications, in *Immunolabelling for Electron Microscopy* (Polak, J. M. and Varndell, I. M., eds.), Elsevier, New York, pp. 143–154.
8. Merighi, A. (1992) Postembedding electron microscopic immunocytochemistry, in *Electron Microscopic Immunocytochemistry* (Polak, J. M. and Priestley, J. V., eds.), Oxford University Press, Oxford, pp. 51–88.
9. Forsgren, A. and Sjöquist, J. (1966) 'Protein A' from *S. aureus*. I. Pseudoimmune reaction with human γ-globulin. *J. Immunol.* **97,** 822–827.
10. Björck, L. and Kronvall, G. (1984) Purification and some properties of streptococcal protein G, a novel IgG-binding reagent. *J. Immunol.* **133,** 969–974.
11. Benhamou, N. (1989) Preparation and application of lectin-gold complexes, in *Colloidal Gold*, vol. 1 (Hayat, M. A., ed.), Academic, New York, pp. 96–145.

12. Bendayan, M. (1984) Enzyme-gold electron microscopic cytochemistry: a new affinity approach for the ultrastructural localization of macromolecules. *J. Electron Micros. Tech.* **1,** 349–372.

13. Reynolds, E. S. (1963) The use of lead citrate at high pH as an electron opaque stain in electron microscopy. *J. Cell Biol.* **17,** 208–210.

14. Berryman, M. A. and Rodewald, R. R. (1990) An enhanced method for post-embedding immunocytochemical staining which preserves cell membranes. *J. Histochem. Cytochem.* **38,** 159–170.

VI

APPLICATIONS FOR NUCLEIC ACID LOCALIZATION BY *IN SITU* HYBRIDIZATION

46

Overview of *In Situ* Hybridization

Roland M. Nardone

1. Introduction

In situ hybridization (ISH) permits the detection and localization of DNA and RNA in a cytological preparation affixed to a microscope slide. Such detection and localization is made possible by hybridization of cellular DNA and/or RNA targets with nucleic acid probes tagged with a signal generating system such as a fluorochrome or enzyme.

1.1. Representative Applications

From its inception, Gall and Pardue *(1)* and John et al. *(2)* emphasized the importance of studying nucleic acid localization within a high-resolution structural framework; thus, nucleic acid detection and distribution was superimposed on cytological detail. This attribute of the technology has resulted in its rapid acceptance and application to diverse biomedical studies including gene expression, gene mapping, genetic disease analysis, cancer diagnosis and prognosis, evolution, and the spread of infectious organisms within a tissue *(3–7)*. (A search of PubMed of the National Library of Medicine for research papers published in 1997 on ISH uncovered 4418 "hits.")

Contributing greatly to the value and acceptance of ISH is the specificity of the core reaction of probe and target (base pairing in nucleic acid hybridization) and the high resolution of signal generating systems that reveal the precise localization of nucleic acid targets. Parallel improvements in photomicrography have made visualization more precise *(8)*.

Many other factors, including applicable advances in recombinant DNA technology and impetus provided by the Human Genome Project, also have contributed to important permutations of methodology and applications, such as polymerase chain reaction- (PCR)-ISH *(6)* and spectral karyotyping *(5)*.

From: *Methods in Molecular Biology, Vol. 115: Immunocytochemical Methods and Protocols*
Edited by: L. C. Javois © Humana Press Inc., Totowa, NJ

2. The Core Method

The seminal aspects on ISH are as follows:

1. The DNA or RNA target is localized within a cellular compartment (i.e., the DNA of chromosomes or mRNA in the cytoplasm) in a cytological preparation.
2. The target as well as the labeled probe that is to be hybridized to the target must be single-stranded, such as an RNA molecule or denatured DNA, thus permitting recognition by base complementarity and subsequent hybridization by hydrogen bonding.
3. A signal generating system, such as a fluorochrome or enzyme, directly or indirectly conjugated to the probe, permits "visualization," that the probe has hybridized to a cellular target and the location of the target is discerned.

2.1. The Typical Protocol

A typical protocol has five major phases:

1. Tissue preparation, including fixation, sectioning, and affixing the sections or the whole cells to a microscope slide.
2. Prehybridization.
3. Hybridization.
4. Posthybridization.
5. Detection.

The preparations most often affixed to silane- *(9)* or poly-L-lysine-coated slides are Carnoy-fixed suspensions of metaphase cells for chromosome studies (*see* Chapter 47); cytospins of cultured cells followed by methanol fixation; sections of formalin-fixed, paraffin-embedded tissues; and frozen sections fixed with acetone or ethanol/acetic acid after cutting. Each of these approaches imparts different intracellular effects that may modify cytological detail and/or hybridization. (*See* Chapters 8 and 9 for a discussion of fixatives and frozen-section preparations.)

Some or all prehybridization steps are optional, depending on the target and specimen. Prehybridization entails soaking the slides containing the specimens in a prehybridization buffer in order to have the specimen rehydrate and equilibrate with the hybridization buffer, to be used later. The prehybridization phase may also entail treatment with proteases or nucleases in order to improve probe access and to reduce background.

The hybridization phase entails thermal denaturation of double-stranded DNA and incubation of the probe with the denatured DNA at a temperature 25°C below the melt temperature. Unless one is interested in partial homology, lower temperatures should be avoided. However, addition of formamide promotes the hybridization, thereby permitting the use of a lower temperature, if it should be required, to prevent structural modification.

The hybridization time is dictated by the size of the probe (the larger the probe, the slower the hybridization rate), the abundance of the target, and the concentration of the probe. Hybridization times are usually longer than for other formats because of the relative inaccessibility of the targets. Often hybridization may be a few hours to overnight.

Commonly used probes include plasmids, cloned sequences of long length, and oligonucleotides. Oligos provide the highest efficiency of hybridization. However, their value is sometimes diminished by their reduced capacity to carry a large number of reporter molecules of a signal generating system. This limitation is diminishing each year as more sensitive reporters are designed.

Posthybridization entails a series of washes with preset stringency conditions to adjust the specificity or the wanted degree of homology and to wash out the unbound probe. Hence, posthybridization washes ordinarily include specified amounts of formamide and sodium chloride and sodium citrate (SSC) stock to adjust for the degree of stringency.

A variety of signal generating systems have been used with ISH. Gall and Pardue *(1)* and John et al. *(2)* used radiolabeled probes and autoradiography. The ensuing three decades have seen a major shift to nonradioactive probes that use enzymes, fluorochromes, or chemiluminescent molecules as the signal generators. Their use and the basis of their detection are discussed in other chapters in this book (*see* Chapters 14 and 23). Suffice it to note that these nonradioactive reporter systems have a sensitivity virtually the same as isotopic methods and, in the case of fluorescent detection, far greater resolution, thereby permitting mapping of probes to specific regions of chromosome bands *(10)*.

2.2. Problems and Their Solutions

The principles that underlie probe design and labeling, hybridization, and detection are largely derived from nucleic acid hybridization studies involving other formats, such as solution-phase hybridization and blotting techniques. Although ISH uses these same principles, there is associated with it an overlay of protocol factors that relate to the requirements of cytological preparations. The most significant problems are:

1. Steps must be taken to ensure adhesion of the specimen to the slide through numerous washes and temperature changes.
2. Access of the target by the probe must be accommodated. The principal barriers to be overcome are the limited porosity of the cell membrane and cross-linked cellular proteins caused by the use of fixatives such as formalin.
3. Inadvertent digestion of RNA targets by cellular or contaminating RNase must be avoided.
4. Interference of RNA detection caused by "background" cellular DNA may occur.

The solutions to these problems are specified in **Table 1**.

Table 1
Protocol Accommodations for *In Situ* Hybridization

Problem	Solution
Sample could be lost as a result of numerous washes and temperature changes.	Pretreat microscope slides with adhesive, such as aminoalkylsilane.
Target accessibility to probe may be restricted because of limited permeability of cell membrane and crosslinking of cellular proteins during fixation.	Alcohol, protease, or lipase treatment may be used to permeabilize cells. Proteolytic digestion is best for negation of crosslinking. DNA is more difficult to completely unmask because of histone complexing.
Nuclease digestion of targets by endogenous RNase or DNase or by contaminating RNase.	Use nuclease-free solutions and glassware. Chelaters, such as EDTA, may be used to reduce nuclease activity by binding divalent cations.
Nonspecific binding of probe to nontarget DNA and RNA. may provide interfering background. Also, cellular electrostatic forces may nonspecifically attract the probe.	Sonicated and denatured salmon sperm DNA (or other anionic macromolecules) may be used to reduce nonspecific probe interaction and electrostatic forces. The latter also may be reduced with dextran sulfate. High-stringency (low-sodium) hybridization ensures that complete complementarity will characterize the probe-target hybrid.

2.3. Permutations of Basic ISH

2.3.1. Multicolor FISH

Two common fluorochromes on different probes and conventional fluorescence microscopy permit an investigator to resolve signals that are very close to each other. When investigating such juxtaposed targets, the probes often are differentially labeled, one with biotin and the other with digoxigenin and with the fluorochromes FITC and rhodamine. A third signal (yellow) can be produced by an overlay of FITC (green) and rhodamine (red) or by the use of a blue coumarine dye. Lichter and Ried *(10)* have used various combinations of three colors to detect seven target areas in chromosome spreads.

Probe 1,	Fluorochrome A
Probe 2,	Fluorochrome B
Probe 3,	Fluorochrome C
Probe 4,	Fluorochrome A + B
Probe 5,	Fluorochrome A + C

Probe 6,	Fluorochrome B + C
Probe 7,	Fluorochrome A + B + C
Counterstain,	Fluorochrome D

The use of a sensitive camera system, such as a cooled charged-coupled device (CCD) camera, facilitates resolution and recording of the image, including the pseudocolors generated by probes 4–7.

A variation of this approach is multicolor spectral karyotyping, which has proved to be effective in unmasking some hidden chromosome abnormalities in hematological malignancies *(5)*. Spectral karyotyping (SKY) involves the hybridization of 24 fluorescently labeled chromosome painting probes and the visualization of the red-blue-green differential color display of all chromosomes. Image requisition entails the use of a SD 200 Spectracube (Spectral Imaging), an optical filter, a microscope, and an interferometer in the optical head. An interferogram is generated at all image points that are deduced from the optical path difference of light which, in turn, depends of the wavelength of the emitted fluorescence. Thus, each chromosome is assigned a pseudocolor which, when combined with chromosome classification based on size and shape, permits accurate identification of translocations and other aberrations.

2.3.2. PCR-ISH

Amplification of low-abundance cellular targets in specimens affixed to a slide, followed by ISH, is the basis of PCR-ISH. The target to be amplified may be DNA or RNA. This approach has proved invaluable for the detection of low-abundance genes, transcripts, and viruses. Often, the detection of mutant genes, abnormal expression, and viruses can be made by PCR-ISH long before traditional pathology analyses can reveal an abnormality *(6)*.

The PCR phase of such a study is very similar to PCR amplification of DNA in a reaction tube placed in a thermocycler. Here, however, the specimen is affixed to a slide; covered with the reactants (buffer, DNTPs, primers, and a thermostabile DNA polymerase); and placed on the heating block of a traditional or modified (for slides) thermocycler programmed to provide the optimum temperatures for denaturation, primer annealing, and extension. After amplification for 20–30 cycles, the specimen is processed for ISH.

Reverse transcription (RT)-PCR ISH is based on RT-PCR. The transcripts in a cell are used to make cDNA by reverse transcription, followed by PCR amplification of the cDNA. Details of these methods and representative applications are covered in Chapter 49.

2.3.3. Combining Immunocytochemistry with In Situ *Hybridization*

Immunocytochemistry, like ISH provides specific localization of target molecules within cells or tissues. Hence, it too can be used to correlate structure with target presence and to discern differences among cells in a heterogeneous population. The targets of immunocytochemistry are protein antigens, and the probes are labeled antibodies; the highest specificity is provided by monoclonal antibodies. Other similarities between the two methods include similarity of signal generating systems (fluors, enzymes, chemiluminescent signals) and similarity of procedures to enhance cell adhesion and cell permeabilization *(11)*.

Specimens to be analyzed may be sequential sections, each one analyzed by one of the methods or the same specimen subjected to a protocol with immunocytochemistry steps followed by ISH steps.

Most often, the ISH phase is done last because some antigens may be destroyed by the high temperature and protein digestion of ISH. Conversely, the antibodies used for immunocytochemistry may have some RNase carryover that will diminish transcript targets. The addition of RNase inhibitors, therefore, may be warranted *(12)*.

3. Controls

ISH has several trouble spots; hence, several issues must be addressed in adopting or establishing a protocol. Often, more than one control is called for. Some controls should relate primarily to the staining and signal generating system response and others to hybridization.

Target access: Try several methods to permeabilize cells, each method having a different basis.

Nonspecific hybridization: Does a noncomplementary probe of identical size and G-C composition hybridize? Does a nuclease-treated specimen show hybridization? Does protease treatment decrease hybridization?

Fidelity of hybridization: Does a reduction in stringency increase hybridization? Does an increase in stringency reduce hybridization? Does competition of hybridization with an unlabeled but identical probe reduce hybridization?

The design and use of positive and negative controls that are specific for the study could be an interesting and important challenge. Success in that regard will make the value of ISH be fully realized.

References

1. Gall, J. G. and Pardue, M. L. (1969) Formation and detection of RNA-DNA hybrid molecules in cytological preparations. *Proc. Natl. Acad. Sci. USA* **63,** 378–383.
2. John, H. A., Birnsteil, M. L., and Jones, K. W. (1969) RNA-DNA hybrids at the cytological level. *Nature* **223,** 582–587.
3. Buckle, V. J. and Craig, I. W. (1986) *In situ* hybridization, in *Human Genetic Diseases: A Practical Approach* (Davies, K. E., ed.), IRL, Oxford, pp. 85–100.
4. Franz, G. D. and Tobin, A. J. (1990) The use of *in situ* hybridization to study gene expression in mouse neurological mutants, in *In Situ Hybridization Histochemistry* (Chesselet, M.-F., ed.), CRC, Boca Raton, FL, pp. 71–110.

5. Veldman, T., Vignon, C., Schrock, E., Rowley, J. D., and Ried, T. (1997) Hidden chromosome abnormalities in hematological malignancies detected by multi-colour spectral karyotyping. *Nat. Genet.* **15,** 406–410.

6. Nuovo, G. J. (1994) *PCR In Situ Hybridization: Protocols and Applications.* 2nd ed. Raven, NY.

7. Lawrence, J. B. and Singer, R. H. (1985) Quantitative analysis of *in situ* hybridization methods for the detection of actin gene expression. *Nucleic Acids Res.* **13,** 1777–1799.

8. Kedershi, N. L. (1991) Photomicrography of immunofluorescently labeled cells. *Amer. Lab.* **23,** 28–32.

9. Van Prooijen-Knegt, A. C. and Raap, A. K. (1983) Spreading and staining human metaphase chromosomes on aminoalkylsilane-treated glass slides. *Histochem. J.* **14,** 333,334.

10. Lichter, P. and Reid, T. (1994) Molecular analysis of chromosome aberrations: in situ hybridization, in *Chromosome Analysis Protocols* (Gosden, J. R., ed.), Humana, Totowa, NJ, pp. 449–478.

11. Zeller, R., Watkins, S., Rogers, M., Knoll, J. H. M., Bagasra, O., and Hanson, J. (1995) *In situ* hybridization and immunohistochemistry, in *Current Protocols in Molecular Biology* (Ausubel, F. M., Brent, R., Kingston, R. E., Moore, D. D., Seidman, J. G., Smith, J. A., and Struhl, K., eds.), Wiley, New York, 14.0.3.

12. Hockfield, S., Carlson, S., Evans, C., Levitt, P., Pintar, J., Silberstein, L. (1994) Combining immunocytochemistry with in situ hybridization, in *Selected Methods for Antibody and Nucleic Acid Probes.* Cold Spring Harbor Laboratory Press, Cold Spring Harbor, NY, pp. 261,262.

47

In Situ Hybridization to Human Chromosomes of an Alkaline Phosphatase-Labeled Centromeric Probe

Roland M. Nardone

1. Introduction

The application of *in situ* hybridization and the underlying rationale are described in Chapter 46. *In situ* hybridization of nucleic acid probes to cellular targets of biological specimens affixed to a slide has benefited greatly from the concepts and strategies so elegantly developed under the umbrella of immuno-cytochemistry. These include direct and indirect affinity assays involving biotinylated nucleic acid probes and streptavidin complexed to an enzyme, indirect immunoassays and indirect immunoaffinity assays involving antibodies that recognize an antigen conjugated to a nucleic acid probe, and direct assays in which enzyme- or fluor-labeled probes are used *(1)*. The selection of a specific probe-detection system is usually dictated by the required sensitivity and resolution. For example, the indirect immunoassay and the indirect immunoaffinity assay use primary and secondary antibodies, thereby facilitating the formation of a branched network of antibodies and an increase in the amount of label that indirectly becomes associated with the probe-target hybrid. Fluorescent tags provide the greatest resolution, thereby optimizing localization. By using two or more fluorochromes simultaneously for two or more cellular targets, it is possible to discern the precise location of multiple targets in the same specimen *(2)*.

Probes tagged with an enzyme signal generating system, such as alkaline phosphatase or horseradish peroxidase, are sometimes used in direct assays. Alternatively, enzymes as reporters may be used with biotinylated probes in a direct affinity assay or in an indirect affinity assay. For the former assay, the

From: *Methods in Molecular Biology, Vol. 115: Immunocytochemical Methods and Protocols*
Edited by: L. C. Javois © Humana Press Inc., Totowa, NJ

reactants are DNA targets to which a biotinylated probe will hybridize. In turn, a streptavidin-enzyme complex will attach to the probe, thereby presenting the enzyme reporter close to the target. An indirect affinity version also uses a biotinylated probe to hybridize to the target. However, after hybridization, streptavidin, with its multiple binding sites, is allowed to react with the biotin of the probe. A biotin-enzyme conjugate is then allowed to react with the streptavidin, thereby forming a bridge between target and signal, the bridge consisting of biotin (on the probe), streptavidin, biotin (conjugated to enzyme), and enzyme as the signal generating system. In the presence of an appropriate substrate, a colored precipitate will form at the target site. This colored precipitate can then be visualized with a standard light microscope.

Hybridization of biotinylated probes and enzyme signal generators have found widespread application in localizing specific base sequences to specific chromosomes and to specific regions of chromosomes, especially when the target is repetitive or high-abundance. The protocol described herein should be viewed as a generic protocol entailing the use of a biotinylated probe and an enzymatic detection system. In this protocol, the enzymatic signal generating system is alkaline phosphatase, and the chromosomal targets are those base sequences that are common to the DNA in the centromeric regions of human chromosomes of a continuous cell line, HeLa. The substrate is tris [hydroxymethyl] aminomethane (T-E), which forms nitroblue tetrazolium/5-bromo—4-chloro—3-indolylphosphate (NBT/BCIP) a black-purple precipitate. The reaction is stopped by PBS/EDTA treatment.

1.1. Generic Review

As described in Chapter 46, virtually all *in situ* hybridization protocols entail the following major steps:

1. Tissue preparation and attachment of specimen to slide.
2. Prehybridization phase.
3. Hybridization.
4. Posthybridization.
5. Detection.

2. Materials

1. Log-phase culture of HeLa or some other rapidly growing mammalian cell line (American Type Culture Collections, Rockville, MD).
2. Colchicine stock solution: 10 μg/mL in complete nutrient medium (e.g., EMEM with 5% calf serum; Sigma, St. Louis, MO).
3. Trypsin-EDTA: 0.05% (Sigma).
4. Calcium-magnesium-free phosphate-buffered saline (CMF-PBS): Dissolve 8 g NaCl, 0.2 g KCl, 1.44 g Na_2HPO_4, and 0.24 g KH_2PO_4 in 800 mL of distilled H_2O. Adjust the pH to 7.4 with HCl. Add H_2O to 1 L, then autoclave.

5. 0.075 M KCl (hypotonic solution).
6. Absolute methanol:glacial acetic acid fixative (3:1; freshly prepared); 40% methanol in distilled water at 4°C: place microscope slides in this solution prior to using.
7. Diff-Quik stain set (Baxter Scientific, Muskegon, MI).
8. Permount.
9. Miscellaneous glass and plasticware, such as pipets, centrifuge tubes, microscope slides, and number-one cover slips.
10. Coplin jars.
11. CMF-PBS with 0.3% Triton-X.
12. Acetylation solution: 0.1 M triethanolamine and 0.25% (v/v) acetic anhydride (*see* **Note 1**).
13. 20X SSC stock solution: 175.3 g NaCl, 88.2 g sodium citrate, and 800 mL distilled water; adjust pH to 7.4 with 10 N NaOH, and then bring final vol to 1 L with dH$_2$O. Autoclave. Use to make 2X and 3X SSC by diluting with distilled water.
14. RNase solution (DNase-free) 100 µg/mL in 2X SSC.
15. Cold ethanol: 70, 80, 85, 95, and 100%.
16. Heat block at 90°C.
17. Hybridization and washing buffer: Hybrisol VII (Oncor, Gaithersburg, MD), which is a mixture of formamide (50%) in SSC (2X).
18. Biotinylated probe: 0.5 µg/mL in hybridization buffer.
19. Microfuge tubes.
20. Blocking buffer: 1X Tris-buffered detergent (TBD: 50 mM Tris, pH 7.5, 200 mM NaCl, and 0.3% Tween-20) with 3% bovine serum albumin.
21. Streptavidin (Boehringer Mannheim, Mannheim, Germany) blocking buffer (1 µg/mL).
22. Biotin-conjugated alkaline phosphatase (Boehringer Mannheim), in blocking buffer, 5 µg/mL.
23. Stain buffer: 100 mM Tris-HCl, pH 9.0, and 100 mM NaCl.
24. NTB/BCIP: 0.33 mg/mL nitroblue tetrazolium salt and 0.16 mg/mL 5-bromo-4-chloro-3-indolylphosphate in stain buffer.
25. T-E, pH 7.4: 10 mM Tris (tris [hydroxymethyl] aminomethane) and 1 mM EDTA in distilled water.
26. Basic Fuchsin: 0.5% in 5 mM Tris-HCl, pH 7.4.

3. Methods

3.1. Preparation of Chromosome Spreads (see Note 2)

1. To a log-phase monolayer culture of HeLa cells, add sufficient stock colchicine solution to give a final concentration of 0.06 µg/mL.
2. Incubate for 4–6 h.
3. Decant the supernatant and rinse with CMF-PBS.
4. Harvest the cells by trypsinization. Trypsinization is accomplished by decanting the medium from a flask of cells and adding 5 mL of trypsin-EDTA. Wait until the cells detach (approx 5 min), then transfer the cell suspension to a centrifuge tube. Pellet the cells by centrifugation at 500g for 5 min.

5. Resuspend the pellet in 5 mL of 0.075 M KCl for 12 min, and pellet again by centrifugation.
6. Decant the supernatant, and without disturbing the pellet, add 5 mL of methanol: glacial acetic acid fixative to it. Fix for at least 20 min.
7. Resuspend the cell pellet and centrifuge. Decant the supernatant, and resuspend the pellet in 5 mL of fresh fixative. Centrifuge again, and resuspend the pellet in 0.5 mL or less of fixative.
8. Remove, but do not dry, alcohol-cleaned slides from a 4°C bath of 40% methanol in distilled water.
9. Load the cell suspension in a Pasteur pipet, and affix the cells to several slides by either of the following methods:
   a. Add a few drops of cell suspension to a slide from a distance of about 12 in and air-dry; or
   b. Hold the slide at a 45° angle, and allow drops of cell suspension to tumble down the slide and air-dry.
10. In order to evaluate the frequency and quality of chromosome spreads, stain one of the air-dried slides following the instructions that accompany the Diff Quik stain set (total time about 1 min) or stain with Giemsa stain *(3)*.
11. Proceed to hybridization protocol using the other air-dried slides.

3.2. In Situ *Hybridization*

The prehybridization steps are best performed in Coplin jars.

1. Soak the slides with cells for 15 min in a solution of CMF-PBS with 0.3% Triton-X in order to rehydrate the cells.
2. Transfer the slides through two 5-min washes in CMF-PBS.
3. Prepare the acetylation solution for immediate and one-time use. Acetylate the slides by soaking for 10 min, and rinse in distilled water for 2 min (*see* **Note 1**).
4. Treat slides with RNase, 100 µg/mL in 2X SSC, for 1 h (*see* **Note 3**).
5. Rinse slides three times, 5 min/rinse, in 2X SSC.
6. Dehydrate the slides through cold (–20°C) ethanol washes, 2 min/wash: 70, 85, and 95%.
7. Place the slides on a 90°C heat block for 5 min to denature the DNA (*see* **Note 4**).
8. Immerse the slides in cold (–20°C) 70% ethanol for 2 min.
9. Dehydrate further in 2-min cold (–20°C) ethanol washes: 80, 95, and 100% (*see* **Note 5**).
10. Add sufficient biotinylated probe (0.5 µg/mL final concentration) to the hybridization buffer in a microfuge tube and heat to 90°C for 5 min (*see* **Note 6**).
11. Cool the tube on ice for 5 min.
12. Cover a region of a slide and a region of a coverslip with 40 µL of hybridization buffer, and mate the two, ensuring that no air bubbles are entrapped.
13. Place covered slide in a humidified chamber at 37°C for 1 h (*see* **Note 7**).
14. Following hybridization, wash slide with gentle agitation in 50% formamide: 2X SSC at 37°C for 12 min. The cover slip should dislodge.
15. Wash the uncovered slides in three changes of 2X SSC at 37°C, 4 min/wash.

3.3. Detection

1. Add three drops of blocking buffer to the slides and incubate at room temperature for 20 min.
2. Wash slides at room temperature by dipping through three changes of 0.3% Tween-20 in 1X TBD.
3. Add two drops of streptavidin (1 µg/mL) in blocking buffer to the slides and incubate at room temperature for 10 min.
4. Repeat **step 2** using fresh wash buffer.
5. Add three drops of biotin-conjugated alkaline phosphatase (5 µg/mL) in blocking buffer to the slides and incubate at room temperature for 10 min.
6. Repeat **step 2** using fresh wash buffer.
7. Add four drops of a freshly prepared solution of NTB/BCIP in stain buffer, and incubate at 37°C. Examine for development of blue stain after 90 min. If color is detected, stop the reaction by rinsing slides in T-E, pH 7.4. If blue color is not detected, resume incubation at 37°C, and reexamine every 10 min.
8. Counterstain with Basic Fuchsin for 5 min.
9. Rinse with distilled water, and air-dry.
10. Scan slides for centromeric staining (blue or purple).
11. Make slides permanent and suitable for viewing with an oil immersion objective by adding Permount and a cover slip.

4. Notes

1. Acetylation of slides prevents nonspecific, electrostatic binding of DNA probe to the cells. It is important that the acetylation solution be prepared immediately before use, with stirring, and that it not be reused. Prepare as follows: add 0.1M triethanolamine to a Coplin jar with a magnetic stirring bar. As stirring proceeds, add sufficient acetic anhydride to give a final concentration of 0.25% (v/v).
2. Having a sufficient number of good chromosome spreads for cytogenetic analysis and for hybridization requires that an actively proliferating cell population be treated with a metaphase blocking agent (colchicine) and with a hypotonic salt solution prior to affixing cells to a slide. The many processing steps for the slide require that the slide be pretreated to enhance attachment of the cells.

   Among the variables found in procedures for making chromosome spreads are the concentration of colchicine, the duration of treatment, the composition of hypotonic salt solution, the approaches to splattering cells on to slides, and the staining procedures. Adjustments in colchicine use may be required for other cell lines that may have a generation time significantly different from that of HeLa (approx 24 h) or that may be less or more sensitive to colchicine. Furthermore, chromosome banding staining usually calls for less-condensed chromosomes and, hence, shorter colchicine treatment. Almost invariable is the use of methanol-glacial acetic acid fixation for chromosome preparations.
3. RNase treatment eliminates RNA sequences that may compete for the probe.
4. In the absence of a heat block, use a treatment that is harsher on the chromosomes. Immerse the slides for 2 min in a solution of 70% formamide in 2X SSC at 70°C.

5. If hybridization is to be postponed, store the slides for up to 2 wk in 95% ethanol at –20°C.
6. The centromeric probe for human chromosomes is a cloned sequence, p82H, of the alphoid-repeated DNA family. The cloning and characterization of this sequence are described in *(4)*.
7. A simple humidified chamber can be constructed by enclosing in a sealable dish a piece of filter paper that has been soaked in 2X SSC. A foil tent covering the slides should be used to prevent condensation from damaging the slides.

Acknowledgments

This protocol is based on one used in the *in situ* hybridization course directed by Jan Kuzava Blancato and held at the Center for Advanced Training in Cell and Molecular Biology. Julie Brent and Dwayne Dexter contributed to development and support services.

References

1. Nardone, R. M. (1997) Nucleic acid *in situ* probing. *Mol. Biotechnol.* **7,** 165–172.
2. Veldman, T., Vignon, C., Schrock, E., Rowley, J. D., and Reid, T. (1997) Hidden chromosome abnormalities in hematological malignancies detected by multicolour spectral karyotyping. *Nat. Genet.* **15,** 406–410.
3. Wang, H. C. and Federoff, S. (1972) Banding in human chromosomes treated with trypsin. *Nature New Biol.* **235,** 52,53.
4. Mitchell, A. R., Gosden, J. R., and Miller, D. A. (1985) A cloned sequence, p82H, of the alphoid repeated DNA family found at the centromeres of all human chromosomes. *Chromosoma* **92,** 369–377.

48

Fluorescence *In Situ* Hybridization Using Whole Chromosome Library Probes

Roland M. Nardone

1. Introduction

Fluorescence *in situ* hybridization (FISH) expands the repertoire of user-friendly (nonradioactive), signal-generating systems, which may be used to identify and characterize chromosomes, interphase nuclei, and other sources of nucleic acids in biological material. FISH is based on the principles of nucleic acid immunocytochemistry described in Chapter 46. The signal-generating system entails the detection of hybridized biotin- or digoxigenin-labeled DNA probes using fluorochrome conjugates. Conventionally, the fluorochrome conjugates are used to label the probe before *in situ* hybridization. PRINS (primed *in situ* labeling), however, can be used to label the probe enzymatically after *in situ* hybridization *(1)*.

1.1. Representative Applications and Probes

The applications of FISH for chromosome-based studies involve the use of chromosome spreads, chromosomes of tissue sections, and interphase nuclei. Other applications include the study of adventitious organisms *(2)* and messenger RNA synthesis *(3)*. The applications, especially to chromosome spreads, are sometimes enhanced by exploiting the opportunity presented by FISH for double or triple labeling.

Table 1 lists representative examples of the application of *in situ* hybridization, including FISH, to studies that use chromosome preparations and interphase nuclei *(4,5)*. FISH is superior to *in situ* hybridization with an enzymatic probe because it provides finer resolution and higher signal intensity. This is especially important when the localization of a gene within a specific chromosome band is to be established.

From: *Methods in Molecular Biology, Vol. 115: Immunocytochemical Methods and Protocols*
Edited by: L. C. Javois © Humana Press Inc., Totowa, NJ

**Table 1
Applications of *In Situ* Hybridization
to Chromosomes and Interphase Nuclei**

Identification of chromosomes (human as well as mouse and hamster)
Determination of ploidy
Detection and characterization of aneuploidy
Gene mapping: localization of specific DNA sequences
Gene mapping: detection of heteromorphisms and determination of parental origin
  of heteromorphisms
Detection of gene mutations
Characterization of structural aberrations
Identification of marker chromosomes
Characterization of tumor cells
Chromosome and karyotype evolution
Spatial topography of chromosomes

**Table 2
Representative Chromosomal Probes**

Total chromosome "painting" probes, each specific for a different pair of homologs
Probes for the centromeric region of all chromosomes
Chromosome-specific centromeric probes (classical, midi-, α-, and β-satellite)
Telomere-specific probes
Probes for gene-specific sequences (e.g., IL-2)

FISH chromosomal probes, which are available commercially, exploit the existence of characteristic highly repetitive base sequences, either in specific regions of chromosomes, such as centromeres, or widely distributed throughout a specific chromosome. The probes may be plasmids or cosmid inserts (up to 40 kb), or specific *Alu* (highly repetitive) sequences that have been copied by polymerase chain reaction (PCR). Some of these probes are used singly or in combination (cocktails) to ensure signal enhancement and/or an increase in target number. An example would be a human chromosome "paint" probe cocktail, which can hybridize to many sequences along the length of a specific human chromosome *(6–8)*. Representative chromosomal probes for use with FISH are listed in **Table 2**.

The protocol described below entails the use of mitogen-activated peripheral leukocytes. The hybridization portion of the protocol can be used with metaphase-arrested HeLa cells (*see* Chapter 47).

1.2. Generic Review of Methods

Table 3 lists the major steps involved in *in situ* hybridization of labeled probes to chromosomal targets in cultured cells. The first and second steps entail the

Table 3
Major Steps for *In Situ* Hybridization to Chromosome

Block cells in metaphase
Use hypotonic treatment/fixation technique, and affix cells to slide
Treat cells with RNase (and/or proteinase K)
Denature DNA of chromosomes by temperature and salt adjustment
Add labeled probe at optimum concentration and at conditions optimum
    for hybridization (temperature, salt concentration); use optimum hybridization time
Remove unhybridized probe by washing
Process for visualization of hybridization
    For nonradioactive probe: fluorescence or enzymatic color development
    For radioactive probe: dip in photographic emulsion, expose, develop, stain
    Microscopic examination

preparation of chromosome spreads on slides. The third and fourth steps are prehybridization steps needed to optimize hybridization by elimination of interfering RNA and protein. The fifth and sixth steps are the hybridization and posthybridization steps, respectively, whereas the seventh step is the immuno-cytochemistry image-development step.

The image resulting from the use of an enzymatic signal-generating system or autoradiography may be visualized (eighth step) by the use of phase and/or bright-field microscopy. Fluorescent preparations require the use of a fluorescent microscope and appropriate filters for excitation and emission. The excitation and emission wavelengths, respectively, for three commonly used fluorochromes are as follows:

DAPI: 365 nm; >420 nm (blue)
Fluorescein: 495 nm; 525 nm (green)
Rhodamine: 552 nm; 570 nm (red)

Cellular autoradiography techniques using radioactive nucleic acid probes have several features in common with nucleic acid immunocytochemistry. The method is based on the hybridization of radioactive probes to cellular targets and the subsequent exposure of photographic emulsion, which, when developed, reveals blackened (exposed) silver grains close to the site of hybridization. Hence, cellular autoradiography techniques permit excellent specificity and localization of the hybridized probe—to 1 μm when tritium is the label used in the autoradiography-based method (*9*).

2. Materials

1. Fresh collection of 3 mL of heparinized blood.
2. Buffy coat (*see* **Note 1**).
3. 25 cm^2 Sterile tissue-culture flask.
4. Ham's F-10 medium supplemented with fetal bovine serum (10%).

5. Phytohemagglutinin (PHA) (Gibco-BRL, Gaithersburg, MD): reconstitute as directed and add sufficient quantity to Ham's F-10 medium (usually about 1–2 mL/100 mL).
6. Colcemid stock solution (Gibco-BRL): 10 µg/mL in phosphate-buffered saline (PBS).
7. Miscellaneous glass and plasticware, such as pipets, centrifuge tubes, microscope slides, and number-one coverslips.
8. 0.075 M KCl.
9. Centrifuge.
10. Methanol:glacial acetic acid fixative (3:1, freshly prepared).
11. Coplin jars.
12. Cold ethanol: 70, 80, and 95%.
13. 20X SSC stock solution: 175.3 g NaCl, 88.2 g sodium citrate, and 800 mL distilled water; adjust pH to 7.4 with 10 N NaOH, and then bring final vol to 1 L with dH_2O. Autoclave. Use to make 2X SSC by diluting with distilled water.
14. Formamide: 70% in 2X SSC; 50% in 2X SSC.
15. Shaker water bath.
16. Biotinylated whole-chromosome probe (Oncor, Gaithersburg, MD) in hybridization solution: For each slide, mix 1.5 µL of probe in 30 µL of 50% formamide in 2X SSC buffer to give a final concentration of 10 ng/µL. Keep at 37°C for 5 min, and then chill quickly in ice bath.
17. Rubber cement.
18. Humidified slide chamber (*see* **Note 2**).
19. PBS: Dissolve 8 g of NaCl, 0.2 g of KCl, 1.44 g of Na_2HPO_4, and 0.24 g of KH_2PO_4 in 800 mL of distilled water. Adjust pH to 7.4 with HCl. Add water to bring vol to 1 L and autoclave.
20. PBS with 0.1% Tween-20.
21. Blocking solution: PBS with dry milk (5%) and Tween-20 (0.1%).
22. Fluorescein isothiocyanate-labeled avidin (Vector Labs, Burlingame, CA): 5 µg/mL in PBS with 1% Tween-20.
23. Antiavidin antibody solution (Vector Labs): 5 µg/mL in PBS with 1% Tween-20.
24. Propidium iodide (Sigma, St. Louis, MO) counter stain: 1 µg/mL in PBS.
25. Antifade mounting medium: 100 mg *p*-phenylenediamine dihydrochloride in 10 mL PBS; adjust to pH 9 (*see* **Note 3**).
26. Nail polish.
27. Fluorescent microscope and filters for fluorescein images.

3. Methods

3.1. Preparation of Leukocyte Chromosome Spreads

1. Collect 3 mL of blood in a heparinized blood collection tube and cool.
2. Transfer the tube to a cold bath, and allow the erythrocytes to settle, or centrifuge the tube at low speed for 5–10 min.
3. Remove about 0.5 mL of the top most layer in the tube (the buffy coat of the plasma), and add it to a tissue-culture flask containing 5 mL of Ham's F-10 medium supplemented with 10% fetal bovine serum and 400 µL of phytohemagglutinin.
4. Incubate for 72 h at 37°C in a carbon dioxide incubator (*see* **Note 4**).

5. One hour before harvest, add sufficient colcemid stock to provide a final concentration of 0.8 µg/mL. Incubate.

6. Collect the contents of the flask, and pellet the cells by centrifugation at 300*g* for 8 min.

7. Decant all but about 0.2 mL of the supernatant, and resuspend the pellet using a Pasteur pipet.

8. Dropwise, and with gentle agitation, add 5 mL of warm 0.075 *M* KCl to the suspended cell pellet.

9. Incubate at 37°C for 10 min.

10. Pellet the cells by centrifugation at 300*g* for 8 min.

11. Discard the supernatant, and add, dropwise and with agitation, 5 mL of methanol–glacial acetic acid fixative.

12. Incubate at room temperature for 20 min.

13. Pellet the cells by centrifugation, decant the supernatant, and add dropwise 5 mL of fixative to the pellet.

14. Refrigerate overnight.

15. Pellet the cells by centrifugation, and decant all but 1.5 mL of supernatant, which is then used to resuspend the cell pellet.

16. Remove cleaned microscope slides from a bath of cold 80% methanol, and to each slide, add several drops of cell suspension (*see* **Notes 5** and **6**).

17. Immediately tilt the slides so that they are at a 45° angle to the ground, and run about 1 mL of fresh fixative down each slide.

18. Dry the slides at room temperature. Do not heat!

19. After drying, further dehydrate each slide for 2 min in a Coplin jar containing cold 70% ethanol. Shake periodically during the dehydration period.

20. Repeat using 80% and then 95% ethanol.

21. Air-dry the slides (*see* **Notes 7** and **8**).

3.2. Denaturation

1. Fill a Coplin jar with a denaturing solution of 70% formamide in 2X SSC, and heat to 70°C. Maintain the temperature.

2. Add slides to the Coplin jar, and incubate for 2 min (with shaking).

3. Without delay, transfer the slides to a Coplin jar containing 70% cold ethanol.

4. Repeat using 80 and 95% ethanol.

5. Allow slides to air-dry.

3.3. Hybridization and Posthybridization Wash

1. Chill the warmed probe–hybridization solution (10 ng/µL) in a 4°C ice bath.

2. Apply 30 µL of probe–hybridization solution to each slide.

3. Quickly overlay the slide with a number-one glass coverslip, and seal the edge with rubber cement.

4. Incubate the slides overnight at 37°C in a humid chamber.

5. Carefully remove the rubber cement and the coverslip from each slide.

6. Place the slides in a Coplin jar of 50% formamide in 2X SSC. Incubate, with shaking, for 15 min.

7. Transfer the slides to a Coplin jar containing 2X SSC at 37°C, and wash for 5 min. Do not allow the slides to dry out after this step.

3.4. Detection with a Fluorophore

1. Immerse slides in PBS containing 0.1% Tween-20.
2. Transfer the slides to the blocking solution (PBS with Tween-20 and 5% dry milk), and incubate for 5 min.
3. Add 60 µL of the fluorescein isothiocyanate-labeled avidin solution to each slide.
4. Incubate the slides for 20 min at 37°C in a humid chamber.
5. Wash the slides three times, 3 min/wash, in the PBS–Tween-20 solution.
6. Incubate slides for 5 min in the blocking solution.
7. Place 60 µL of antiavidin antibody solution on each slide.
8. Incubate for 20 min at 37°C in a humid chamber.
9. Repeat **steps 3–5** (*see* **Note 9**).
10. Counterstain the DNA of the interphase and mitotic cells by immersing the slides in the propidium iodide counterstain for 5 min at room temperature.
11. Wash the slides by immersing in distilled water for 5 min.
12. Air-dry the slides.
13. Mount the slides in antifade mounting medium.
14. Secure the corners of the coverslips with nail polish.
15. Air-dry for 1 h, and examine using a fluorescent microscope equipped with filters for fluorescein excitation and emission.

4. Notes

1. The buffy coat is the top most layer of plasma after erythrocyte sedimentation. It has a higher leukocyte:erythrocyte ratio than other regions of the plasma. Nevertheless, whole blood, without erythrocyte sedimentation, also may be used.
2. A simple humidified chamber can be constructed by enclosing in a sealable dish a piece of filter paper that has been soaked in 2X SSC. A foil tent covering the slides should be used to prevent condensation from damaging the slides.
3. Antifade mounting medium prevents quenching, and is especially useful when bright light is needed for photography or long-term examination of the specimen. A 9:1 mixture of PBS and glycerol also can serve as a mounting medium. However, its quenching properties are limited.
4. The mitogenic action of phytohemagglutinin on "resting" peripheral leukocytes is best seen about 72 h after addition of the mitogen. However, significant increases in mitotic cells can be seen at 48 h.
5. Glass slides are cleaned with a solution of 1% HCl in absolute ethanol and dried with a lint-free fabric.
6. Many variations exist regarding how the cell suspension is dropped on the slide. The variations include the height of the pipet and the angle of the slide.
7. When limited hybridization is expected, the chromosome preparations may be pretreated with RNase and/or proteinase K to eliminate interfering RNA or protein, respectively (*see* Chapter 47).

8. Other modifications that may be introduced here relate to preparation of chromosomes for banding. Procedures for specific banding techniques should be consulted.
9. The purpose of the repetition of these steps is to provide additional cycles of amplification of the signal.

Acknowledgment

This protocol is based on one used in the *in situ* hybridization course directed by Jan Kuzava Blancato and held at the Center for Advanced Training in Cell and Molecular Biology.

References

1. Koch, J., Hindkjaer, J., Mogensen, J., Kolvraa, S., and Bolund, L. (1991) An improved method for chromosome-specific labeling of alpha satellite DNA *in situ* using denatured double stranded DNA probes as primers in a PRimed IN Situ labeling (PRINS) procedure. *GATA* **8(6),** 171–178.
2. Harper, M. E., Marselle, L. M., Gallo, R. C., and Wong-Stahl, F. (1986) Detection of lymphocytes expressing human T-lymphotropic virus type III in lymph nodes and peripheral blood from infected individuals by *in situ* hybridization. *Proc. Natl. Acad. Sci. USA* **83,** 772–776.
3. Dirks, R. W., Van Gijlswijk, R. P. M., Vooijs, M. A., Smit, A. B., Bogerd, J., Van Minnen, J., Raap, A. K., and Van der Ploeg, M. (1991) 3'end fluorochromized and haptenized oligonucleotides as *in situ* hybridization probes for multiple simultaneous RNA detection. *Exp. Cell Res.* **194,** 310–315.
4. Harper, M. E., Ullrich, A., and Saunders, G. F. (1981) Localization of the human insulin gene to the distal end of the short arm of chromosome 11. *Chromosoma* **83,** 431–439.
5. Burns, J., Chan, V. T. W., Jonasson, J. A., Fleming, K. A., Taylor, S., and McGee, J. O. D. (1985) Sensitive system for visualizing biotinylated DNA probes hybridized *in situ*: rapid sex determination of intact cells. *J. Clin. Pathol.* **38,** 1085–1092
6. Cremer, T., Lichter. P., Borden, J., Ward, D. C., and Manuelidis, L. (1988) Detection of chromosome aberrations in metaphase and interphase tumor cells by *in situ* hybridization using chromosome specific library probes. *Hum. Genet.* **80,** 235–246.
7. Lichter, P., Cremer, T., Borden, J., Manuelidis, L., and Ward, D.C. (1988) Delineation of individual human chromosomes in metaphase and interphase cells by *in situ* suppression hybridization using recombinant DNA libraries. *Hum. Genet.* **80,** 224–234.
8. Pinkel, D., Landegent, J., Collins, C., Fuscoe, J., Segraves, R., Lucas, J., and Gray, J. W. (1988) Fluorescence *in situ* hybridization with human chromosome specific libraries: detection of trisomy 21 and translocations of chromosome 4. *Proc. Natl. Acad. Sci. USA* **85,** 9138–9142.
9. Choo, K. H., Brown, R. M., and Earle, E. (1991) *In situ* hybridization of chromosomes, in *Protocols in Human Molecular Genetics* (Mathew, C. G., ed.), Humana, Totowa, NJ, pp. 233–254.

49

In Situ RT-PCR and Hybridization Techniques

Vishakha Thaker

1. Introduction

In situ polymerase chain reaction (PCR) is a very powerful tool, which enhances our ability to detect minute quantities of a rare, single copy number, target nucleic acid sequences in freshly frozen or paraffin-embedded intact cells or tissue sections *(1–10)*. In 1986, the introduction of PCR methods opened new horizons and revolutionized research in all areas of molecular biology *(11,12)*. Dr. Hasse and his coworkers in 1990 used multiple primers and successfully amplified the target nucleic acid sequences in intact cells by combining a traditional *in situ* hybridization protocol with a powerful PCR technology *(13)*.

The recent introduction of new enzymes *(14,15)*, sealing reagent, and automatic *in situ* PCR machines have made this technique much easier and less time-consuming. This technique has been improved enormously in the past two years. It is possible to obtain reproducible results under carefully designed reaction conditions with the proper selection of reagents and equipment. *In situ* PCR technology has a tremendous potential for its applications in diagnostic histopathology, viral diseases, study of *in situ* gene expression, regulation, and mutation *(16–21)*.

The following protocol and discussion contains fundamental principles and as much detail as possible but remains a general outline of the procedures and practical considerations. Each individual experiment must be well planned, with sufficient theoretic contemplation given to the unique characteristics of the study target and experimental materials. Whenever possible, practical procedural tips have been included in an attempt to save time, trouble, and materials.

There are two different approaches to performing the *in situ* reverse transcription (RT)-PCR technique *(22–25)*.

From: *Methods in Molecular Biology, Vol. 115: Immunocytochemical Methods and Protocols*
Edited by: L. C. Javois © Humana Press Inc., Totowa, NJ

1. Direct *in situ* RT-PCR: Direct detection of the amplified products by incorporating digoxigenin-labeled 11-dUTPs during the PCR reaction, followed by binding with the Fab fragment enzyme-conjugated (alkaline phosphatase) anti-digoxigenin antibody and subsequent staining of the complex with the specific substrate Nitroblue-tetrazolium and 4-bromo-5-chloro-3-indolylphosphate (NBT-BCIP).
2. Indirect *in situ* RT-PCR: Indirect detection of the PCR amplified signal by hybridization with a digoxigenin-labeled oligo probe specific for the target followed by binding with the alkaline phosphatase-conjugated antidigoxigenin antibody and detection of the hybridized complex by staining with the specific substrates (NBT-BCIP).

2. Materials
2.1. Equipment and Glassware

1. Microtome.
2. 65°C oven.
3. *In situ* PCR machine (Thermal Cycler PTC-100™ M. J. Research, Watertown, MA).
4. Microcentrifuge.
5. Benchtop centrifuge.
6. Vortex machine.
7. pH meter.
8. Magnetic stirrer.
9. Hybridization oven.
10. A chemical hood.
11. A laminar flow or a tissue culture hood.
12. UV chamber or DNA station (M. J. Research).
13. Incubator.
14. Water bath.
15. Refrigerator.
16. Freezer.
17. A light microscope.
18. Autoclave and a baking oven for glassware.
19. Humidity chamber (Shandon Lipshaw, Pittsburgh, PA; or to make your own, *see* **Note 1**).
20. Thermometer.
21. Forceps, scalpels, labeling tapes.
22. Slide holders.
23. Glass and plastic Coplin jars or staining dishes (Fisher Scientific, Pittsburgh, PA).
24. 20 × 30-mm glass cover slips (Bellco Glass, Co. Vineland, NJ).
25. Plastic cover slips (PGC Scientific, Gaithersburg, MD).
26. Silanated slides (Digene Diagnostics, MD).
27. Disposable sterile polypropylene tubes (15- and 50-mL capacity).
28. RNase-free Eppendorf tubes (*see* **Note 2**).
29. Micropipetters.

30. Microtips (range 0.5–1000 µL).
31. Tube racks.
32. Measuring cylinders.
33. Funnels.
34. Ice buckets.
35. Slide holders.
36. Aluminum foil.

2.2. Reagents

1. Diethyl pyrocarbonate (DEPC) (Fluka Chemical Corp., Ronkonkoma, NY).
2. DEPC-treated water: 800 µL DEPC/L of distilled water; stir overnight or at least 1 h; autoclave (*see* **Note 3**).
3. 10% buffered formalin.
4. 4% paraformaldehyde (Fluka): heat 400 mL DEPC-treated 1X PBS to 65°C; add 16 g paraformaldehyde; stir with a magnetic stirrer until it dissolves (approx 1–2 h); adjust pH to 7.5 with 10 N NaOH (*see* **Note 4**).
5. RNASE Zap (Ambion, Inc., Austin, TX).
6. Xylene.
7. Absolute alcohol.
8. Graded alcohols: 50% ethanol: mix DEPC-water and absolute ethanol 1:1; 70% ethanol: mix DEPC-water and absolute ethanol 1:3; 95% ethanol: mix DEPC-water and absolute ethanol 5:95.
9. DEPC-treated PBS (10X, 3X, 1X); for 1L DEPC-treated 10X PBS: 23.5 g Na_2HPO_4, 4.5 g NaH_2PO_4, 87.6 g NaCl, 0.8 mL DEPC solution, 1000 mL dH_2O; autoclave for 20 min. Dilute with DEPC treated-dH_2O to make 3X and 1X.
10. UV-irradiated double-distilled water (UV-ddH_2O) (*see* **Note 5**).
11. Proteinase K (Fluka) or Pepsin (Boehringer Mannheim Biochemicals, Indianapolis, IN); for stock solution: 10 mg Proteinase K, 10 mL 1X DEPC-treated PBS; mix and store 1-mL aliquots in RNase-free Eppendorf tubes at –20°C.
12. 0.1 M glycine in 1X PBS: mix 7.5 g glycine/L of 1X DEPC-treated PBS.
13. 0.3% hydrogen peroxide (H_2O_2) solution in methanol: 1.0 mL 30% H_2O_2 and 99.0 mL absolute methanol.
14. Methanol.
15. 0.02 N HCl: mix 0.2 mL 12 N HCl and 119.8 mL DEPC-dH_2O (*see* **Note 11**).
16. Reverse transcription reagents (Life Technologies, Inc., Germantown, MD): antisense primer; ultrapure 4(dNTPs), 100 mM each; dithiothreitol (DTT); oligo dT; first-strand reaction buffer, 50 mM Tris-HCl, pH 8.3, 5 mM KCl, 3 mM $MgCl_2$; Superscipt II (Reverse Transcriptase); RNasin (Promega, Madison, WI); UV-irradiated dH_2O.
17. PCR reaction reagents (Perkin-Elmer and Life Technologies, Inc., Foster City, CA): Ultrapure 100 mM 4(dNTP) stock solution; UV-irradiated dH_2O; Digoxigenin 11-dUTP (Boehringer Mannheim); PCR cocktail, 10X PCR buffer, $MgCl_2$, a set of sense and antisense primers, BSA, Taq polymerase-Taq-start antibody complex (*see* **Note 6**).

18. Taq start antibody (CLONTECH Laboratories, Palo Alto, CA).

19. Digoxigenin-11-ddUTP (Boehringer Mannheim).

20. Hybridization solution (Boehringer Mannheim): Deionized formamide; 20X SSC; 100X Denhardt solution; 10 mg/mL Salmon sperm DNA; 10% SDS.

21. 2X SSC: 300 mM NaCl, 30 mM sodium citrate.

22. Detection buffers: 1 L Buffer 1: combine 12.11 g Tris; 5.84 g NaCl; 0.40 g MgCl$_2$; 30.0 g BSA; adjust to pH 7.5. 1 L Buffer 2: combine 12.11 g Tris; 5.84 g NaCl, 10.0 g MgCl$_2$; adjust to pH 9.5. 1 L Buffer 3: combine 2.42 g Tris; 1.86 g EDTA; adjust to pH 7.5.

23. Anti-digoxigenin antibody, a Fab fragment (Boehringer Mannheim).

24. 0.1 M Tris-HCl, pH 7.5: Combine 12.11 g Tris/L of DEPC-water and adjust the pH with pure concentrated 12 N HCl.

25. 0.1 M Tris/50 mM EDTA, pH 8: Combine 12.11 g Tris, 18.61 g EDTA/L of DEPC-water and adjust pH.

26. Nitroblue tetrazolium (NBT).

27. 4-bromo-5-chloro-3-indolylphosphate (BCIP).

28. Mounting medium (Aqua mount, Biomeda c/o Fisher Scientific, Inc., Pittsburgh, PA) (*see* **Note 17**).

3. Overview of Protocol and Fundamental Principles (*see* Notes 18 and 19)

3.1. Fixation and Sample Preparation

Experimental samples are mainly derived from tissue culture cells, laboratory animals, or human tissues collected from hospitals after surgical biopsies and autopsies. With human and animal tissue specimens, it is important to arrest metabolic processes within 5–10 min of collection in order to preserve mRNAs from degradation by internal enzymatic reactions (*26,27*). Most hospitals use 10% buffered formalin as a tissue fixative. Subsequently, each tissue slice is trapped in a paraffin block. Series of 4–5-µm-thick sections are cut and mounted on silanated slides. Formalin-fixed archival tissues have been successfully used in *in situ* PCR and *in situ* hybridization protocols (*28–32*). However, the procedure for RNA protection is not always followed. It is often difficult to alter or control the routine procedures of hospitals for the required protection of mRNAs in surgically removed human tissues.

Proper fixation is one of the most critical steps in an *in situ* RT-PCR or *in situ* hybridization experiment, because each tissue type must have optimized fixation conditions. Particularly archival tissues may require individual specific treatment in order to meet with *in situ* experimental requirements. Errors in fixation will only be discovered after the entire hybridization process has been completed (*see* **Note 10**).

When selecting the appropriate fixative, its possible effect on tissue morphology, target signal retention, and the PCR process (with particular regards to temperature sensitivity) must be carefully considered. You may choose between crosslinking or precipitating fixative types according to the tissue sample being used. You will know that you have achieved the best fixative results when the probe and reagents have sufficiently penetrated your tissue sample and have provided strong signal results, yet all of the target DNA or RNA has been retained and the morphology of the tissue sample is intact *(33)* (*see* **Note 11**).

Although the best probe penetration might be achieved with precipitating fixatives such as acetic acid or ethanol, imperfect tissue morphology and a loss of target signal may also result. Improved RNA retention and tissue morphology can be achieved with the use of aldehyde (cross-linking) fixatives and using a 4% paraformaldehyde solution is a common compromise *(34)*. The proper balance between target retention and tissue permeability must be determined by trial and error according to the target selected. Consideration should be given to the fact that mRNA is degraded enzymatically, whereas DNA is more stable. RNA studies require that the tissue be fixed or frozen within 10 min of collection in order to retain sufficient message, and the time elapsed between sample collection and fixation should always be accounted for when the results are interpreted.

The *in situ* RT-PCR process subjects tissue samples to various severe chemical and enzymatic reactions and temperature fluctuations. For that reason, silanated or positively charged slides must be used that are able to retain the tissue sample throughout the process.

Tissue culture cells are easy to handle under laboratory conditions. It is necessary to fix these cells immediately after cytospin or before paraffin embedding. Different fixatives are used depending on the goal of the experiment.

One should avoid the use of highly cross-linking fixatives, such as mercuric chloride, glutaraldehyde, modified formalin, and picric-acid-based fixatives. These extensively cross-linking fixatives render the tissue virtually impermeable to the RT-PCR reaction components and to the probe.

3.1.1. Preparation of Paraffin-Embedded Tissue Culture Cell Sections

For *in situ* mRNA detection:

1. Grow cells in five T-150 tissue culture flasks cells either in suspension or in monolayer.
2. When growth reaches 60–70% confluency, harvest cells by trypsinizing and spinning the cells at 500g in 15-mL centrifuge tube for 10 min.
3. Wash the cell pellet 3X with DEPC-treated 1X PBS.
4. After carefully aspirating the supernatant, add 10% buffered formalin or fix it in freshly prepared 4% paraformaldehyde for 4–16 h. After this step follow the standard guidelines for paraffin-wax embedding as described in **Subheading 3.1.2.**

3.1.2. Paraffin-Wax Embedding

1. Place tissues from laboratory animals or surgically removed human tissue samples in freshly prepared 4% paraformaldehyde for 4–16 h at room temperature.
2. Transfer the tissues in 0.5M sucrose in 1X PBS at room temperature for 8–12 h.
3. Process in the following:
   a. Normal saline: (0.85% NaCl), 15 min, two times.
   b. Normal saline: ethanol (1:1), 30 min, two times.
   c. Ethanol 70%, 30 min, two times.
   d. Ethanol 85%, 30 min, one time.
   e. Ethanol 95%, 30 min, one time.
   f. Ethanol 100%, 30 min, two times.
   g. Xylene, 20 min, three times.
   h. Xylene: paraffin (1:1), 20 min, two times.
   i. Paraffin, 20 min, three times.
4. Embed tissue samples in paraffin wax in embedding cassettes.
5. Place the embedded tissues at 4°C for 2–12 h.

3.2. Preparation of Positive and Negative Controls

For positive controls, it is extremely important to have a tissue culture cell line with an abundant amount of target nucleic acid. Varieties of tissue culture cell lines are available from American Tissue Culture Center, Rockville, MD. Rapidly growing tissue culture cell lines transfected with the target nucleic acid could be successfully used for positive controls. For example, formalin-fixed paraffin-embedded foreskin fibroblast tissue culture cell-line FS4 can be used as a positive control for human lysyl oxidase mRNA detection and c-H-*ras* transformed RS-485 cell line for normal *ras* message detection. Thin sections (4–5 μm) of paraffin-embedded positive control cell lines could be placed simultaneously near the side of the experimental human or animal tissue. Similarly, one can also use freshly cut animal or human tissue (with the target sequences) preserved, sectioned, and mounted under standard laboratory conditions. For negative controls, choose a cell line (or a tissue) that completely lacks the target nucleic acid (*see* **Note 12**).

3.3. Glass Slide Preparation

While performing the protocol of *in situ* RT-PCR on the tissue sections, each tissue undergoes a variety of relatively harsh chemical and enzymatic treatments and temperature shocks. Hence, the tissue or cells should be mounted on slides that can best retain and hold the specimen under relatively harsh chemical and temperature treatment.

Varieties of pretreated precoated slides are commercially available. Your choice of coating on the slide can play an important role. I started my work with Teflon-coated three-well slides but soon realized that the area available for mounting the tissue was much smaller than the tissue sizes from the human tissue blocks. It was also very time-consuming and inconvenient to seal three-well Teflon-coated slides with a clear nail polish. I personally use commercially made silanated slides from Digene Diagnostics, Silver Spring, MD. As an alternative, Dr. Bagasra's protocol for silanating plain slides works very well *(35)*.

3.3.1. Specimen Mounting

1. Cut formalin- or paraformaldehyde-fixed paraffin-embedded tissue into 4–5-μm-thick sections on a microtome.
2. These tissue sections are floated in a nuclease-free clean waterbath filled with DEPC-treated water.
3. Place each section on a silanated slide by scooping the slide under the floating tissue section and lifting the slide with the attached tissue section up and out of the waterbath.
4. With a soft paint brush gently remove water bubbles and foldings of the tissue. After mounting sections to the slides, allow to air-dry.
5. The day before the experiment, bake slides at 65°C overnight.

3.4. Pretreatment/Proteinase K Digestion

Tissues fixed with crosslinking fixatives require extensive pretreatment to permeabilize the tissue in order to allow the reverse transcription and polymerase chain reaction components access to the target nucleic acid sequences. The success of the experiment depends on the careful optimization of permeabilization and proteinase K digestion *(36)*. Overdigestion with proteinase K can cause breakage in the tissue morphology, and tissues might become fragmented and partially or completely fall off the slide. Excessive pretreatment can also cause a leakage of the amplicon (PCR-amplified product) into other areas of the tissue, causing ambiguous results *(1,27)*.

1. Refix the tissue in freshly prepared 4% paraformaldehyde for 1–4 h after xylene treatment and before proteinase K digestion (*see* **Note 13**).
2. For optimizing pretreatment conditions, treat a series of tissues (sectioned from the same tissue block) with different concentrations of proteinase K (5, 10, 15, 20, 25, 30 μg/mL) and incubate for 15 min at 37°C in the humidity chamber.
3. At the end of incubation, observe the integrity of the tissue morphology under the light microscope. One should choose the highest concentration of proteinase K that can retain the intact tissue morphology. Overdigested tissues that show distortion, fragmentation, wavy texture, or a loss of tissue morphology because of partial detachment of the tissue from the slide should be discarded (*see* **Note 14**).

3.5. DNase Digestion

It is necessary to treat the tissues with highly pure RNase-free DNase to destroy all the endogenous DNA in the cells and tissues so that only RNA remains available for cDNA synthesis and amplification. This avoids problems associated with the incorporation of labeled nucleotides into DNA through repair mechanism during the reaction cycles of *in situ* RT-PCR and prevents nonspecific amplification of DNA gene sequences by Taq DNA polymerase *(37)*.

3.6. Primer Design

A set of sense and antisense primers should be selected to synthesize specific cDNAs and also to detect the amplified messages of the genes, complementary to their specific gene sequences. It is important to consider the following points while designing the primers for reverse transcriptase and polymerase chain reaction:

1. Length of sense and antisense strands should be between 20 and 30 bp.
2. At the 3' ends of both primers, there should be at least one CG-type basepair in any combination to facilitate complementary strand formation.
3. The GC content of the primers should be between 45 and 55%.
4. 3' ends of primers should not be complementary to avoid primer dimer formation.
5. Reverse transcription primers should be designed so that they do not contain secondary structures *(38)*.

3.7. Reverse Transcription Reaction (cDNA Synthesis)

In the cells and tissues, low- or single-copy number viral RNA or messenger RNA can be detected by *in situ* RT-PCR. Templates of mRNAs are transcribed to the first strand complementary DNAs (cDNAs) by incubating the tissue sections with the specific RT-reaction mixture. A variety of reverse transcriptase enzymes are available commercially. The first strand synthesis protocol varies depending on the type of RT enzyme and its specific reaction conditions. One should carefully follow the instructions recommended by the enzyme manufacturing companies. The optimum temperature for the first strand cDNA synthesis ranges between 42 and 55°C *(39)*. The cDNA is synthesized by reverse transcriptase in the reaction mix containing either antisense highly specific primer or random hexamers and free nucleotides (dNTPs), at approx 42°–50°C for approx 60 min. Subsequent heating at 70°C for 10 min and rapid cooling for 1 min keeps the cDNA products in a linearized form. The newly formed cDNA becomes the template for the polymerase chain reaction (*see* **Notes 15** and **16**).

3.8. The Polymerase Chain Reaction

The cDNAs, which are assumed to be synthesized at the end of the RT reaction, are subjected to amplification reaction by following two methods. The first, "Indirect *in situ* PCR" method, involves the incorporation of unlabeled nucleotides *(40)*. In the indirect method, the target cDNA is amplified by unlabeled nucleotides. This is immediately followed by the hybridization step with the digoxigenin-labeled oligo probe. The resulting digoxigenin-labeled hybridized product is then detected by using the target detection method. In the second, "Direct *in situ* PCR" method, digoxigenin-labeled 11-dUTPs are added into the cocktail of the regular dNTPs, and their ratio is maintained as recommended by the manufacturers (Boehringer Mannheim). This amplification does not require the subsequent hybridization step by labeled probe *(41)*. The digoxigenin-labeled amplified target can be directly detected by following the signal detection protocol. This protocol is relatively short and easy to perform. However, it takes a long time to optimize reaction conditions. There are more chances to get false-positive signals by the repair processes of nicked and damaged DNA. It is also less specific because it lacks the step of subsequent hybridization of the amplicon with a highly specific labeled probe *(42,43)*.

3.8.1. The Hot-Start Strategy

Initial annealing between the primers and the target sequence determines the amplification specificity. "Hot start" entails the initiation of a primer-target annealing step at a higher temperature, which significantly reduces the possibility of mispriming and thereby improves the specificity of subsequent PCR *(44)*. Under the *in situ* conditions, the "hot-start" could be performed by adding into the PCR reaction system either the enzyme AmpliTaq Gold (specially designed for "hot start" Perkin-Elmer) or the complex of Taq polymerase enzyme-Taqstart™ antibody (PT1576-1, CLONTECH). Each step of the protocol should be handled carefully (*see* **Notes 17** and **18**).

3.9. Hybridization

For indirect detection of unlabeled RT-PCR products, slides are hybridized with a digoxigenin-labeled probe. The probe should be chosen so that it spans an exon-intron junction, ensuring that the hybridization occurs only with message-derived products and not with fragments of genomic DNA. Synthetic oligo probe detects only one of the two amplified target cDNA sequences. I have used 3' end digoxigenin-labeled synthetic oligonucleotide probe in my indirect *in situ* RT-PCR experiments.

3.9.1. Preparation of the Probe

In the tissue sections, a hybridization is performed with a highly specific probe in order to detect the target (either amplified by indirect *in situ* RT-PCR or unamplified). For the preparation of nonradioactive digoxigenin-labeled probes, one should refer to the guidelines of the company manuals: Genius™ System User's guide for membrane hybridization and Nonradioactive *in situ* hybridization application manual (2nd ed., Boehringer Mannheim).

For labeling synthetic oligonucleotides with digoxigenin, there are three methods developed by Boehringer Mannheim: 3' end labeling, 3' end tailing, and 5' end labeling. The PCR-generated digoxigenin-labeled probes are also being successfully used in hybridization-detection methods. I have used both 3' end labeled synthetic oligonucleotide probe (24 bp) as well as PCR-generated double-stranded cDNA probe (238 bp) to detect lysyl oxidase message in human prostate tissues. The gel- or HPLC-purified oligonucleotide probe is labeled at the 3' terminus with digoxigenin-11-ddUTP, using terminal transferase, but only one molecule of digoxigenin becomes attached. However, it is highly specific, and sometimes the signal develops after 12–24 h of incubation with the substrate. The hybridization with PCR-generated double-stranded cDNA probes (usually 200–300 bp) provides a higher concentration of digoxigenin to bind with the enzyme (alkaline phosphatase, horseradish peroxidase) or fluorescein (FITC)-conjugated antidigoxigenin antibodies. The intensified signal can be detected within 10 min to 2 h.

3.9.2. Posthybridization

After the hybridization step, the posthybridization wash conditions should be optimized for each type of probe, whereby an optimized stringent posthybridization wash should be balanced so that the much weaker and fewer hydrogen bonds between the probe and nontarget molecules are disrupted but enough below the melting point temperature (T_m) of the probe/target complex *(1)*. Under low-stringency wash conditions, oligo probes tend to produce more background than larger PCR-generated cDNA probes or genomic probes. Another difference between oligo probes and larger genomic probes relates to the actual conditions of the posthybridization wash. It is easy to achieve a temperature with larger probes for the posthybridization wash in which the specific signal persists and the background signal is lost. However, these conditions could be too stringent with the oligo probe. One must reduce the stringency until one reaches the narrow window above the Tm of the background hybridization yet below the T_m for the oligo probe-target annealing (*see* **Note 19**).

3.10. Signal Detection

Indirect *in situ* RT-PCR protocol requires a hybridization step for the detection of the amplified target cDNA products. Boehringer Mannheim color detection kits or separate detection reagents are available for direct or indirect detection of digoxigenin-labeled cDNA targets. The sample must first be blocked using blocking agent (either BSA or a sheep serum) to block any protein binding sites to which the antibody conjugate may bind. This is followed by incubation of a tissue sample using alkaline phosphatase- or horseradish peroxidase-conjugated antidigoxigenin antibodies. A freshly prepared substrate solution with NBT and BCIP in alkaline buffer solution is then added on the top of each tissue, and the chamber is placed in the dark for the development of the purple-blue-colored signal. The reaction is stopped by transferring the slides into a solution containing Tris-EDTA after observing under the microscope for the proper intensity of the colored signal. Excessive background can be removed by dipping the slides for 5–7 min in 95% methanol (*see* **Notes 20** and **21**). Finally, the slides are mounted with a mounting media (Aquamount/Crystal mount, Biomeda) (*see* **Notes 17**, **22**, and **23**).

4. Method

For normal and neoplastic human breast and prostate tissues.

4.1. Day 0

Wipe the entire work area with RNase Zap to destroy RNases.

Clean staining dishes, forceps, and slide holders; wrap them with an aluminum foil, and bake them at 350–400°C for approx 4–6 h.

Prepare following solutions:

1. 4% paraformaldehyde in 1X PBS (DEPC-treated). Store at 4°C.
2. 0.1*M* glycine in 1X PBS (DEPC-treated).
3. DEPC-treated deionized distilled autoclaved water (approx 4–5 L).
4. Deionized, distilled sterile water (3–4 L).
5. Deionized, distilled sterile double-filtered UV-irradiated water (5 mL).
6. 10X PBS, 3X PBS, 1X PBS (DEPC-treated).
7. Graded alcohols: Dilute 100% alcohol with DEPC-treated water to make 50, 70, and 95% alcohols.
8. Proteinase K solution: 1 mg/mL stock solution.
9. 100 m*M* Tris-HCl.
10. Program and clean *in situ* PCR machine, set oven at 65°C, clean waterbaths, and set the oven's temperature at 37°C.
11. Check the stock of reagents for reverse transcriptase and PCR reaction, RNase-free DNase, Kim wipes, parafilms, plastic coverslips.
12. Take 4 (minimum) to 16 (maximum) paraffin-embedded tissue sections mounted on the silanated slides, and place them into the 65°C oven, and melt the paraffin wax overnight.

4.2. Day 1

1. Increase the temperature of the oven from 65°C to 80°C to melt the remaining paraffin wax from the slides. Heat the slides at 80°C for 1 h.
2. Take two baked staining dishes. Place them in a chemical hood; pour xylene (sufficient enough to cover the slides) into the first dish and put a label on it. Add 100% absolute alcohol into the second dish and put a label on it.

3. At the end of the 1-h heat treatment, carefully transfer all slides from the oven, and place them in a slide holder.
4. Put slides in xylene for 7–8 min.
5. Dry slides in absolute alcohol for 5 min.
6. Repeat **steps 4** and **5**.
7. Pour freshly made 4% paraformaldehyde in a clean baked dish and fix the slides for 2–4 h.
8. Wash slides twice (5 min each) by gently moving the slide holder up and down in 3X PBS.
9. Wash slides twice (5 min each) in 1X PBS.
10. Dehydrate slides in graded alcohols: 50, 70, 95, and 100% for 2 min each.
11. At this point, slides could be stored at –70°C in a desiccator for at least 6–12 mo, or label the slides with appropriate labels and continue the experiment.

4.2.1. Proteinase K Digestion

1. In a clean, RNase-free humidity chamber, add some sterile DEPC-treated water or wet some paper towels and place them on the bottom of the humidity chamber.
2. Place the humidity chamber in a 37°C water bath or in an incubator to bring the temperature to equilibrium.
3. Rehydrate slides in graded alcohols: 100, 95, 70, and 50% for 2 min each.
4. Place a clean sheet of aluminum foil on a bench top work area. Carefully remove each slide from the slide holder. Wipe remaining alcohol around the periphery of the tissue with Kim-wipe and place it on the clean sheet of aluminum foil.
5. Place all the slides in a humidity chamber at 37°C for 10 min.
6. Thaw the stock solution of proteinase K. Make the following concentrations of proteinase K in different tubes: 5, 10, 15, 20, 25, 30, and 50 µg/mL.
7. Optimize conditions for proteinase K digestion by adding 250 µL of different concentrations of the enzyme on different slides of the same tissue block. Incubate for 15 min at 37°C.
8. Remove the humidity chamber from the incubator, drain off the enzyme by tilting the slide on a paper towel. Inactivate the enzyme activity by putting the slides into a staining dish with 0.1 *M* glycine solution for 20 min.
9. Wash slides for 5 min in DEPC-treated 1X PBS.
10. Dehydrate in graded ethanols: 50, 70, 95, and 100% for 2 min each.
11. Observe these slides under the light microscope to check integrity of tissue morphology, fragmentation, breakage, or distortion of the tissue section.
12. Make a judgment about the optimum proteinase K concentration for different tissue sections, and select that particular concentration for a future final experiment.
13. Sometimes it is a good approach to check the quality of tissue fixation at this stage (*see* **Note 24**).
14. Select the following tissue slides (sectioned from one tissue block) after proteinase K digestion: one with suboptimal digestion; one with optimum digestion; and one with over digestion.

15. Add 100 µL of 1X PCR reaction buffer on the top of the each slide, cover the tissue with either a clean plastic or a glass cover slip, and seal with rubber cement (*see* **Notes 18** and **25**).
16. Place these slides in a Thermal Cycler and heat them for 10 min at 95°C.
17. Peal off the rubber cement, remove the cover slip, stain the slides for 5 min with 1% Eosin, and observe them under the light microscope. An intact tissue morphology will confirm the quality of slides and fixation (*see* **Note 11**).

4.2.2. DNase Digestion

1. Select slides with optimized proteinase K digestion and follow the steps below for DNase digestion. Prepare in an Eppendorf tube: 1.0 µL RNasin (RNase inhibitor), 11.5 µL UV-irradiated dH_2O, 37.5 U DNase (RQ1 RNase free DNase, Promega Inc., Madison, WI)/(1 U/µL); Total (per slide): 50.0 µL.
2. Depending on the size of the tissue sections, add RQ1 RNase free DNase (600–750 U/mL dH_2O) to the sections (50–100 µL).
3. Immediately cover the tissue with a rectangular piece of parafilm (slightly larger than a size of a tissue), with the unexposed side down.
4. Incubate the slides overnight (12–16 h) at room temperature (25–28°C) in the humidity chamber.

4.3. Day 2

1. Following the incubation, carefully remove the parafilm coverslip from the top of the each tissue section with nuclease-free, sterile forceps.
2. Wash the slides twice with DEPC-treated 1X PBS and dehydrate in graded ethanols: 50, 70, 95%, and absolute ethanol, for 2 min each.

4.3.1. Reverse Transcriptase Reaction

1. Mark control slides and keep them in Coplin jar with 100 mM Tris-HCl, pH 8.0 until the PCR step.
2. Add to each of the other sections 65 µL H_2O + 5 µL oligo (dt). Cover with parafilm and incubate for 10 min at 70°C in the Thermo Cycler.
3. Incubate 1 min on ice.
4. Prepare RT reaction mix as follows: 250–500 ng Anti-sense primer, 0.5 mM dATP, 0.5 mM dCTP, 0.5 mM dGTP, 0.5 mM dTTP, 0.01 M Dithiothreitol, 20 U (Promega) RNasin, 50 mM, pH 8.3 Tris-HCl, 75 mM KCl, 3 mM $MgCl_2$, 400 U Superscript II; UV-irradiated dH_2O to total 60 µL.
5. Place 60 µL of the RT reaction mix onto each tissue slide.
6. Seal tissue sections with clean baked cover slips (20 × 30 mm) using rubber cement (*see* **Note 18**).
7. Leave sealed slides on the bench top for 5–7 min to dry the rubber cement.
8. Insert slides into a Thermal Cycler and run a program for RT reaction.
9. Incubate slides at 42°C for 1 h immediately followed by raising the temperature to 70°C for 10 min in a Thermal Cycler.

10. At the end of the incubation, quickly remove all slides from the Thermal Cycler and place on ice for 1 min.
11. Remove cover slips with forceps, wash briefly in DEPC-treated 1X PBS and dehydrate in graded ethanols.

4.3.2. Polymerase Chain Reaction

The conditions for this step have to be set up for every primer by standard tube (in vitro) PCR. The concentration of the different components and the temperature cycle should be the same (or very similar) for both techniques. The number of cycles is lower for *in situ* PCR, because otherwise one would obtain strong staining that could prevent visualization of the fine morphological details of signal localization.

4.3.2.1. INDIRECT METHOD

1. Preparation of Taq polymerase-Taqstart antibody complex (*see* **Note 26**): In 0.5-μL Eppendorf tube, mix the following (for 10 PCR reactions, 50 μL each): 4.4 μL Taqstart-antibody, 17.6 μL Dilution Buffer, 4.0 μL Taq polymerase. Incubate the mixture at 22°C for 5 min.
2. Preparation of 10-mM dNTP stock solution: Mix in an Eppendorf: 60.0 μL UV-irradiated dH$_2$O, 10.0 μL dATP (100 mM stock), 10.0 μL dCTP (100 mM stock), 10.0 μL dGTP (100 mM stock), 10.0 μL dTTP (100 mM stock), for a total volume of 100 μL.
3. PCR cocktail for indirect *in situ* PCR: 10.0 μL 10X PCR buffer (100 mM Tris-HCl, pH 8.3, 500 mM KCl), 3.0 μL MgCl$_2$ (50 mM stock), 100 pmol Sense strand primer, 100 pmol Antisense primer, 2.5 μL dNTP mix (10 mM stock), 2.0 μL BSA (0.02% stock), 10 U Taq polymerase-Taqstart antibody complex, UV-irradiated filtered dH$_2$O to make up a total volume of 100 μL.
4. Cover the sections with the PCR reaction cocktail (quantity of the reaction mix should be proportional to the size of the tissue).
5. Carefully place nuclease-free cover slip (siliconized) on the top of the tissue (*see* **Note 17**).
6. Carefully seal the edge of the cover slip with rubber cement and leave slides on bench top for 5–10 min to dry the rubber cement (*see* **Note 15**).
7. Prepare following control PCR reaction mix for the control slides:
   a. Omit reverse transcriptase reaction.
   b. Omit primers in PCR reaction mix.
   c. Omit Taq polymerase.
   d. Add only one primer and another unrelated primer.
   e. Use RNase-treated tissue section for PCR.
   f. RT positive, PCR positive, and hybridize with an unrelated probe.
   g. RT negative, PCR negative, and hybridize with a specific sense probe.
8. Cover each control slide with an appropriate reaction mix and seal with rubber cement and a coverslip.
9. Place all slides into the Thermal Cycler.

10. Run the PCR program (*see* **Note 19**).
11. Program the Thermal Cycler to keep the slides at 22–25°C at the end of the PCR cycles (if you plan to leave the slides overnight in the machine).

4.3.2.2. DIRECT METHOD

In the case of direct *in situ* RT-PCR, digoxigenin-labeled 11-dUTP is added in the 4(dNTP) mix. The recommended concentration ratio of digoxigenin 11-dUTP to dTTP is 1:19 (Boehringer Mannheim).

4.4. Day 3

1. Carefully remove the rubber cement and cover slip from each slide.
2. Heat the slides at 92°C for 1–2 min either in a Thermal Cycler or on a heat block in order to immobilize the signal.
3. Place them in a slide holder and treat with DEPC-treated 1X PBS for 5 min.
4. Dehydrate in graded alcohol: 50, 70, 95, and 100% for 2 min each.
5. Add 2 ng/µL digoxigenin-labeled oligo probe into the following hybridization solution: 50% deionized formamide, 10 mL 20X SSC, 10X (final concentration) 100X Denhardt solution, 1 mg/mL (final concentration) 10 mg/mL salmon sperm DNA, 1% (final concentration) 10% SDS.
6. Add 30–40 µL of probe containing hybridization solution to each tissue section treated with the *in situ* RT-PCR reaction protocol. Cover with siliconized glass coverslips, seal with rubber cement.
7. Heat the slides at 95°C for 10 min in a Thermal Cycler.
8. Incubate slides overnight (for 8–12 h) at 37–42°C in a humidity chamber.

4.5. Day 4—Signal Detection

1. Remove coverslip and wash each slide in 2X SSC at 48°C for 20 min.
2. Soak slides in buffer 1 at room temperature for 10 min.
3. Wipe the slide dry around the tissue but keep the tissue area wet. Put onto the each section 200 µL of antidigoxigenin antibody (a Fab fragment), diluted 1:200 in buffer 1. Incubate at least for 2 h at room temperature.
4. Wash twice in buffer 1 (5 min each).
5. Equilibrate the sections in buffer 2 for 10 min at room temperature.
6. Prepare substrate solution: 10.0 mL buffer 2, 27.7 µL Nitro-blue tetrazolium (NBT), 22.5 µL BCIP, 6.0 drops Levamisole (1 mM).
7. Carefully wipe the area around each tissue section, place the slides flat in the humidity chamber, and check the horizontal level of the chamber.
8. Add 250 µL substrate solution to each tissue section and incubate in the dark. Check under the microscope until the development of the color (bluish purple) is complete. It might take from a few hours to 24–48 h.
9. Stop the reaction by immersing in buffer 3 for 5 min.
10. Mount the slides in a water soluble mounting medium (e.g., Aqua mount).
11. Observe under the light microscope (*see* **Fig. 1**).

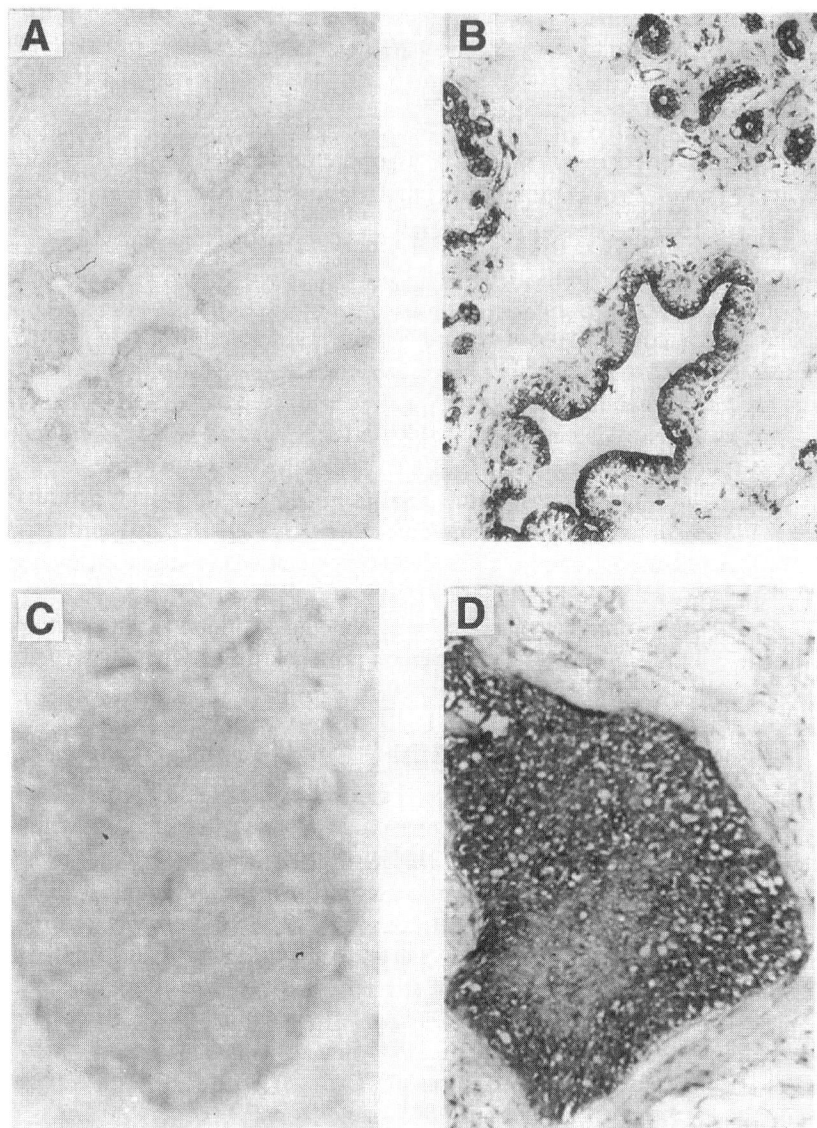

Fig. 1. Indirect *in situ* RT-PCR for *ras* mRNA in normal and neoplastic human breast tissues. **(A)** and **(C)** Negative controls for *ras* message. **(B)** Detection of *ras* mRNA signal in normal epithelial cells of mammary duct and lobules. **(D)** Detection of *ras* mRNA signal in neoplastic ductal epithelial cells in breast tissue with ductal carcinoma *in situ* (DCIS).

5. Notes

1. To make your own humidity chamber:
    a. Take a large (8" × 10") tightly sealed plastic box.
    b. Wipe thoroughly with RNase Zap solution to destroy RNases and rinse with DEPC-treated distilled water.
    c. Cut two or three filter papers or clean paper towels according to the size of the bottom of the plastic box and place flat on the bottom of the box.
    d. Pour DEPC-treated water uniformly to cover the entire surface.
    e. Cut several plastic pipets according to the length of the box and place them on the top of the wet filter paper in such a way that they can support approx 20 horizontal glass slides.
    f. Cover with the lid and wrap the entire box with plastic wrap.
2. RNase-free Eppendorf tubes are available commercially. Alternatively, soak regular Eppendorf tubes overnight in DEPC-containing distilled water. Next day, drain the water and place the tubes in a beaker or a glass jar and autoclave. Place the jar in an oven dryer at 80°C for 4–5 h.
3. **Caution: DEPC is highly carcinogenic!** Wear gloves and facial mask before handling. It is safer to purchase it in a solution form (Fluka Chemical Corp., Ronkonkoma, NY) and use a chemical hood while pipeting the aliquot.
4. Make a fresh solution for every new procedure, using a chemical hood. Do not use after 15–20 d.
5. UV-irradiated, double-distilled water (UV-ddH$_2$O): Filter autoclaved double-distilled water through 0.22-μm filter. Transfer 1 mL to 1.5-mL capacity RNase-free Eppendorf tubes. Without closing, expose them to UV irradiation for approx 30 min. Close the tubes inside the UV station. Store in –20°C freezer (make approx 20 tubes).
6. In the case of direct *in situ* PCR, digoxigenin-labeled 11-dUTP is added to the above 4(dNTP) mix. The recommended concentration ratio of digoxigenin 11-dUTP to dTTP is 1:19 (Boehringer Mannheim).
7. One should check the solubility of signal and counterstain in the mounting medium before selecting it. When water-soluble medium is used (e.g., Aqua mount or Crystal mount), many times, bubbles appear during long storage because of excessive drying. After mounting the slide with aqua mount, seal the edges of the cover slip with a colorless nail polish. This will protect the slide from drying and bubble formation.
8. Before starting experiments with human or animal tissue samples, it is extremely important to optimize in vitro experimental conditions. With a purified template nucleic acid, standardize RT and PCR conditions. Check the specificity and crossreactivity of primers and probes. Sometimes it is necessary to alter MgCl$_2$ concentration under *in situ* reaction conditions. The blocking reagent for filter hybridization could be different than the *in situ* protocol. (I use 1% purified casein solution for filter hybridization and 3% BSA for *in situ* signal detection.)
9. **Caution:** All specimen handling and subsequent procedures must be carried out using ultra pure, diethylpyrocarbonate-treated deionized distilled auto-

claved water (DEPC-water). All solutions, equipment, and glassware used for the pretreatment, RT-PCR protocols, and hybridization stages should be appropriately treated and baked to destroy nucleases, particularly ribonuclease, which is heat-stable.

10. If the goal is to detect mRNA, then optimize all positive control reaction conditions with a tissue that is fixed with special care for mRNA protection. This helps in optimizing subsequent reaction conditions and reproducing the data.

11. If the tissues come off the slide:
    a. Examine the original tissue block. It may have dried out. If possible, cut new tissue sections from a better block, or try to use tissue sections derived from the inner part of the tissue block.
    b. After xylene and alcohol treatment, keep the slides in 4% paraformaldehyde for 4 h or overnight at 4°C in 2% paraformaldehyde.
    c. Postfix the tissue for 10 min in 4% paraformaldehyde after proteinase K digestion.
    d. Check on the quality of the silanated slides.

12. In formalin-fixed paraffin-embedded tissues, severe nicking of nucleic acid targets may occur during fixation, chemical processing, or from nuclease activity prior to the tissue fixation. Apoptosis could also cause nicks in the DNA. The polymerase action of Taq polymerase enzyme is capable of repairing nicks *(1,23)*. When using "Direct *in situ* RT-PCR" protocol, detection of false-positive signal is probably a result of the incorporation of labeled nucleotides during nicked DNA repair by Taq polymerase. The appearance of false-positive signal may also be because of a detection system artifact, particularly in those tissues that have high levels of endogenous enzymes, such as alkaline phosphate or peroxidase. Negative results could result from the failure of the *in situ* RT-PCR reaction or overfixation of the tissue, which could have rendered target nucleic acid inaccessible to the reagents or the probe. Therefore, a large number of controls must be considered for every tissue under investigation *(1,23,45)*.

13. A noncrosslinking Streck's fixative is commercially available from Streck Laboratories (Omaha, NE). This fixative is especially formulated for *in situ* applications.

14. If the primer-negative control slides show intensive nonspecific positive signal, it could be because of the following: When the negative control tissue sections are inadequately digested with proteinase K, all DNA within the tissue does not get completely exposed to the subsequent step of overnight digestion with concentrated RNase-free DNase. Residual DNA that is protected underneath the undigested protein can provide sites for Taq polymerase for DNA synthesis. Repeat the experiment with optimum proteinase K digestion followed by adequate DNase digestion.

15. In 1995, Perkin-Elmer introduced a new enzyme rTth-DNA polymerase with a dual activity. It can perform both RT and PCR in the presence of manganese acetate buffer, sense and antisense primers, and nucleotides. This protocol is easier to perform and reduces total *in situ* RT-PCR reaction time *(46)*.

16. When the signal is detected partially on one part of the tissue, or there is no uniformity in the signal amplification pattern, check the following:

a. PCR or RT reaction mix might not have covered the entire area of the tissue.

b. There might be bubbles underneath the coverslip. When tissues are sealed with a light plastic coverslip, bubbles are formed during the heating steps of PCR cycles. As a result, the reaction mix does not act on or reach uniformly to the entire area of the target tissue. I recommend using thick (minimum 1-mm thickness) siliconized glass coverslips to seal the slides during PCR cycles. If you already have nuclease-free plastic coverslips, seal them with rubber cement, and on the top of the plastic coverslip put a good-size glass coverslip to create some weight in order to prevent bubble formation during PCR cycles.

c. Check the level of the humidity chamber. If the antidigoxigenin antibody solution or a substrate solution is not evenly distributed over the tissue, intensity of the signal will differ in the different areas of the tissue.

d. A significant delay in fixation after the surgery, and an improper handling of the tissue, can cause partial enzymatic degradation of mRNAs in the tissues.

17. Make sure that there are no air bubbles. I usually place 40–50 µL PCR reaction mix on the top of the coverslips. Invert one slide at a time, touch it to the drop of reaction mix, and quickly lift the slide. The coverslip should remain adhered to the slide because of surface tension. Flip the slide, center the coverslip with a clean pipet tip, and gently remove the air bubbles.

18. The sealed glass coverslip prevents evaporation of reaction mixture during PCR cycling. For sealing:

a. Buy rubber cement in tube form.

b. Remove the cap, and insert the mouth of the tube into the small piece of flexible plastic or transparent rubber tube (with same width and approximately 1" long).

c. Insert 1-mL pipet tip and fix it properly inside the flexible tube.

d. For sealing slides, place reaction mix onto the top of the coverslips, invert one slide at a time, and touch it to the drop of reaction mixture on the coverslip (the coverslip will remain adhered to the slide), then gently flip the slide.

e. Squeeze rubber cement tube very gently around the entire edge of coverslip.

f. Let it air-dry completely before placing slide into the PCR machine. Improper sealing can ruin the experiment.

19. An example of a PCR cycle would be:

a. 96°C. for 2 min.

b. 94°C for 1 min.

c. 55–60°C for 1 min.

d. 72°C for 0.45 s.

e. 20–25 cycles.

f. 72°C for 10 min.

20. Dirty patches or chromogen precipitates on the tissue:

a. If, while viewing the slides, you notice deposits of pale yellow crystals or patches over the entire tissue, it may be precipitates of partially dissolved bovine serum albumin from the blocking solution. I prefer to make concentrated blocking solution (10X) and store it in 15-mL tubes. Two hours before the blocking step, thaw two tubes in a 37°C waterbath and vortex several times before adding it to buffer 1.

b. Sometimes, the substrate reagents (buffer 2, pH 9.5 with NBT-BCIP) form dark blue-black precipitates during the detection step. Buy the product from a different company. (I have used Enzo Diagnostics, Inc. [Farmingdale, NY], Vector Laboratories, Inc. [Burlingame, CA], and Boehringer Mannheim's substrates). Filter and prewarm the substrate buffer solution at 37°C after the addition of NBT-BCIP.

21. If there is too much signal on the entire slide, check the following:

a. Concentration of the probe (it should be 2 ng/μL). Under standardized posthybridization wash conditions, a very highly concentrated unbound probe will not be washed away. Recalculate the probe concentration and increase the stringency of wash conditions.

b. Do not exceed the incubation time (>2 h) with diluted (1:200) antidigoxigenin antibody solution.

c. Destroy endogenous alkaline phosphatase by treating the slides with either 0.02 N HCl at room temperature for 8 min or 20% acetic acid for 20–30 s at 4°C for 20 s.

22. When control slides show optimum signal, but there is poor or no signal in the experimental slides, this may be because of:

a. Degradation of the target nucleic acid sequences during initial fixation.

b. Overfixation with 4% paraformaldehyde. Check the concentration of paraformaldehyde, reduce the fixation time, and reoptimize the proteinase K digestion.

c. Poor sealing of slides causing the evaporation of RT and PCR reagent mix from the slide during reaction cycles.

d. Improper preparation of reaction mixtures or suboptimal concentration of $MgCl_2$.

e. Inactive or old batch of enzymes.

f. Extremely stringent wash conditions. (Oligoprobes can be washed away under very high stringency.)

g. Drying of the antidigoxigenin antibody. (Always check the surface level of the humidity chamber before the addition of antidigoxigenin antibody solution.)

h. Selection of a wrong batch of tissues.

23. To remove coverslips from the slides:

a. If permount is used as a mounting medium, a slide should be soaked in xylene for 10–15 min. The treatment with xylene dissolves permount and the coverslip comes off.

b. When slides are permanently mounted with Aquamount, it is necessary to soak slides in hot water (85°C) for 30–40 min to remove the coverslip.

24. It is important to proceed further only if tissue retains its morphology intact after proteinase K digestion. Discontinue the experiment if the tissue is distorted. This saves your valuable time and reagents.

25. The size of coverslip should be bigger than the tissue and its width should be less than the width of the glass slide. I prefer 20 × 30-mm or 20 × 40-mm glass coverslips. Proper airtight sealing prevents evaporation of reagents and gives uniform signal over the entire surface of the tissue.
26. Preparation of Taq polymerase-Taq-start antibody complex: I mix 1:1 Taq-start antibody: Taq polymerase enzyme and incubate at 22°C for 10 min.

References

1. Nuovo, G. J. (1997) PCR in situ hybridization: protocols and applications, 3rd ed. Lippincott-Raven, Philadelphia, PA.
2. Komminoth, P., Long, A. A., Ray, R., and Wolfe, H. J. (1992) In situ polymerase chain reaction detection of viral DNA, single copy genes and gene rearrangements in cell suspensions and cytospins. *Diag. Mol. Pathol.* **1,** 85–97.
3. Ray, R., Komminoth, P., Machado, M., and Wolfe, H. J. (1991) Combined polymerase chain reaction and in situ hybridization for the detection of single copy genes and viral genomic sequences in intact cells. *Mod. Pathol.* **4,** 124A.
4. Nuovo, G. J., Gallery, F., MacConnell, P., Becker, J., and Bloch, W. (1991) An improved technique for the in situ detection of DNA after polymerase chain reaction amplification. *Am. J. Pathol.* **139,** 1239–1244.
5. Gosden, J. and Hanratty, D. (1993) PCR *in-situ* a rapid alternative to *in-situ* hybridization for mapping short low copy number sequences without isotopes. *Biotechniques* **15,** 78–80.
6. Martinez, A., Miller, M. J., Quinn, K., Unsworth, E. J., Ebina, M., and Cuttitta, F. (1995) Non-radioactive localization of nucleic acids by direct *in-situ* PCR and *in-situ* RT-PCR in paraffin-embedded sections. *J. Histochem. Cytochem.* **43,** 739–747.
7. Chieu, K. P., Cohen, S. H., Morris, D. W., and Jordan, G. W. (1992) Intracellular amplification of proviral DNA in tissue sections using the polymerase chain reaction. *J. Histochem. Cytochem.* **40,** 333–341.
8. Spann, W., Pachmann, K., Zabnienska, H., Pielmeier, A., and Emmerich, B. (1991) *In-situ* amplification of single copy gene segments in individual cells by polymerase chain reaction. *Infection* **19,** 242–244.
9. Bagasra, O., Sechamma, T., and Promerantz, R. J. (1993) Polymerase chain reaction *in-situ*: intracellular amplification and detection of HIV-1 proviral DNA and other specific genes. *J. Immunol. Meth.* **158,** 131–145.
10. Komminoth, P. and Long, A. A. (1993) In situ polymerase chain reaction. An overview of methods, applications and limitation of a new molecular technique. *Virchows Arch.* **64,** 67–73.
11. Rodu, B. (1990) The polymerase chain reaction: the revolution within. *Am. J. Med. Sci.* **299,** 210–216.
12. Eeles, R. A., Warren, W., and Stamps, A. (1992) The PCR revolution. *Eur. J. Cancer* **28,** 289–293.
13. Haase, A. T., Retzel, E. F., and Staskus, K. A. (1990) Amplification and detection of lentiviral DNA inside cells. *Proc. Natl. Acad. Sci. USA* **87,** 4971–4975.

14. Zevallos, E., Anderson, V., Novakoski, M., Bard, E., and Gu, J. (1994) Detection of human retinoblastoma gene expression by rTth-driven reverse transcribed in situ PCR. *Cell Vision* **1,** 88.

15. Kellogg, D. E., Rybalkin, Y., and Chen, S. (1994) Taq-start antibody: hot start PCR facilitated by a neutralizing monoclonal antibodies directed against Taq DNA polymerase *Biotechniques* **16,** 1134–1137.

16. Simsor, S. and Nuovo, G. J. (1995) PCR and RT-PCR in situ hybridization: application in viral detection. *Trends Biotechnol.* **11,** 13–23.

17. Embertson, J., Zupanic, M., Beneke, T., Till, M., Wolinsky, S., Ribas, J. L., Burke, A., and Haase, A. T. (1993) Analysis of human immunodeficiency virus-infected tissues by amplification and in situ hybridization reveals latent and permissive infections at single-cell resolution. *Proc. Natl. Acad. Sci. USA* **90,** 357–361.

18. Nuovo, G. J., Becker, J., Simsir, A., Morgiotta, M., Khalife, G., and Shevchuk, M. (1994) HIV-1 nucleic acids localize to the spermatogonia and their progeny: a study by polymerase chain reaction in situ hybridization. *Am. J. Pathol.* **144,** 1142–1148.

19. Kato, J., Hirata, S., Hagihara, K., Osada, T., Hirai, M., and Ikagami, J. (1992) Gene expression of steroid hormone receptors in brain and peripheral tissues relative to reproduction. *Acta Histochem. Cytochem.* **25,** 667–680.

20. Lev-Lehman, E., Ginzberg, D., Hornreich, G., Ehrlich, G., Meshorer, A., Eckstein, F., Soreq, H., and Zakut, H. (1994) Antisense inhibition of acetylcholinesterase gene expression causes transient hematopoietic alterations in vivo. *Gene Therapy* **1,** 127–135.

21. Shroyer, K. R., Brooks, C. G., Markham, N. E., and Shroyer, A. L. (1995) Detection of human papillomavirus in anorectal squamous cell carcinoma. *Am. J. Clin. Pathol.* **104,** 299–305.

22. Kumazaki, T., Hamada, K., and Mitsui, Y. (1994) Detection of mRNA expression in a single cell by direct RT-PCR. *Biotechniques* **16,** 1017–1019.

23. Chiocchia, G. and Smith, K. A. (1997) Highly sensitive method to detect mRNAs in individual cells by direct RT-PCR using Tth DNA polymerase. *Biotechniques* **22,** 312–314.

24. Greenwood, D., Yao, Wp, and Housley, G. D. (1997) Expression of the P2X2 receptor subunit of the ATP-gated ion channel in the retina. *Neuroreport.* **8,** 1083–1088.

25. Sanno, N., Jin, L., Qian, X., Osamura, R. Y., Scheithauer, B. W., Kovacs, K., and Lloyd, R. V. (1997) Gonadotropin-releasing hormone and gonadotropin-releasing hormone receptor messenger ribonucleic acids expression in nontumorous and neoplastic pituitaries. *J. Clin. Endocrinol. Metab.* **82,** 1974–1982.

26. Amarsham International Plc. (1994) A guide to radioactive & non-radioactive *in situ* hybridization systems.

27. Giang, Gu. (1995) *In situ* PCR and related technology. Eaton Publishing Co.

28. Isaacson, S. H., Asher, D. M., Gibbs, C. J., and Gajdusek, D. C. (1994) In situ RT-PCR amplification in archival brain tissue. *Cell Vision* **1,** 25–28.

29. Sebbelov, A. M., Svendsen, C., Jensen, H., Kjaer, S. K., and Norrild, B. (1994) Prevalence of HPV in premalignant and malignant cervical lesions in Greenland and Denmark: PCR and in situ hybridization analysis on archival material. *Res. Virol.* **145,** 83–92.

30. Drut, R. M., Day, S., Drut, R., and Meisner, L. (1994) Demonstration of Epstein-Barr viral DNA in paraffin-embedded tissues of Burkitt's lymphoma from Argentina using the polymerase chain reaction and in situ hybridization. *Ped. Pathol.* **14,** 101–109.

31. Nuovo, G. J. and Silverstein, S. J. (1988) Comparison of formalin, buffered formalin, and Bouin's fixation on the detection of human papillomavirus DNA from genital lesions. *Lab. Invest.* **59,** 720–724.

32. O'Leary, J. J., Browne, G., Landers, R. J., Crowley, M., Healy, I. B., Street, J. T., Pollock, A. M., Murphy, J., Johnson, M. I., Lewis, F. A., Mohamdee, O., Cullinane, C., and Doyle, C. T. (1994) The importance of fixation procedures on DNA template and its suitability for solution-phase polymerase chain reaction and PCR in-situ hybridization. *Histochem. J.* **26,** 337–346.

33. Greer, C. E., Lund, J. K., and Manos, M. M. (1991) PCR amplification from paraffin-embedded tissues: recommendation on fixatives for long term storage and prospective studies. *PCR Meth. Appl.* **95,** 117–124.

34. Greer, C. E., Peterson, S. L., Kiviar, N. B., and Manos, M. M. (1991) PCR amplification from paraffin-embedded tissues: effects of fixative and fixative times. *Am. J. Clin. Pathol.* **95,** 117–124.

35. Bagasra, O., Seshamma, T., Romerantz, R., and Hansen, J. (1995) *In situ* PCR* Hybridization to detect low-abundance nucleic acid targets. *Current protocols in molecular biology*, *vol. 2*, sections 14.8.1–14.8.23 (Suppl. 31).

36. Chen, R. H. and Fuggle, S. V. (1993) In situ cDNA polymerase chain reaction: a novel technique for detecting mRNA expression. *Am. J. Pathol.* **143,** 1527–2534.

37. Teo, I. A. and Shaunak, S. (1995) PCR in situ: aspects which reduce amplification and generate false-positive results. *Histochem. J.* **27,** 660–669.

38. Pallansch, L., Beswick, H., Talian, J., and Zelenka, P. (1990) Use of an RNA folding algorithm to choose regions for amplification by the polymerase chain reaction. *Anal. Biochem.* **185,** 57–62.

39. Freeman, W. M., Vrana, S. L., and Vrana, K. E. (1996) Use of elevated reverse transcription reaction temperatures in RT-PCR. *Biotechniques* **20,** 782–783.

40. Long, A. A., Komminoth, P., Lee, E., and Wolfe, H. J. (1993) Comparison of indirect and direct *in situ* polymerase chain reaction in cell preparations and tissue sections. Detection of viral DNA gene rearrangements and chromosomal translocations. *Histochemistry* **99,** 151–162.

41. Salstrom, J., Alemi, M., and Wilander, E. (1993) Pitfalls of in situ PCR using direct incorporation of labeled nucleotides. *Anti-Cancer Res.* **13,** 1153–1155.

42. Zebhe, I., Sällström, J. F., Hacker, G. W., Hauser-Kronberger, C., Rylander, E., and Wilander, E. (1994) Indirect and direct on situ PCR for the detection of human papillomavirus. An evaluation of two methods and a double staining technique. *Cell Vision* **1,** 163–167.

43. Teo, I. A. and Shaunk S. (1995) PCR in situ: an appraisal of an emerging technique. *Histochem. J.* **27,** 647–659.
44. Bassam et al. (1993) Automated hot start PCR by mineral oil and paraffin wax. *Biotechniques* **14,** 30–34.
45. Nuovo, G. J., Gallery, F., Hom, R., MacConnell, P., and Bloch, W. (1993) Importance of different variables for enhancing in situ detection of PCR amplified DNA. *PCR Meth. Appl.* **2,** 305–312.
46. Myers, T. W. and Gelfand, D. H. (1991) Reverse transcription and DNA amplification by a thermus thermophilus DNA polymerase. *Biochemistry* **30,** 7661–7666.

VII

SPECIAL APPLICATIONS
FOR THE CLINICAL LABORATORY

50

Overview of the Clinical Immunohistochemistry Laboratory: Regulations and Troubleshooting Guidelines

Patricia A. Fetsch and Andrea Abati

1. Introduction

Immunohistochemical procedures have become an integral part of the clinical laboratory routine, evolving from a research tool to a diagnostic necessity in pathology. As a specifically defined laboratory section, the Immunohistochemistry Laboratory must meet federally mandated standards of operation as defined in the Clinical Laboratory Improvement Amendments of 1988 (CLIA-88) *(1)*. The published guidelines for the practice of pathology do not include regulations devoted exclusively to immunohistochemistry; however, within the Federal Register (Section 493.1259, p. 7170) are CLIA regulations governing the use of "special stains" in the practice of pathology that can be applied to the immunoperoxidase procedures.

Discussed herein are the general requirements of the immunohistochemistry laboratory for compliance with CLIA-88. In addition, a troubleshooting guide for immunoperoxidase procedures is included.

2. Clinical Laboratory Regulations

Congress passed the CLIA-88 to set criteria for improving the quality of clinical laboratory services. The goal of this law was to standardize laboratory testing across the United States in all sites conducting testing on human specimens for health assessment or for the diagnosis, prevention, or treatment of disease. Failure to comply with these requirements may result in sanctions by the Health Care Financing Administration (HCFA), whose task is that of implementing CLIA-88. These sanctions may include changes in specific aspects of the laboratory operation, the suspension of part or all of Medicare payment for

From: *Methods in Molecular Biology, Vol. 115: Immunocytochemical Methods and Protocols*
Edited by: L. C. Javois © Humana Press Inc., Totowa, NJ

services, or even a complete shutdown of a facility. The regulations as printed in the Federal Register of Feb. 28, 1992 (vol. 57, pp. 7001–7288) can be found in many regional, university, law, and reference libraries. In addition, the College of American Pathologists (CAP) provides a highly regarded laboratory accreditation program that may be of benefit prior to federal inspection *(2)*. The guidelines discussed in this chapter provide a general reference for CLIA-88.

2.1. Procedure Manual

Procedures should be organized and indexed in the form of manuals and must be written in compliance with National Committee for Clinical Laboratory Standards (NCCLS) GP2-A2 *(3)*. Technical procedures designed for use at the bench should be complete, easy to follow, and readily available to testing personnel. These procedures must contain the following information for each assay performed:

1. Principle of the test.
2. Specimen requirements, including specimen collection, processing, storage, preservation, and the criteria for specimen rejection.
3. Reagents, standards, and controls used in the assay and instructions on their preparation, storage, and shelf life.
4. Instrumentation/calibration.
5. Step-by-step directions for all methods and antibodies currently in use, including fixatives used and any pretreatment protocols required.
6. Calculations.
7. Frequency and tolerance of controls and the corrective action to be taken if controls fail to meet the laboratory criteria for acceptability.
8. Expected values and reporting of results.
9. Procedure notes, as well as a description of the course of action to be taken in the event that a test system becomes inoperable.
10. Limitations of method (e.g., interfering substances).
11. Pertinent literature references.
12. Effective date and schedule for review.
13. Distribution supported by documentation that the procedure has been reviewed by all testing personnel.
14. Name of author.

Technical approaches must be scientifically valid and clinically relevant, and documentation that the laboratory director has reviewed all procedures on an annual basis is required. A copy of discontinued procedures is to be maintained for 2 yr thereafter, recording initial date of use and retirement date.

2.2. Reagents

All reagents must be properly labeled and dated as to content, date prepared or received, when placed in service, expiration date, and storage requirements. Routine monitoring of buffer pH is documented.

Reagent performance and adequacy are verified before placing the material in service. This can be accomplished through direct analysis of the new reagent or by parallel testing with reagents currently in use.

Antibody titration records are required for new antibody lots or when new antibodies are introduced into the laboratory. This is accomplished by running the assay at various antibody dilutions, using known positive and negative controls to determine the appropriate concentration for maximum sensitivity and specificity. Since incubation times, buffers, specimen processing, and fixation will affect the dilution used, optimal dilutions must be determined by each laboratory under its own special conditions.

Antibodies and other reagents are disposed of after the manufacturer's expiration date.

2.3. Equipment Maintenance

The daily monitoring of temperatures on various types of laboratory equipment is critical. Tissue-processing temperatures (water baths, ovens) may affect the quality of immunohistochemical staining and thus need to be verified each day. In addition, refrigerators and freezers require daily temperature checks, as the storage conditions of immunoreagents must be optimal for maintaining expected shelf life. If a "frost-free" freezer is in use, there must be assurance that the specimens, tissues, and reagents that are stored in that freezer are not damaged by the cycle of freezing, thawing, and refreezing.

A mercury-based maximum-minimum thermometer may be used to monitor temperatures during periods when the laboratory is not staffed. Large-volume refrigerators/freezers can be monitored with an upper- and lower-limit alarm system or a recording thermometer. All thermometers are to be calibrated with an appropriate National Institute of Standards and Technology (NIST) thermometer when put in service and every 6 mo thereafter.

Scheduled preventive maintenance is performed to prevent breakdowns or malfunctions, to prolong the life of an instrument, and to maintain optimum operating characteristics. For automated immunostainers, the performance and documentation of maintenance/function checks should be done as defined by the manufacturer with (at least) the frequency as specified. In general, common laboratory equipment such as pipets, centrifuges, and balances need to be serviced or calibrated twice yearly.

All instrument maintenance, service, and repair records should be available to, and usable by, the technical staff operating the equipment. These records are to be retained for the life of the instrument. This information can be invaluable for troubleshooting purposes.

Specific guidelines for laboratory information systems include regulations for computer configuration, procedure manuals, system security, data entry/reports/retrieval, hardware and software, and system maintenance.

2.4. Quality Control (QC)

Controls are used to ensure proper technique and the specificity of the stain. The use of similarly processed positive and negative staining controls is essential for interpretation of immunohistochemical reactions and must be done for each antibody. These controls may be commercially prepared, previously tested patient samples or proficiency testing specimens for which results have been confirmed.

The use of separate positive control tissues known to contain the antigen being evaluated must be included for each antigen in a run. It is most cost-effective to use large in-house tissues that have been fixed with the regular workday's surgical cases for this type of control. In general, autopsy tissues tend to exhibit autolysis in some organs and may not be optimal for use as control material. For cytology, cell culture material, effusions, and fine needle aspiration material can be prepared in the form of cytospins and stored (unfixed) at $-20°C$ for 3–6 mo for use as controls on similarly processed material. Cell blocks can also be used for cytology control material when appropriate. All controls should be fixed and prepared with the same protocol as the patient's slide material.

Internal or built-in controls are present when the specimen contains the target marker, not only in the tumor to be identified, but also in adjacent normal tissue. The evaluation of internal controls (when present) can be used as an indication of appropriate immunoreactivity. For ubiquitous antigens, internal controls are acceptable for use as a positive control, but the laboratory manual must clearly state, on a case-by-case basis, the manner in which internal positive controls are used for quality assurance.

A negative control (or specimen "blank") is used to assess nonspecific background staining of the specimen. In lieu of primary antibody, the use of nonimmune IgG from the same species as the primary antibody on one slide of each particular specimen is tested along with the rest of the case slides. Isotype-matched immunoglobulin negative controls, while optimal, are not always feasible and therefore are not essential. In addition, if a pretreatment step is performed (i.e., microwave, protease/trypsin digestion), a separate negative control must be run for that particular immunostaining protocol on one case slide.

The quality control program must clearly define goals for monitoring performance, procedures, policies, tolerance limits, corrective action, and related information. Records must be maintained regarding the reactivity of controls on a daily basis, along with an ongoing mechanism to evaluate the corrective actions taken when a control is unacceptable. These quality control records must be maintained for 2 yr.

2.5. Storage of Slides

The immunostained slides are filed with the remainder of slides from the case. Slides (and reports) are kept for a minimum of 10 yr.

2.6. Quality Assurance

An active program of surveillance of the quality of the immunostains produced must be defined. The primary elements of such a quality assurance (QA) program include procedures and policies for patient test management, quality control, proficiency testing, comparison of test results, relationship of clinical information to patient test results, personnel assessment, communications, complaint investigations, QA review with staff, and QA records. The documentation and review by the laboratory director of all QA procedures is imperative and cannot be overstressed. A brief explanation of each of the QA elements is as follows:

1. A patient test management system must assure optimum specimen integrity and identification from the pretesting to the post-testing process. Criteria must be established for patient preparation, specimen collection, labeling, preservation, and transportation. An appropriate specimen identification and accessioning system is in place to minimize sample mixups. Turnaround time (i.e., the interval between specimen receipt by laboratory personnel and reporting of results) for each test is defined and adhered to. Accuracy and reliability of test reporting systems, appropriate storage of records, and prompt retrieval of test results should be demonstrated (*see* **Note 1**).

2. The laboratory must have an ongoing mechanism to evaluate corrective actions and review their effectiveness when quality control is unacceptable.

3. Successful participation in a CLIA-88 approved proficiency testing program is mandated. An example of this type of external audit system is the College of American Pathologists (CAP), MK series, which provides two sets of challenges per year. There must be evidence of the identification and review of problems discovered through the use of this program and the documentation of corrective actions taken. If external proficiency testing is not available, blind testing of specimens with known results, exchange of specimens with other laboratories, or an equivalent system can be used to assess the reliability of test procedures on a semiannual basis.

4. When the laboratory uses different methodologies or instruments, or performs testing at multiple testing sites, a system is to be in place that evaluates and verifies the comparability between these test results. For example, correlation studies must ensure that manual and automated methods of immunostaining within a laboratory are in agreement. This must be documented biannually. In addition, any reference laboratories utilized must be CLIA-88-certified, and the lab director must monitor the quality of test results received from these outside sources.

5. A mechanism must be in place to evaluate immunohistochemical results that are inconsistent with clinicopathologic studies. This evaluation should be performed and recorded by a laboratory physician.

6. Personnel qualifications for high-complexity testing are stated in the CLIA-88 regulations. The director of the laboratory must be a qualified physician or a doctoral-level clinical scientist. Detailed job descriptions are required as well

as a system of documenting that all analysts are knowledgeable about the contents of procedure manuals relevant to the scope of their testing activities. Continuing education programs are an essential part of the laboratory quality improvement plan.

7. Documentation of problems because of breakdowns in communication, complaints reported to the laboratory, and records of the corrective action taken should be available. QA meetings with the staff are necessary to discuss identified problems and corrective actions taken to prevent reoccurrences.

2.7. Safety/Environment

The work area should be well-lighted with sufficient space and have the necessary water, air, gas, and electrical outlets.

Water quality (reagent grade II is appropriate for immunohistochemical procedures) must be regularly monitored as to bacterial contamination, resistivity, and silica content. If glassware is washed in the laboratory, a specific procedure is followed that requires rinsing with deionized or distilled water prior to drying as well as a determination that glassware is free of cleaning agents prior to use.

Temperature and humidity is controlled to minimize evaporation of reagents and to keep performance of electronic equipment optimal. Ventilation is adequate for the removal of noxious fumes and odors. Formaldehyde and xylene vapor concentrations must be below maximum permissible levels. For formaldehyde, this level is 0.75 ppm for an 8-h time-weighted average, or 2.0 ppm for a 15-min short-term exposure. For xylene, the level is 100 ppm for an 8-h time-weighted average and 200 ppm for a 15-min short-term exposure. The monitoring of the work area and employees can be performed on a yearly basis. Chemical and biological safety cabinets are checked for proper airflow on a yearly basis.

Waste disposal of infectious specimens, contaminated materials, and chemicals must be in compliance with local, state, and federal (EPA) regulations. Flammable safety cabinets are required for storage of alcohols, xylenes, and other combustible materials.

Safety policies and procedures are documented, and Material Safety Data Sheets are provided for all chemicals used in the laboratory. In addition, a chemical hygiene plan that defines the safety procedures for all hazardous chemicals is written in detail. All laboratory personnel must review these policies on an annual basis.

Universal precautions training that complies with the OSHA standard on occupational exposure to bloodborne pathogens, as well as a fire training program, should be provided on an annual basis for all laboratory employees. Personnel are required to use proper personal protective devices when handling corrosive, flammable, biohazard, or carcinogenic substances. Eye wash stations should be readily accessible and tested regularly for proper performance. (*See* **Notes 2** and **3** for additional resources.)

3. Troubleshooting of Immunoperoxidase Assays

When analyzing an immunostained specimen, deposits of the colored chromogen indicate the presence of the antigen and represent specific positive staining. The pattern of staining in the cells can be cytoplasmic, nuclear, membranous, or surface; focal or diffuse. This section addresses some of the most common problems encountered in immunoperoxidase procedures and the appropriate solutions to correct them (*see* Chapters 23–25). (*See* **Note 4** for additional troubleshooting resources.)

3.1. Absence of Staining

1. Procedure not followed: Staining steps were not performed in the correct order. Specimen pretreatment protocols such as digestion or microwave heating were not followed. Review manufacturer's package inserts.
2. Sodium azide present in buffers: The presence of sodium azide will prevent the development of the peroxidase color reaction.
3. Improper fixation and processing: Proper antigen fixation is the cornerstone of immunoperoxidase techniques, for without it, results will be poor. Some markers may be destroyed by certain fixatives; overfixation may mask certain antigens (*see* Chapter 8) (*4*). Review manufacturer's package inserts for appropriate fixation techniques. Paraffin-embedded tissue should never be exposed to temperatures >60°C as this can destroy some antigens.
4. Drying out of specimens during staining: Samples must be kept moist by applying sufficient reagent to prevent evaporation and by using a humidity chamber. When wiping off excess liquid, process only a few slides at a time. Repeated drying of specimens will result in poor morphology and staining.
5. Improper concentration of hydrogen peroxide in the substrate solution.
6. Specimens improperly counterstained: This problem may occur when using an alcohol-soluble chromogen such as 3-amino-9-ethylcarbazole (AEC). Counterstains containing alcohol, dehydration steps, or xylene/toluene-based mounting media will dissolve soluble colored precipitates.

3.2. Weak Staining

1. Too much buffer left on slides: After rinsing, as much liquid as possible should be wiped from around the specimen to avoid the dilution of antibody.
2. Use of old substrate: Substrate with hydrogen peroxide added to the chromogen must be made up fresh immediately before use.
3. Incubation times too short.
4. Antibody too dilute.
5. Improper storage of reagents or expired reagents: Some antibodies can be aliquoted and frozen to extend their shelf life.

3.3. Excess Background Staining

Positive staining that is not a result of antigen-antibody binding is termed nonspecific background stain. The most common cause of this is the attachment of protein to highly charged collagen and connective tissue elements of the specimen. Therefore, it is imperative to consistently include a negative control slide on all test samples for background staining assessment. There are a number of conditions that can contribute to excess background:

1. Endogenous peroxidase activity not removed: In this instance, staining will be observed in red and white blood cells.
2. Nonspecific binding of protein to the specimen: Use nonimmune serum from same animal species as the secondary antibody to reduce nonspecific binding. A higher concentration of salt in the buffer solutions, such as use of a 1:10 Tris:saline solution may also aid in the reduction of nonspecific binding.
3. Improper antibody dilutions: Use of concentrated antibody solutions can cause high background. This is especially common when changes are made in the incubation times of a procedure. For example, a 2-h incubation at room temperature is lengthened to 18 h at 4°C. It may also be observed when different specimen types (paraffin vs frozen section) are being tested.
4. Improper fixation: Poorly fixed specimens exhibit intense nonspecific staining. If specimen is too thick to permit complete fixative penetration, inadequate preservation will result.
5. Paraffin incompletely removed: The residue causes overall background staining that extends beyond the borders of the specimen.
6. Improper rinsing of slides: Slides should be thoroughly rinsed in three buffer baths after each incubation, the only exception being after incubation with the blocking serum.
7. Overdevelopment of substrate reaction: This may be because of excess chromogen in the solution, a high concentration of antigen in the specimen, or increased temperatures causing an accelerated reaction. Substrate reactions may be monitored microscopically for optimal staining.
8. Excessive application of tissue adhesive, gel-coated slides, or the use of albumin when preparing cytospins may contribute to background stain. The use of charged slides (Fisher Superfrost/Plus, Fisher Scientific, Pittsburgh, PA) can eliminate this problem.
9. Increased thickness of specimen: Tissue sections should be cut 4–5 µm thick, and cell smears should be spread as thinly as possible. Cytospins should be only a monolayer in thickness.
10. Endogenous biotin: In certain tissues, most commonly kidney and liver, background staining can be high because of the presence of endogenous biotin. This can be minimized by preincubation of the sections with a dilute avidin solution, followed by incubation with a dilute biotin solution (Vector Laboratories, Burlingame, CA) before the application of the primary antibody.
11. Microwave heating: Nonspecific granular cytoplasmic staining pattern is often accentuated when paraffin sections are microwave-pretreated (5). A biotin block

step may reduce this artifact. Careful attention to the biotin-blocked negative control in conjunction with the sample in question should eliminate potential false-positive results.

12. Necrotic tissue.

3.4. Positive Control Acceptable But Specimen Stains Weakly

1. Improper fixation and processing of unknown: This emphasizes the need for controls and unknowns to be processed in an identical manner.
2. Antigen present in low concentration.
3. Excess buffer or nonimmune serum allowed to remain on specimen.
4. Antigen partly destroyed or masked by fixation: If the antigen was masked as a result of overfixation in formalin, the use of digestion or microwave heating prior to the application of the primary antibody may increase staining intensity (*see* Chapter 13).

3.5. Artifacts

1. Undissolved granules of chromogen or counterstain: Precipitates are not confined to cells but spread randomly across the specimen. This can be corrected by filtering the chromogen or counterstain.
2. Incomplete dezenkerization of B5-fixed sections: This is observed as a black precipitate that is spread randomly across the specimen. It can be corrected by removing the mercury from the specimen prior to immunostaining.
3. Bacterial or yeast contamination.
4. Pigments such as melanin and hemosiderin: These differ in texture and color from the chromogen but may be difficult in interpretation. However, the negative control will demonstrate their true character, and by using a chromogen of a contrasting color (i.e., AEC stains positive cells a red color), this problem can be minimized.

4. Notes

1. Instruments, kits, or test systems labeled "for research use only" or "for investigational use only" have not been cleared by the United States Food and Drug Administration for sale to diagnose disease or other conditions. These products cannot be used as a diagnostic tool unless diagnosis is confirmed with a medically accepted test or procedure. Therefore, the laboratory must clearly state in the report that the test results are not to be used for diagnostic or treatment purposes.
2. Additional resources for assistance in meeting CLIA-88 regulations can be obtained from:
   a. National Committee for Clinical Laboratory Standards (NCCLS), 771 East Lancaster Avenue, Villanova, PA 19085-1596. 610-525-2453. FAX: 610-527-8399.
   b. College of American Pathologists (CAP), 325 Waukegan Rd., Northfield, IL 60093-2750. 800-323-4040. FAX: 800-289-1815.

    c. Health Care Financing Administration (HCFA), PO Box 26687, Baltimore, MD 21207-0487. 410-290-5850.

    d. Occupational Safety and Health Administration (OSHA), Dept. of Labor, 200 Constitution Ave NW, N3647, Washington, DC. 202-219-8151.

3. To order a copy of the Clinical Laboratory Improvement Amendments of 1988, refer to stock no. 069-001-00042-4 (price $4.50/copy) and mail request to: Government Printing Office, PO Box 371954, Pittsburgh, PA 15250-7954. 202-783-3238. FAX 202-512-2250.

4. Additional resources for troubleshooting assistance can be found in:

    a. Bourne, J. (1983) Handbook of immunoperoxidase staining methods. Dako Corp.: Carpinteria, CA.

    b. Wordinger, R., Miller, G., and Nicodemus, D. (1987) Manual of immunoperoxidase techniques, 2nd ed. ASCP (Chicago, IL).

References

1. Department of Health and Human Services, Health Care Financing Administration (1992) *Clinical Laboratory Improvement Amendments of 1988; Final Rule.* Federal Register **57**, 7001–7288.

2. College of American Pathologists Laboratory Accreditation Program (1997) *Standards for Laboratory Accreditation, Commission on Laboratory Accreditation Inspection Checklist*, Sections 1 and 8.

3. National Committee for Clinical Laboratory Standards (1992) *Clinical Laboratory Technical Procedure Manuals, Approved Guideline GP2-A2*, 2nd ed. NCCLS, Villanova, PA.

4. Battifora, H. (1991) Assessment of antigen damage in immunohistochemistry: the vimentin internal control. *Anat. Pathol.* **96**, 669–671.

5. Rodriguez-Soto, J., Warnke, R., and Rouse, R. (1997) Endogenous avidin-binding activity in paraffin-embedded tissue revealed after microwave treatment. *Appl. Immunohistochem.* **5**, 59–62.

51

Overview of Immunocytochemical Approaches to the Differential Diagnosis of Tumors

Dennis M. Frisman

1. Introduction

In 1974, the presence of intracytoplasmic immunoglobulins in plasma cells was demonstrated in routinely processed paraffin tissue (the mainstay of diagnostic pathology) using an immunoperoxidase technique (*1*). This new capability has introduced the surgical pathologist to a technique that can aid in reaching a diagnosis on already available material, including cytological smears and touch preps, frozen sections, and paraffin-embedded tissues fixed in formalin, ethanol, and other fixatives (*see* **Table 1**; Chapters 9–12). This novel approach of combining the sensitive and specific antibody-antigen reaction with the morphologic interpretation rendered by the experienced trained eye of the pathologist marks probably the most important advance in the world of diagnostic pathology in the last 20 yr.

Currently, the pathologist can use immunocytochemical procedures in tissue diagnosis in three different ways:

1. Identification of antigens that indicate tissue derivation of poorly differentiated neoplasms.
2. Identification of infectious organisms.
3. Identification of antigens that may aid in distinguishing malignant from benign processes, such as monoclonal immunoglobulin light chain expression in malignant lymphomas or aberrant proto-oncogene/suppressor gene expression in various malignant tumors.

There are now hundreds of antigens identifiable by this technique. Hence, the pathologist must use a concise practical approach to tailor his or her selection of immunostains to help narrow the differential diagnosis. Before the

From: *Methods in Molecular Biology, Vol. 115: Immunocytochemical Methods and Protocols*
Edited by: L. C. Javois © Humana Press Inc., Totowa, NJ

Table 1
Pathology Preparations for Immunocytochemistry[a]

Source	Preparation	Fixation
Tissue	Paraffin-embedded block	Formalin, ethanol, B5, Bouin's, or Carnoy's
	Frozen sections	Air-dried-acetone
Cytologic fine needle aspirates, brush, wash, body fluids	Smear/filter	Ethanol
	Cytospin	Air-dried-acetone
	Cell block	Formalin or ethanol
Sputum/gynecologic PAP smear	Smear	Ethanol with detergent
Imprints (touch preps), squash preps, scrape preps	Smear	Ethanol or air-dried-acetone

[a]Sources of specimens, the type of preparation, and the fixation needed for immunocytochemistry. Almost any type of specimen preparation or fixative can be used. The gentlest fixation is air-dried-acetone, which consists of first air-drying the slide followed by a brief acetone postfixation (*see* Chapters 9–12).

pathologist is ready to wade through the myriad of immunostains that needs to be ordered on a particular case, he or she needs to understand about controls, antibody selection, the types of antigens to assay, and how to interpret the slides as provided in the following discussion.

1.1. Controls

A negative immunostaining result should not always be regarded as a lack of antigen expression. It may also be a result of possible failure in antigen detection from poor fixation or an insensitive or faulty technique. Positive controls should always be evaluated alongside case slides. External positive controls consist of additional slides containing tissue that is preferably fixed the same as the case tissue and is known to contain the antigen being assayed. Internal positive controls are the most valuable, since the fixation is guaranteed to be identical to the tumor being evaluated. Internal positive controls consist of normal structures found on the case slide being evaluated that are known to express the antigen being assayed and are present adjacent to the tumor. An example is the normal epidermis adjacent to a skin tumor being evaluated for carcinoma by using antikeratin antibody. Unfortunately, slides containing both the tumor being evaluated and the corresponding normal structures that serve as internal controls are not always available, especially for antigens such as proto-oncogene products. One must then rely exclusively on external positive controls. If the positive control (either external or internal) is negative for the antigen being assayed, the technique should be considered invalid.

Even with the appropriate reactive positive controls, a negative result still does not absolutely rule out a particular differentiation state. Some tumors may lose their ability to express some antigens (dedifferentiation), although they can still be recognized to be of a particular differentiation by other methods (morphology, electron microscopy, and so forth). Because of this reason, negative results can never be as meaningful as positive results.

The converse of the above point should also be considered: A positive immunostaining result does not always indicate that the antigen is expressed. Other causes of a positive reaction could be because of background from fixed serum proteins within the tissue or faulty technique such as inadequate washes, wrong titers, overdigestion with proteases, or air-drying artifact. Unexpected crossreactions may also occur. To control for these possible causes of misinterpretation, negative controls should be run alongside the case slides. A negative control consists of a slide containing tissue identical to the case slide (preferably a section cut adjacent to the case slide from the paraffin block). This slide goes through the exact same procedure as the case slide, except a nonsense antibody that is known not to react with the tissue is substituted for the primary reagent. In addition, one should check the case slide for positivity in unexpected structures, such as keratin positivity in blood vessels. Positivity seen on the negative control slide or in the wrong areas on the case slide may invalidate the procedure.

1.2. Antibody Selection

The antibodies available to detect tissue antigens can be either polyclonal antisera, usually made in rabbits, or mouse monoclonal antibodies. Special considerations are warranted when using polyclonal antisera. To reduce non-specific background staining, antisera must often be "cleaned up" by absorption or column chromatography (*see* Chapters 2–5). In addition, there can be tremendous lot-to-lot variation in the reactivity of the antisera, even variation in the same lot over time. Therefore, the use of mouse monoclonal antibodies is preferable. It must be remembered, however, that a monoclonal antibody recognizes only one epitope on an antigen molecule. There is a risk of obtaining a false negative reaction from either an unstable epitope in an otherwise intact antigen or variations in conformation or primary structure of the antigen. To work around this problem, one can use an antibody cocktail composed of several monoclonal antibodies to different epitopes on the same antigen molecule. These monoclonal antibody cocktails, although the most ideal reagents, take time and effort to develop in the laboratory. However, there are commercially available cocktails for some antigens, including pan-reactive cytokeratins and CD45.

Another way of increasing sensitivity and minimizing false negatives besides monoclonal antibody cocktails is to use antibody panels. Rather than test for just one epithelial marker such as cytokeratin to diagnose carcinoma, it may be prudent to run other additional epithelial-specific markers, such as epithelial membrane antigen (EMA) and BER-EP4. To determine whether a tumor has neuroendocrine differentiation, rather than just test for chromogranin expression, test also for synaptophysin, CD57 (Leu 7), and/or γ-γ neuron-specific enolase (NSE) expression. Just be aware that a large antibody panel may show redundancies of reactivity and will certainly be costly.

Currently, antibodies reactive to the antigen to be assayed may be commercially available from several sources. There may be as many as 100 different antibodies available for some antigens such as cytokeratin. Unfortunately, not all antibodies available have shown the same desired reactivity in practice. To aid in the selection of an antibody, one should review the literature not only to find those that have been put to successful use in other laboratories but also to note additional procedures that may be required to enhance the antibody reactivity, such as the need lo use a microwave antigen retrieval system (*see* Chapter 13) or protease digestion. The antibody still needs to be performance-tested in one's own laboratory before it can be relied on diagnostically. This includes testing the antibody against a panel of normal and malignant tissues to assure that the antibody is reactive to the antigen being assayed, and that there are no unexpected cross reactions, and by titering the antibody to the appropriate concentration, which is often different from that suggested by the manufacturer. Although this testing is time-consuming and expensive, it will help ensure against possible future diagnostic misinterpretations.

1.3. The Antigens: Differentiation Markers vs Prognostic Markers

The following is a discussion of the antigens to be assayed. Assays for infectious agents will not be dealt with in this chapter. Most of the antibodies available today are directed against differentiation antigens. Their expression by a particular cell tells of the cell's functional state or embryologic derivation. Both benign and malignant counterparts express the same differentiation antigens, although there may be a quantitative difference in expression. An example would be S100 protein expression in both benign melanocytic nevi and malignant melanomas (2). Hence, these antigens should not be used to differentiate between benign and malignant tumors. There are, however, two conditions where the expression of differentiation antigens may be able to aid the pathologist in distinguishing benign from malignant states:

1. The detection of monoclonality in immunoglobulin-producing neoplasms as mentioned above: The presence of either, but not both, κ or λ light chain in lymphocytic proliferations may indicate the presence of a malignant neoplasm such as B-cell lymphoma or myeloma/plasmacytoma.
2. Metastasis: Sometimes cells of a particular origin can be determined to be malignant just by virtue of being detected in sites other than where they are expected to occur.

Hence, just knowing the differentiation state of the cells (by the use of immunocytochemistry) is often enough for the pathologist to diagnose malignancy. An example is finding glandular antigens (that are not crossreactive with mesothelial antigens) in cells from the pleural cavity. This alone is usually sufficient to diagnose metastatic adenocarcinoma, since glandular cells of any type are not normally found in the pleural space. Another example is finding prostatic specific antigen (PSA) in cells found in a lymph node. PSA is highly specific for the glandular cells of the prostate, whether benign or malignant. PSA-positive glandular cells in a lymph node are diagnostic of malignancy, whereas PSA-positive glandular cells in the prostate are not.

The evaluation of differentiation antigens usually requires only a qualitative interpretation. It is often enough just to note whether an antigen is present or absent in order to make a diagnosis, although sometimes noting the strength of the reaction (weak vs strong) or the proportion of cells showing positive may be useful. However, this semiquantitative evaluation may be more a function of fixation than of tumor behavior.

In addition to the differentiation antigens, immunocytochemistry can also identify antigens that may aid in the separation of benign from malignant processes and also relate information about the prognosis of a malignant tumor. These antigens fall into three categories:

1. Cell cycle-dependent antigens.
2. Hormone receptor antigens.
3. Proto-oncogene/suppressor antigens.

Unlike differentiation antigens, prognostic markers are usually evaluated semiquantitatively, both in terms of the proportion of tumor cells showing positivity and the strength of the reaction.

Cell-cycle-dependent antigens are proteins involved with DNA replication, such as DNA polymerases. Although all tumors, whether benign or malignant, express these antigens, certain malignant tumors may show increased proportions of positivity. Hence, quantifying their expression will give an index of proliferation akin to measuring S-phase fraction in DNA analysis by flow cytometry (*see* Chapter 34). The most popular antibody that has been examined with immunocytochemistry is Ki-67 on frozen sections. Recently, an antibody (MIB1) raised against the recombinant Ki-67 molecule has become available that can be applied to formalin-fixed paraffin-embedded tissues *(3)*. Ki-67 has been shown to have independent prognostic value when examining malignant melanomas, breast carcinomas, and non-Hodgkin's lymphomas *(4)*. Another commonly used antibody (PC10) recognizes proliferating cell nuclear antigen (PCNA), also known as cyclin, which also works on paraffin tissues and has been shown to be correlated with Ki-67 reactivity *(5)*.

The hormone receptors most studied in the immunocytochemistry of tumors are the estrogen (antibodies Abbott H222-ER-ICA *[6]* and Dako 1D5 *[7]*) and progesterone (antibodies JZB39 *[8]* and PgR-1A6 *[9]*) receptors on breast carcinoma. The presence of these receptors is a crucial favorable prognostic indicator, and it behooves the pathologist to measure these in every new case of breast cancer. Traditionally, the pathologist was required to freeze a portion of the cancer tissue from each case so that the receptors could be measured by the dextran-coated charcoal biochemical technique—the gold standard. However, it has now been shown that the immunocytochemistry technique is just as useful, if not better, than the gold standard, so now the pathologist can get the answers from the paraffin blocks, including those from which the tissue was insufficient in quantity to freeze for the biochemical assay.

The value of the detection of the overexpression of proto-oncogenes and suppressor genes by immunocytochemistry as a prognostic indicator is still in its infancy, although immunohistochemical markers for these antigens have been available for many years. Part of the problem is the inability to correlate the protein expression detected by immunocytochemistry with the DNA amplification or RNA production as detected by *in situ* hybridization and Southern and Northern blotting assays. Antibodies reactive with c-erb-B2 (Her-2/neu) *(10)* and c-myc p62 (Myc1-9E10) *(11)* proteins in paraffin-embedded tissues are available, although the jury is still out regarding whether their expression is an independent poor prognostic indicator in cancer, especially breast carcinoma. Interestingly, rather than having prognostic value in themselves, antibodies directed against c-erbB-2 may be a predictor of those tumors that are more likely to respond to adjuvant chemotherapy for breast cancer *(12)*. Abnormalities in expression of the suppressor gene product, p53, is one of the most ubiquitous findings in neoplasia. Antibody PAb1801 *(13)* can detect this antigen in paraffin-embedded tissues. Another antibody, RB, which is directed against the retinoblastoma suppressor gene product, is now available for use on paraffin-embedded tissue *(14)*.

1.4. Interpretation

Before embarking on the immunocytochemical approach to diagnosis, some points concerning interpretation of the immunostained slides should be made. Even after controlling for nonspecific reactions and increasing the sensitivity of the technique, interpretation may still prove to be difficult. Endogenous pigments such as melanin, hemosiderin, bile, lipofuscin, formalin pigment, and mercury from tissues that are not dezenkerized rarely can be confused with the red-brown pigment of the horseradish-peroxidase substrate, DAB. If there is confusion, switching to another substrate, such as AEC (red) or 4CN (blue) may be helpful (*see* Chapter 23). To evaluate tumors heavily pigmented with melanin, one can either bleach out the melanin *(15)* or the hematoxylin counterstain can be replaced with azure B that stains melanin green-blue *(16)*.

The pattern of staining is very important. Antigens have a characteristic distribution in the cell. Hence, immunostains should reflect this distribution. Most antigens as defined by currently available antibodies are present in the cytoplasm and exhibit cytoplasmic staining with the nucleus remaining unstained. The intermediate filaments, HMB-45, hormones, and neuroendocrine markers (chromogranin, synaptophysin) are cytoplasmic markers. The S-100 protein antigen is unique in that it appears in both the nucleus and the cytoplasm. Occasionally, some cytoplasmic markers will show a localized staining pattern for certain tumors, such as the perinuclear staining of cytokeratin in Merkel cell tumors and Ki-1 on Reed-Sternberg cells of Hodgkin's Disease. Many lymphoid markers, such as CD45, CD20 (L26), and CD45RO (UCHL-1) stain cellular membrane antigens and appear as a distinctive ring around cells. Antibodies available against many viral antigens (such as HPV, adenovirus, and Hepatitis B core antigen) will characteristically stain only the nucleus. Special caution must be taken when interpreting uncharacteristic staining patterns, such as cytokeratin staining the nucleus, as specific.

The chromagen precipitate should appear crisp and granular in sections containing the antigen being assayed. Be careful not to interpret hazy, amorphous-appearing precipitate as specific. This nonspecific pattern may arise artifactually from "laking" of staining reagents on portions of tissue not well adherent to the slide or crossreaction with serum proteins that are fixed within the tissue. With all of the above points as background, the following will show the practical approach the pathologist can use in diagnosing tumors, starting with the poorly differentiated tumors.

2. Initial Approach Used
to Classify Poorly Differentiated Tumors:
Three Antibody Panel

Even with well-stained, well-fixed H&E sections, the pathologist will be confronted with poorly differentiated neoplasms that defy classification using sound morphologic criteria. After resorting to a battery of histochemical stains (PAS, mucin, silver stains, reticulum stains, PTAH, etc.) the epithelial, neural, hematopoietic, or mesenchymal nature of certain tumors may still remain elusive. So often the pathologist is left with a tray of "special-stained" slides demonstrating the same tumor in many different color combinations, yet with no more clues to the tissue derivation of the tumor than with the original H&E. In many of these cases, the "true colors" of the tumor can be elucidated using a screening panel of monoclonal antibodies in an immunohistochemical assay.

An abbreviated panel of antibodies directed against three antigens—keratin, S-100 protein, and CD45 (leukocyte common antigen—LCA)—can separate

Table 2
Diagnostic Categories as Determined by a Panel of Antibodies

| Dx category | Antigen positivity | | | Diagnoses | |
	Ker.[a]	S100	CD45	Main	Other considerations
A	+	–	–	Carcinoma	Mesothelioma
					Synovial sarcoma
					Epithelioid sarcoma
	+	+	–	Carcinoma: breast, salivary, sweat gland	Chordoma
					Mixed tumor
B	–	+	–	Neural tissue/ melanoma	Liposarcoma
					Chondrosarcoma
C	–	–	+	Hematopoietic	None
D	–	–	–	Sarcoma	All

[a]Since carcinomas vary in their composition of keratins, it is important for the keratin-detecting antibody to be broad spectrum, i.e., able to react with both high- and low-molecular-weight species of keratin. This can be achieved by using a cocktail of commercially available monoclonal antibodies. The use of antibodies specific for different keratins in subtyping carcinomas will be discussed later. Monoclonal antibodies against S-100 protein and CD45 are readily available from many commercial sources.

poorly differentiated tumors into four diagnostic categories: carcinoma, neural tissue tumor/melanoma, hematopoietic tumor, and sarcoma, as shown in **Table 2**.

The following is a discussion of further immunopathology workup once a diagnostic category is established for a tumor. A few additional points need to be addressed before continuing.

2.1. The Use of Intermediate Filaments to Determine Cell Lineage

The five intermediate filaments and their respective tissues are listed in **Table 3**. Only the intermediate filament, cytokeratin, is selected as useful in the initial classification of tumors. The other intermediate filaments can cause diagnostic confusion because (1) they are usually not expressed in their poorly differentiated counterparts (especially GFAP, NFP, and Desmin) and (2) they are often coexpressed on many types of tumors. Poorly differentiated neuroectodermal tumors may often express more than two intermediate filaments. Vimentin demonstrates the most lineage infidelity.

2.2. Failure of All Three Antibodies to React with the Tumor

Although supportive of a sarcoma diagnosis, lack of expression of the antibodies may be secondary to a fixation or preservation artifact as well as dedif-

Table 3
Tissue Localization of Intermediate Filaments

Intermediate filament	Tissue
Keratin	Epithelia
NFP	Neural
GFAP	Glial
Vimentin	Mesenchyma
Desmin	Muscle

ferentiation of the tumor. The demonstration of vimentin (the intermediate filament expressed in mesenchymal cells) in these tumors by an antibody that some may want to include in a screening panel supports the diagnosis of sarcoma and, at the same time, reassures to a certain degree that some antigens are still preserved in the preparation. However, tumors from all other diagnostic categories may express vimentin (vimentin is found in all melanomas, 50% of carcinomas, and in some lymphomas). This caveat must be kept in mind when including this antibody in the initial screen.

In fact, because of the ubiquitous expression of vimentin in tissues and its partial susceptibility to formalin fixation, some authors advocate including vimentin routinely as an internal positive control on all cases *(17)*. Since many useful molecules also show sensitivity to fixation or processing for histology, vimentin positivity may be used as a gauge of the general preserved antigenicity of the tissue being examined.

2.3. Failure of Hematopoietic Tumors to Express CD45

Some lymphomas are unreactive with antibodies to CD45, such as lymphomas showing plasmacytoid differentiation or anaplastic large-cell lymphomas.

2.4. Expression of Keratin by Sarcomas

Some sarcomas besides epithelioid sarcoma or synovial sarcoma show coexpression of vimentin and keratin *(18)*. Examining these tumors for the expression of other epithelial markers will help clarify their true nature.

2.5. Differential Diagnosis of Small Cell Tumors

The initial panel of antibodies described above is also useful in discriminating among small cell tumors, except that a few additional markers are needed. A panel of neuroendocrine markers (chromogranin A, synaptophysin, and CD57) should be used to differentiate neuroendocrine tumors. The use of a panel rather than just one of these markers is recommended to maximize the ability to detect poorly differentiated neuroendocrine tumors that lose the expression of one or more antigens. The marker, CD99, that detects p30/32mc2 glycoprotein *(19–21)* is very useful for distinguishing Ewing's sarcoma and primitive neuroectodermal tumor from other small cell tumors.

3. Subclassification of Tumors by Diagnostic Category

3.1. Carcinomas (Keratin-Positive Tumors)

Many antibodies are available that react differentially to various epithelial tumors and thus are useful in establishing the tissue of origin. Following is a list of some of these antibodies and how they can be used to subclassify carcinomas.

3.1.1. Variation in Cytokeratin Expression

3.1.1.1. Cytokeratin Subtyping

The intermediate filament, cytokeratin,, represents a family of proteins containing at least 19 members that are distinguishable by their molecular weights and acidic properties (pI). Moll and colleagues *(22)* have created a system of cataloging these proteins (numbered from 1 to 19) based on these properties. Cytokeratin is found in the cytoplasm of epithelial cells as pairs consisting of an acidic and basic member that vary in epithelium from different sites. Generally, complex (stratified squamous) epithelium contains higher molecular-weight species of cytokeratin (classically types 5 and 14 for basal cells and 1,2, and 10 for suprabasal cells), while simple glandular epithelium contains lower molecular-weight species (classically types 8 and 18). Tumors tend to express the same cytokeratins as their normal counterparts, and hence their tissue derivation can be determined by using rnonoclonal antibodies against specific keratin subtypes. **Table 4** shows a list of tumors with the keratin (by Moll cytokeratin number) that they express *(23)*.

Many cytokeratin-specific antibodies are available and have been useful in the distinguishing of carcinomas. For example, an antibody that recognizes cytokeratin 7 or cytokeratin 19 will react with cholangiocarcinomas and not with hepatocellular carcinoma. Two low-molecular-weight cytokeratins have been shown to be useful in distinguishing various metastatic adenocarcinomas: CK7 (Clone Ks20.8) and CK20 (Clone OVTL 12/30) *(24,25)*. **Table 5** shows how different tumors can be distinguished by using these two antibodies. In addition, CK20 has shown the ability to differentiate Merkel cell carcinoma of the skin from other small cell carcinomas *(26)*.

3.1.1.2. Keratin Pattern of Expression

Sometimes the distribution of keratin staining in a particular cell may aid in determining its origin. As mentioned before, Merkel cell and other neuroendo-crine carcinomas characteristically have a perinuclear ball-like reactivity.

Table 4
Cytokeratin Expression in Various Epithelial Tumors[a]

Tumor	Cytokeratin[b]
Squamous cell	
Skin, tongue, vocal cord	5, 6, 14, 16, 17
Esophagus	6, 14, 17, 19
Cervix	5, 6, 13, 17, 19
Adenocarcinoma	
Pancreas, gallbladder, cervix, lung, breast, ovary, endometrium	7, 8, 18, 19
Gastric, colon, kidney	8, 18, 19
Liver	
Hepatocyte	8, 18
Bile duct	7, 8, 18, 19
Transitional cell bladder	5, 7, 8, 13, 17, 18, 19
Neuroendocrine	8, 18

[a]Modified from Nagle (*16*).
[b]By Moll cytokeratin number.

Table 5
CK7 and CK20 Expression in Epithelial Tumors

Tumor source	CK7 positive	CK20 positive
Stomach	+	+
Pancreas	+	+
Bladder	+	+
Ovary—mucinous	+	+
Ovary—nonmucinous	+	−
Lung—nonsmall cell	+	−
Breast	+	−
Endometrium	+	−
Mesothelium	+	−
Liver (bile duct)	+	±
Colon	−	+
Lung—small cell	−	−
Kidney	−	−
Liver (hepatocyte)	−	−
Prostate	−	−
Squamous cell	−	−

Table 6
CEA Expression in Various Epithelial Tumors

CEA reactivity	Epithelial tumor
Positive	Gastrointestinal, pancreas, biliary, lung, transitional cell, sweat glands, mucosal squamous cell, mucinous carcinomas of female GU tract, medullary carcinoma of thyroid
Positive or negative	Breast, squamous cell, endometrioid, Brenner tumor
Negative	Renal cell, hepatocellular, prostate, follicular thyroid, adrenal cortical, serous carcinomas of female GU tract, embryonal, yolk sac, mesothelioma

Adenocarcinomas have a tendency to show more peripheral staining of keratin, whereas mesotheliomas show more perinuclear.

3.1.2. Carcinoembryonic Antigen (CEA)

CEA is a heterogeneous glycoprotein of molecular weight 180,000 kDa. It was originally considered to be specific for colon carcinoma; however, it is now known to be expressed in many different tumors and fetal tissues as well as adult colonic mucosa *(27)*. **Table 6** shows a summary of CEA reactivity in epithelial tumors.

Many polyclonal antisera and some monoclonal antibodies available to CEA have been shown to crossreact with a variety of CEA-related proteins, such as nonspecific crossreactive antigen (NCA), meconium antigen (NCA-2), biliary glycoprotein (BGP), granulocyte antigen (NCA-160), pregnancy specific glycoprotein (PSG), and so on. Several monoclonal antibodies are available that do not recognize these nonspecific antigens. The lack of crossreactivity with NCA-160, that is, lack of granulocyte activity, is the single best criterion used to select monoclonals with a high degree of specificity for adenocarcinomas from the gastrointestinal tract, such as colon or gastric carcinomas. These highly specific antibodies, however, lack activity to normal colon and adenocarcinomas from other sites. Therefore, they may be useful for differentiating metastatic adenocarcinomas from unknown primary tumors. However, to differentiate adenocarcinomas from other tumors, such as mesotheliomas in pleural fluids, a polyclonal or wide-spectrum CEA may be more useful for two reasons: A wider range of tumors will be reactive, and CEA-related antigens are not expressed by mesothelial cells.

3.1.3. Polymorphic Epithelial Mucins (PEM)

These are high-molecular-weight (>100 kDa) sialomucins, consisting of a core protein with a tandem repeat of 20 amino acids and heavy glycosylations

(28). Most antibodies available are reactive with the core protein. The abnormal or decreased glycosylations characteristically seen in tumors unmasks epitopes on the core protein, giving rise to heterogeneous reactivity with different antibodies. To date, none of the antibodies directed to these antigens are specific for tumors or epithelial tumors of particular sites.

3.1.3.1. EPITHELIAL MEMBRANE ANTIGEN (EMA)/MILK FAT GLOBULE PROTEIN (MFG)

Antibodies to these PEMs *(29,30)* are directed to the apical membrane portions of secretory mammary cells surrounding milk fat globules (mol wt >250 kD). They react with a variety of epithelial tumors, including mesotheliomas. Like keratin, these antibodies are also expressed on synovial sarcomas, epithelioid sarcomas, and chordomas. They are not as reactive with as wide a variety of tumors as antikeratins. For instance, they do not react with many endocrine tumors. However, they may recognize some renal cell carcinomas and small cell anaplastic carcinomas that are often unreactive with keratin antibodies. EMA is also expressed by many nonepithelial elements, such as plasma cells, Reed-Sternberg cells, T-cell lymphomas, malignant histiocytes, and plasmacytomas. This may limit the usefulness of antibodies to EMA as epithelial markers.

3.1.3.2. B72.3

B72.3 *(31)* is a monoclonal antibody that recognizes a very high-molecular-weight mucin (TAG-72, mol wt >1000 kD) present on cells showing glandular differentiation. Originally thought to be present only on malignant cells (e.g., metastatic adenocarcinomas in pleural fluids), it is now known to be expressed on normal epithelial cells (breast glands showing apocrine metaplasia). When B72.3 positive cells are found in pleural fluids, one should be highly suspicious of the presence of adenocarcinoma, because there are no other reasons for cells showing glandular differentiation to be in the pleural space. However, when B72.3 positive cells are found in peritoneal fluids or fine needle aspirates where there are reasons for the presence of glandular cells (e.g., endometriosis), benign processes should be considered in the diagnosis.

3.1.3.3. OTHER PEMs

Other PEMs include DF3, F36/22, NCRC11, LICR LON/M8, OM-1, SM-3, BC1, and BA-Br-3. Their usefulness in the differential diagnosis of glandular tumors is currently under investigation.

3.1.4. BER-EP4

The monoclonal antibody, BER-EP4 *(32)*, recognizes a cell membrane glycoprotein consisting of two polypeptide chains (mol wt 34 kD and 39 kD). Its

Table 7
Specific Antigens Expressed by Various Epithelial Tumors

Tumor	Antigens
Prostate adenocarcinomas	Prostate specific antigen (PSA) (25)
	Prostatic acid phosphatase (PAP)
Breast carcinomas	Estrogen receptor protein
Hepatocellular carcinomas	Alpha-fetoprotein
Thyroid neoplasms	
Follicular, papillary	Thyroglobulin
Medullary carcinoma	Calcitonin
Neuroendocrine tumors	
General	Synaptophysin, chromagranin, CD57(26) γ-γ NSE (27)
Specific hormones	Insulin, glucagon, somatostatin, gastrin, pancreatic polypeptide, serotonin, vasoactive intestinal peptide, cholecystokinin
Pituitary	ACTH, LH, PRL, TSH, FSH
Germ cell tumors	Chorionic gonadotropin, alpha-fetoprotein, placental lactogen, placental alkaline phosphatase

antigen is expressed on a wide variety of epithelial tumors, including small-cell undifferentiated carcinomas and neuroendocrine tumors. It is not found on nonepithelial nor mesothelial tumors.

3.1.5. CU18

CU18 (33) recognizes the antigen BCA-225 (225 kD glycoprotein), which is present on various carcinomas but is most characterized on breast carcinoma.

3.1.6. GCDFP-15

Gross cystic disease fluid protein-15 (34) (also known as BRST-2) recognizes a protein originally derived from breast carcinoma but is also found in other carcinomas, such as from sweat glands and prostate.

3.1.7. Tissue Site-Specific Antigens

Antibodies are available that recognize only particular tissues. A partial list is presented in **Table 7** (35–37), showing the most specific markers that can assure the diagnoses in most cases, including PSA reactivity in prostate cancer and calcitonin in thyroid medullary carcinoma. Estrogen and progesterone receptor proteins are also prognostic markers as discussed before. They are also found in other gynecologic neoplasms, as well as other tumors, such as meningiomas.

Table 8
Panel of Antibodies to Differentiate Metastatic Adenocarcinoma: Females

Tumor	CK7	CK20	CEA	ERP	S-100	GCDFP
Colon	–	+	+	–	–	–
Other GI	+	±	+	–	–	–
Lung	+	–	+	–	–	–
Breast	+	–	±	±	±	±
Ovary/endometrium	+	–	±	±	+	–

Table 9
Panel of Antibodies to Differentiate Metastatic Adenocarcinoma: Males

Tumor	CK7	CK20	CEA	PSA
Colon	–	+	+	–
Other GI	+	±	+	–
Lung	+	–	+	–
Prostate	–	–	–	+

3.1.8. Summary of Use of Epithelial Markers— Differentiation of Metastatic Adenocarcinomas

Tables 8 and **9** show how some of the previously mentioned antibodies can distinguish between metastatic adenocarcinomas of unknown primaries in females and males, respectively. The panels are similar except for ERP (estrogen receptor protein), GCDFP-15, and S-100, which are used in females for their specificity for gynecologic tumors; and PSA (prostatic specific antigen), used in males for its specificity for prostatic adenocarcinoma. These antibodies cannot absolutely distinguish among the tumor sources listed. However, they can guide pathologists and clinicians on which organs need to be further examined to determine the primary source of a metastatic tumor.

3.2. Neural Tissue Tumor/Melanoma (S-100-Positive Tumors)

3.2.1. S-100 Subtyping

S-100 protein *(38)* is a 20-kDa calcium-binding protein composed of two subunits, S-100α and S-100β, which are differentially expressed by individual human tissues. For example, S-100(α,α) is found in myocardial and skeletal muscle cells; S-100(α,β) is found in glial cells, melanocytes, chondrocytes, and adnexal glands of the skin; and S-100(β,β) is found in Schwann cells and Langerhans cells of the skin *(39)*. There are currently monoclonal antibodies

Table 10
Antigen Expression in S-100 Positive Tumors

Tumor	MBP[a]	GFAP	NFP	HMB45	EMA
Neuroma	+	–	+	–	–
Pheochromocytoma		–	+	–	–
Neurofibroma	+	–	–	–	–
Schwannoma	+	–	–	–	–
Granular cell tumor	+	–	–	–	–
Astrocytoma	–	+	–	–	–
Glioblastoma	–	+	–	–	–
Ependymoma	–	+	–	–	–
Gliosarcoma	–	+	–	–	–
Malignant melanoma	–	–	–	+	–
Clear cell sarcoma				+	–
Meningioma	–	–	–	–	+
Chordoma		–	–	–	+
Other S-100+:					
Liposarcoma	–	–	–	–	–
Chondrosarcoma	–	–	–	–	–
Histiocytosis X	–	–	–	–	–

[a]Abbreviations as in text.

available that can differentiate between the two subunits and, hence, may be useful in differentiating tumors.

3.2.2. Other Antigens

Table 10 lists S-100 positive tumors and their positivity with additional markers. Myelin basic protein (MBP) *(40)* is a single-chain polypeptide that associates with the central and peripheral nervous system. Glial fibrillary acidic protein (GFAP) and neurofilament proteins (NFP) are intermediate filaments associated with glial cells and neural cells, respectively. HMB-45 *(41)* is a mouse monoclonal antibody that reacts with an antigen present in premelanosomes. It is highly sensitive and specific for melanomas and certain nevi (junctional, congenital, and blue nevi). EMA has been discussed in **Subheading 3.1.3.1.**

3.3. Hematopoietic Tumors (CD45-Positive Tumors)

The benefit of the immunoperoxidase technique is best exhibited by the ability to detect neoplastic B cells. By demonstrating the presence of κ or λ (but not both) in lymphoid or plasma cells, one can establish a diagnosis of lymphoma or plasmacytoma. Unfortunately, in routinely processed tissue, only immunoglobulins in the cytoplasm and not on the cell surface membrane can be detected.

Table 11
Antibody Reactivity in Hematopoietic Tumors

Tumor	Antibodies
Non-Hodgkin's Lymphoma	Immunoglobulins, CD20 (L26), CD45RA (MB1),
B cell	MB2, CD75 (LN1), CD74 (LN2)
T cell	CD3, CD45RO (UCHL-1), CD43 (LEU22, MT1),
	β-F1
Anaplastic large cell	CD30
Hodgkin's disease	CD30, CD15
Histiocytic tumors	Lysozyme, α-1-antitrypsin, α-1-antichymotrypsin,
	MAC-387, CD68
Myeloid tumors	CD15, NP57

Hence, one is successful in demonstrating monoclonal B-cell populations only in lymphomas showing plasmacytoid differentiation, some B-cell immunoblastic lymphomas, and plasmacytomas. To make matters worse, the demonstration of specific staining for immunoglobulin (κ, λ, IgG, IgM, IgA) can be difficult because of high background staining resulting from the presence of these proteins in normal serum. Another immunostain is available that can help distinguish nodular lymphomas from reactive follicular hyperplasia. BCL-2, a marker whose presence indicates cells that escape apoptosis, is expressed on most nodular lymphomas but not on germinal centers *(42,43)*. Currently, there are no immunostains available to demonstrate monoclonality in T-cell neoplasms.

Although small-cell lymphomas are generally considered low grade, one small-cell lymphoma (mantle cell lymphoma) is important to distinguish because of its higher grade behavior and different chemotherapeutic regimen requirement. An antibody directed against cyclin D1 (BCL-1) is useful in distinguishing this malignancy from other lymphomas *(44)*.

Table 11 lists the reactivity of antibodies to various lymphoreticular neoplasms in formalin- or B5-fixed tissues. Many of the monoclonal antibodies available to B- or T-cell lymphomas are not strictly lineage specific (MB1, MB2, LN1, MT1). CD20 (L26) *(45)* for B-cells and CD45RO (UCHL-1) *(46)* for T-cells has been shown to be fairly specific. UCHL-1 is not present on all T-cells, therefore one may want to include another T-cell marker to detect T-cell lymphomas. β-F1 *(47)*, which recognizes a framework epitope on the T-cell β chain antigen receptor, and anti-CD3 *(48)* may be particularly promising in specifically detecting T-cell lymphomas.

Anti-CD30 (Ki-1) *(49)* and Anti-CD15 (Leu-M1) *(50)* both react with Reed-Sternberg cells of Hodgkin's Disease. Leu-M1 also reacts with myeloid cells and is a very specific, though not very sensitive, marker for adenocarcinomas.

Lysozyme, α-1-antitrypsin, and α-1-antichymotrypsin are all present in histiocytes but may be difficult to detect because their presence as a normal component of serum, and like immunoglobulins, may cause high background. MAC-387 *(51)* is present only on a subset of histiocytic cells. Anti-CD68 (KP-1) *(52)* recognizes a lysosomal membrane protein and, hence, may be the most useful in detecting histiocytic tumors. NP57 *(53)*, which recognizes neutrophil elastase, may prove to be very useful in detecting acute myeloid leukemias in tissue sections.

The plethora of antibodies available to detect lymphoreticular antigens on frozen or fresh tissue has necessitated the creation of an international workshop to characterize them (hence the development of the CD or cluster designation nomenclature). Although the number of antibodies available to lymphoreticular antigens on routinely processed tissue is greatly lagging behind, the number of new ones is increasing dramatically, and keeping track of their reactivities may prove difficult.

3.4. Sarcomas (Tumors Negative for Keratin, S-100, and CD45)

Fewer antibodies are available to characterize these tumors than those from the other categories. **Table 12** lists the most useful antigens that can be detected in routinely processed tissues. Antibodies to desmin, an intermediate filament, are very specific for muscle tumors. Because of its sensitivity to fixation and its absence in the early differentiation of muscle cells, desmin is often difficult to detect in muscle sarcomas.

CD31 *(54)*, Factor VIII-related antigen *(55)*, and CD34 *(56)* cannot often be demonstrated in poorly differentiated vascular tumors. The use of the lectin Ulex europaeus agglutinin 1 (UEA-1) *(57)* is a more sensitive method to detect endothelial cells. Like other lectins, however, UEA-1 reacts with many other cell types.

Malignant fibrous histiocytomas (MFH), often a wastebasket diagnosis in histology, cannot be established by immunohistochemistry. The expression of the histiocyte markers and FXIIIa (originally described as a fibrohistiocytic marker that is also on dermal dendrocytes) *(58)* listed in **Table 12** in the malignant cells of MFH remains controversial. In addition, these markers are also found in other sarcomas.

4. Conclusion

If the pathologist follows the general guideline above, he or she will discover making many more specific diagnoses than by using morphologic criteria alone. However, there will be cases where the immunocytochemical results may confound the diagnosis further. Even when confronted with an inconsistent immunostaining result, the pathologist should not be swayed too far from the original differential diagnosis for a particular tumor. It must be remem-

Table 12
Antigen Expression in Sarcomas

Tumor	Antigen
Muscle	Desmin, muscle-specific actin
Rhabdomyosarcoma	Myoglobin
Leiomyosarcoma	Smooth muscle actin
Angiosarcoma	CD31 (JC70/A), factor VIII-related antigen, CD34 (QBEND/10)
Malignant fibrous histiocytoma	α-1-antitrypsin, α-1-antichymotrypsin, CD68, FXIIIa

bered that tumor cells, like most normal diploid cells in the human body, have the complete complement of information to produce almost any protein. An aberrant expression of an antigen different from that expected for the lineage of the tumor should not be too surprising in a neoplastic cell.

A true positive immunostain means that the tumor is expressing a specific antigen. Since the tumor has siphoned energy away from its machinery geared for reproduction and invasion to make this antigen, its presence may mean something important about the biology of the tumor and, hence, have prognostic and therapeutic implications. In this light, immunostains should never be treated as "just another" special histochemical stain.

References

1. Taylor, C. R. and Burns, J. (1974) The demonstration of plasma cells and other immunoglobulin containing cells in formalin-fixed, paraffin-embedded tissue using peroxidase labeled antibody. *J. Clin. Pathol.* **27,** 14.
2. Hachisuka, H., Sakamoto, F., Nomura, H., Mori, O., and Sasai, Y. (1986) Immunohistochemical study of S-100 protein and neuron specific enolase (NSE) in melanocytes and the related tumors. *Acta Histochem.* **80,** 215–223.
3. Cattoretti, G., Becker, M. H., Key, G., Duchrow, M., Schlueter, C., Galle, J., and Gerdes J. (1992) Monoclonal antibodies against recombinant parts of the Ki-67 antigen (MIB 1 and MIB 3) detect proliferating cells in microwave-processed formalin-fixed paraffin sections. *J. Pathol.* **168,** 357–363.
4. Gerdes, J. (1990) Ki-67 and other proliferation markers useful for immunohistological diagnostic and prognostic evaluations in human malignancies. *Semin. Cancer Biol.* **1,** 199–206.
5. Hall, P. A., Levison, D. A., Woods, A. L., Yu, C. C., Kellock, D. B., Watkins, J. A., Barnes, D. M., Gillett, C. E., Camplejohn, R., Dover, R., et al. (1990) Proliferating cell nuclear antigen (PCNA) immunolocalization in paraffin sections: an index of cell proliferation with evidence of deregulated expression in some neoplasms. *J. Pathol.* **162,** 285–294.

6. McCarty, K. S., Jr., Miller, L. S., Cox, K. B., Konrath, J., and McCarty, K. S., Sr. (1985) Estrogen receptor analyses. Correlation of biochemical and immunohistochemical methods using monoclonal antireceptor antibodies. *Arch. Pathol. Lab. Med.* **109,** 716–721.

7. Goulding, H., Pinder, S., Cannon, P., Pearson, D., Nicholson, R., Snead, D., Bell, J., Elston, C. W., Robertson, J. F., Blamey, R. W., et al. (1995) A new immunohistochemical antibody for the assessment of estrogen receptor status on routine formalin-fixed tissue samples. *Hum. Pathol.* **26,** 291–294.

8. Ozello, L., DeRosa, C., Habif, D. V., and Greene, G. L. (1991) An immunohistochemical evaluation of progesterone receptor in frozen sections, paraffin sections, and cytologic imprints of breast carcinomas. *Cancer* **67,** 455–462.

9. Battifora, H. and Mehta, P. (1993) Comparison of antibodies to progesterone receptor. *Appl. Immunohistochem.* **1,** 83,84.

10. Slamon, D. J., Godolphin, W., Jones, L. A., Holt, J. A., Wong, S. G., Keith, D. E., Levin, W. J., Stuart, S. G., Udove, J., Ullrich, A., et al. (1989), Studies of the HER-2/neu proto-oncogene in human breast and ovarian cancer. *Science* **244,** 707–712.

11. Spandidos, D. A., Pintzas, A., Kakkanas, A., Yiagnisis, M., Mahera, H., Patra, E., and Agnantis, N. J. (1987) Elevated expression of the myc gene in human benign and malignant breast lesions compared to normal tissue. *Anticancer Res.* **7,** 1299–1304.

12. Muss, H. B., Thor, A. D., Berry, D. A., Kute, T., Liu, E. T., Koerner, F., Cirrincione, C. T., Budman, D. R., Wood, W. C., Barcos, M., et al. (1994) c-erbB-2 expression and response to adjuvant therapy in women with node-positive early breast cancer. *N. Engl. J. Med.* **331,** 211.

13. Kerns, B. J., Jordan, P. A., Moore, M. B., Humphrey, P. A., Berchuck, A., Kohler, M. F., Bast, R. C., Jr., Iglehart, J. D., and Marks, J. R. (1992) p53 overexpression in formalin-fixed, paraffin-embedded tissue detected by immunohistochemistry. *J. Histochem. Cytochem.* **40,** 1047–1051.

14. Shi, S. R., Cote, R. J., Yang, C., Chen, C., Xu, H. J., Benedict, W. F., Taylor, C. R. (1996) Development of an optimal protocol for antigen retrieval: a "test battery" approach exemplified with reference to the staining of retinoblastoma protein (pRB) in formalin-fixed paraffin sections. *J. Pathol.* **179,** 347–352.

15. Alexander, R. A., Hiscott, P. S., Hart, R. L., and Grierson, I. (1986) Effect of melanin bleaching on immunoperoxidase, with reference to ocular tissues and lesions. *Med. Lab. Sci.* **43,** 121–127.

16. Kamino, H. and Tam, S. T. (1991) Immunoperoxidase technique modified by counterstain with azure B as a diagnostic aid in evaluation heavily pigmented melanocytic neoplasms. *J. Cutaneous Pathol.* **18,** 436–439.

17. Battifora, H. (1991) Assessment of antigen damage in immunohistochemistry. The vimentin internal control. *Am. J. Clin. Pathol.* **96,** 669–671.

18. Swanson, P. E. (1991) Heffalumps, jaguars, and cheshire cats: a commentary on cytokeratins and soft tissue sarcomas. *Am. J. Clin. Pathol.* **95(suppl. 1),** S2–S7.

19. Weidner, N. and Tjoe, J. (1994) Immunohistochemical profile of monoclonal antibody O13: antibody that recognizes glycoprotein P30/32MIC2 and is useful in

diagnosing Ewing's sarcoma and peripheral neuroepithelioma. *Am. J. Surg. Pathol.* **18,** 486–494.

20. Ambros, I. M., Ambron, P. F., and Strehl, S. (1991) MIC2 is a specific marker for Ewing's sarcoma and peripheral primitive neuroectodermal tumors. *Cancer* **67,** 1886–1893.

21. Stevenson, A. J., Chatten, J., Bertoni, F., and Miettinen, M. (1994) CD99 (P30/32MIC2) neuroectodermal/Ewing's sarcoma antigen as an immunohistochemical marker. Review of more than 699 tumors and the literature experience. *Appl. Immunohistochem.* **2,** 231–240.

22. Moll, R., Franke, W. W., Schiller, D. L., Geiger, B., and Krepler, R. (1982) The catalog of human cytokeratins: patterns of expression in normal epithelia, tumors, and cultured cells. *Cell* **31,** 11–24.

23. Nagle, R. B. (1988) Intermediate filaments: a review of the basic biology. *Am. J. Surg. Pathol.* **12(Suppl. 1),** 4–16.

24. Wang, N. P., Zee, S., Zarbo, R. J., Elacchi, C. E., and Gown, A. M. (1995) Coordinate expression of cytokeratins 7 and 20 defines unique subsets of carcinomas. *Appl. Immunohistochem.* **3,** 99–107.

25. Loy, T. S. and Calaluce, R. D. (1994) Utility of cytokeratin immunostaining in separating pulmonary adenocarcinomas from colonic carcinomas. *Am. J. Clin. Pathol.* **102,** 764–767.

26. Miettinen, M. (1995) Immunohistochemical marker for gastrointestinal, urothelial, and Merkel cell carcinoma. *Mod. Pathol.* **8,** 384–388.

27. Sheahan, K., O'Brien, M. J., Burke, B., Dervan, P. A., O'Keane, J. C., Gottlieb, L. S., and Zamcheck, N. (1990) Differential reactivities of carcinoembryonic antigen (CEA) and CEA-related monoclonal and polyclonal antibodies in common epithelial malignancies. *Am. J. Clin. Pathol.* **94,** 157–164.

28. Layton, G. T., Devine, P. L., Warren, J. A., Birrell, G., Xing, P., Ward, B. G., and McKenzie, I. F., (1990), Monoclonal antibodies reactive with the breast carcinoma-associated mucin core protein repeat sequence peptide also recognise the ovarian carcinoma-associated sebaceous gland antigen. *Tumour Biol.* **11,** 274–286.

29. Pinkus, G. S. and Kurtin, P. J. (1985) Epithelial membrane antigen—A diagnostic discriminant in surgical pathology: immunohistochemical profile in epithelial, mesenchymal, and hematopoietic neoplasms using paraffin sections and monoclonal antibodies. *Hum. Pathol.* **16,** 929–940.

30. Battifora, H. and Kopinski, M. I. (1985) Distinction of mesothelioma for adenocarcinoma—an immunohistochemical approach. *Cancer* **55,** 1679–1685.

31. Johnston, W. W., Szpak, C. A., Lottich, S. C., Thor, A., and Schlom, J. (1986) Use of a monoclonal antibody (B72. 3) as a novel immunohistochemical adjunct for the diagnosis of carcinomas in fine needle aspiration biopsy specimens. *Hum. Pathol.* **17,** 501–513.

32. Sheibani, K., Shin, S. S., Kezirian, J., and Weiss, L. M. (1991) BER-EP4 antibody as a discriminant in the differential diagnosis of malignant mesothelioma versus adenocarcinoma. *Am. J. Surg. Pathol.* **15,** 779–784.

33. Mesa-Tejada, R., Palakodety, R. B., Leon, J. A., Khatchenan, A. O., and Greaton, C. J. (1988) Immunocytochemical distribution of the breast carcinoma associated glycoprotein identified by monoclonal antibodies. *Am. J. Pathol.* **130,** 305–314.

34. Wick, M. R., Lillemoe, T. J., Copland, G. T., Swanson, P. E., Manivel, J. C., and Kiang, D. T. (1989) Gross cystic disease fluid protein-15 as a marker for breast cancer: immunohistochemical analysis of 690 human neoplasms and comparison with alpha-lactalbumin. *Hum. Pathol.* **20,** 281–287.

35. Allsbrook, W. C. and Simms, W. W. (1992) Histochemistry of the prostate. *Hum. Pathol.* **23,** 297–305.

36. Gray, M. H., Smoller, B. R., McNutt, N. S., and Hsu, A. (1990) Neurofibromas and neurotized melanocytic nevi are immunohistochemically distinct neoplasms. *Am. J. Dermatopathol.* **12,** 234–241.

37. Pahiman, S., Esscher, T., and Nilsson, K. (1986) Expression of gamma-subunit of enolase, neuron-specific enolase, in human non-neuroendocrine tumors and derived cell lines. *Lab. Invest.* **54,** 554–560.

38. Moore, B. W. (1965) A soluble protein characteristic of the nervous system. *Biochem. Biophys. Res. Commun.* **19,** 739–744.

39. Takahashi, K., Isobe, T., Ohtsuki, Y., Akagi, T., Sonobe, H., and Okuyama, T. (1984) Immunohistochemical study on the distribution of alpha and beta subunits of S-100 protein in human neoplasm and normal tissues. *Virchow Arch. B Cell. Pathol.* **45,** 385–396.

40. Bansal, G., Martenson, R. E., Leveille, P., and Campagnoni, A. T. (1987) Characterization of a novel monoclonal anti-myelin basic protein antibody: use in immunoblotting and immunohistochemical studies. *J. Neuroimmunol.* **15,** 279–294.

41. Gown, A. M., Vogel, A. M., Hoak, D., Gough, F., and McNutt, M. A. (1986) Monoclonal antibodies specific for melanocytic tumors distinguish subpopulations of melanocytes. *Am. J. Pathol.* **123,** 195–203.

42. Wood, B. L., Bacchi, M. M., Bacchi, C. E., Kidd, P., and Gown, A. M. (1994) Immunocytochemical differentiation of reactive hyperplasia from follicular lymphoma using monoclonal antibodies to cell surface and proliferation-related markers. *Appl. Immunohistochem.* **2,** 48–53.

43. Utz, G. L. and Swerdlow, S. H. (1993) Distinction of follicular hyperplasia from follicular lymphoma in B5-fixed tissues: comparison of MT2 and BCL-2 antibodies. *Hum. Pathol.* **24,** 1155–1158.

44. Swerdlow, S. H., Yang, W., Zukerberg, L. R., Harris, N. L., Arnold, A., and Williams, M. E. (1995) Expression of Cyclin D1 protein in centrocytic/mantle cell lymphoms with and without rearrangement of the BCL1/Cyclin D1 gene. *Hum. Pathol.* **26,** 999–1004.

45. Cartun, R. W., Coles, F. B., and Pastusrak, W. T. (1987) Utilization of monoclonal antibody L26 in the identification, and confirmation of B-cell lymphomas: a sensitive and specific marker applicable to formalin- and B5-fixed, paraffin-embedded tissues. *Am. J. Pathol.* **129,** 415–421.

46. Clark, J. R., Williams, M. E., and Swerdlow, S. H. (1990) Detection of B- and T-cells in paraffin-embedded tissue sections: diagnostic utility of commercially obtained 4KB5 and UCHL-1. *Am. J. Clin. Pathol.* **93,** 58–69.

47. Said, J. W., Shintaku, I. P., Parekh, K., and Pinkus, G. S. (1990) Specific phenotyping of T-cell proliferations in formalin-fixed paraffin-embedded tissues. Use of antibodies to the T-cell receptor beta-F1. *Am. J. Clin. Pathol.* **93,** 382–386.

48. Anderson, C., Rezuke, W. N., Kosciol, C. M., Pastuszak, W. T., and Cartun, R. W. (1991) Identification of T-cell lymphomas in paraffn-embedded tissues using polyclonal anti-CD3 antibody: comparison with frozen section immunophenotyping and genotypic analysis. *Mod. Pathol.* **4,** 358–362.

49. Stein, H., Mason, D. Y., Gerdes, J., O'Connor, N., Wainscoat, J., Fallesen, G., Gatter, K., Falini, B., Delsol, G., Lemke, H., Schwarting, R., and Lennert, K. (1985) The expression of the Hodgkin's Disease associated antigen Ki-1 in reactive and neoplastic lymphoid tissue: evidence that Reed-Sternberg cells and histiocytic malignancies are derived from activated lymphoid cells. *Blood* **66,** 848–858.

50. Hanjan, S. N. S., Kearney, J. F., and Cooper, M. D. (1982) A monoclonal antibody (MMA) that identifies a differentiation antigen on human myelomonocytic cells. *Clin. Immunol. Immunopathol.* **23,** 172.

51. Hall, P. A., D'Ardenne, A. J., and Stansfeld, A. G. (1988) Paraffin section immunohistochemistry. I. Non-Hodgkin's lymphoma. *Histopathol.* **13,** 149–160.

52. Pulford, K. A., Rigney, E. M., Micklenn, K. J., Jones, M., Stross, W. P., Gatter, K. C., and Mason, D. Y. (1989) KP1: a new monoclonal antibody that detects a monocyte/macrophage associated antigen in routinely processed tissue sections. *J. Clin. Pathol.* **42,** 414–421.

53. Ralfkiaer, E., Pulford, K. A. F., Lauritzen, A. F., Avnstrom, A., Guldhammer, B., and Mason, D. Y. (1989) Diagnosis of acute myeloid leukaemia with the use of monoclonal anti-neutrophil elastase (NP57) reactive with routinely processed biopsy samples. *Histopathol.* **14,** 637–643.

54. DeYoung, B. R. (1993) CD31. An immunospecific marker for endothelial differentiation. *Appl. Immunohistochem.* **1,** 97–100.

55. Leader, M., Collins, M., Patel, J., et al. (1986) Staining for factor VIII related antigen and Ulex europaeus agglutinin 1 (UEA-1) in 230 tumors: an assessment of their specificity for angiosarcoma and Kaposi's sarcoma. *Histopathol.* **10,** 1153–1162.

56. Nickoloff, B. J. (1991) The human progenitor cell antigen (CD34) is localized on endothelial cells, dermal dendritic cells, and perifollicular cells in formalin-fixed normal skin, and on proliferating endothelial cells and stromal spindle-shaped. *Arch. Dermatol.* **127,** 523–529.

57. Alles, J. U. and Bosslet, K. (1988) Immunocytochemistry of angiosarcomas. A study of 19 cases with special emphasis on the applicability of endothelial cell specific markers to routinely prepared tissues. *Am. J. Clin. Pathol.* **89,** 463–471.

58. Reid, M. B., Gray, C., Fear, J. D., and Bird, C. C. (1986) Immunohistological demonstration of factors XIIIa and XIIIs in reactive and neoplastic fibroblastic and fibro-histiocytic lesions. *Histopathol.* **10,** 1171–1178.

52

Overview of Automated Immunostainers

Gilbert E. Herman, Edna A. Elfont, and Alton D. Floyd

1. Introduction

A significant consideration in the use of immunoreagents for microscopic slide staining is their cost. For many antibodies, fractions of a milliliter of fluid may be very expensive, particularly in the undiluted form. Traditional staining methods frequently use more reagent than allowed for by supply budget constraints. Additionally, routine immunostaining requires several dozen changes of reagent and is therefore labor-intensive, particularly when large numbers of slides must be stained. Using manual staining methods, wash procedures are such that they may result in slide to slide variations. Drying artifact is difficult to avoid, and occasionally, a slide may not even receive a reagent.

Although a number of approaches have been used to automate traditional immunostaining, each of these involves a mechanical device that sequentially brings slides into contact with various solutions and simultaneously conserves reagent usage. The automated stainers that have been introduced use primarily two methods for accomplishing this task: a capillary gap into which reagents flow or a variation of the droplet on a horizontally held slide. This chapter surveys the four immunostainers in widest use in the United States at the time of this study: Instrumentation Laboratory's (Lexington, MA) Code-On Immuno/DNA Slide Stainer, Shandon's (Pittsburgh, PA) Cadenza, Leica Jung's (Deerfield, IL) Histostainer Ig, and Ventana's (Tuscon, AZ) 320.

1.1. Potential Advantages of Automation

A common rationale for the acquisition of an automated immunostainer is that there will be a significant reduction in skilled technicians' time. The true value of automation, however, is improved reproducibility and reliability of staining. Intra- and interlaboratory standardization of reagents and automated

From: *Methods in Molecular Biology, Vol. 115: Immunocytochemical Methods and Protocols*
Edited by: L. C. Javois © Humana Press Inc., Totowa, NJ

immunostaining procedures allows direct comparison of staining results from run to run. Among the potential benefits derived from automating immuno-staining are the following:

1. Automated immunohistochemistry stainers (AIHS) repetitively and reliably dispense and allow for the use of microliter quantities of reagents for individual slide reactions, thus saving expensive immunologic reagents.
2. AIHS perform reagent additions in proper sequence as dictated by computer instructions, eliminating the possibility of "skipping" reagents.
3. AIHS employ uniform washing procedures between reagent steps, which are usually more efficient than manual methods.
4. AIHS save skilled technologists' bench time and decrease turnaround times by eliminating or minimizing repetitive pipeting and incubation procedures. Some instruments allow a decrease in the incubation time of immunologic reagents by using more favorable physicochemical parameters that speed antigen–antibody reactions, and others allow overnight immunostaining in the absence of technologists.
5. AIHS may minimize the exposure of technologists to potentially dangerous chemicals.
6. AIHS may control immunologic and enzymatic reaction conditions by providing uniform temperature and/or humidity, incubation times, and buffer washes.
7. AIHS may decrease the amount of technical training needed to perform immunostaining as compared to manual methods by using preprogrammed computer-driven procedures.
8. AIHS may, through the use of computers and attached printers, record key information about the immunostaining procedure so staining runs can be verified, an important consideration when an aberrant reaction occurs.

1.2. Problems Not Corrected by Automation

The above list of potential advantages of automated immunostainers summarizes their strengths. Automation cannot, however, correct faulty cell or tissue preparative techniques, which would lead to the failure of any immunostaining method. Automated staining also cannot correct for the improper selection of tissue to be examined, or for selection of antisera and associated reagents that produce nonoptimal or inconclusive results. Automation is no panacea for laboratories that have poor quality control procedures, since good laboratory practices are required to operate automated stainers successfully.

1.3. Performance Criteria
to Be Met by an Automated Immunostainer

There is a certain level of performance that should be expected from all AIHS. The performance features that are of particular concern to our group are listed below.

1. A reproducible quantity of reagent must be dispensed on the slide or move into the gap with no greater than a 10% error.
2. No greater than a 10% evaporative loss should occur from either the capillary gap or the reagent container in the course of a run.
3. The reagent should rise in the gap to a reproducible height from slide set to slide set and from run to run with no greater than a 10% error.
4. The recommended buffer washes must completely eliminate each reagent from a capillary gap or from the surface of a horizontally held slide.
5. The instrument should need minimal technician's attention during a run.
6. There must be no significant carryover of reagents.
7. The antibodies must not adsorb to the surfaces of the reagent containers.
8. There should be reproducible antibody reaction intensities from slide to slide and run to run using the same tissue and same antibody.
9. Ideally, a stainer should allow use of multiple antibodies and detection systems during one run.
10. Ideally, the stainer's software should allow for flexible programming and be user-friendly.
11. The stainer must have adequate error tracking programs and ways of printing the programs for each run.
12. The trouble-shooting manual and customer support services must be available to prevent excessive down time. There should be a 24-h or less response time by the company.
13. The stainer must not require major repair within the first 90 d.
14. The stainer must not generate excessive heat or noise, or cause disruption of other laboratory activities.
15. Reagents should have reasonable shelf life with no availability problems. Quality control of reagent lots must be available if questions of sensitivity or specificity should arise.
16. Slide breakage must not occur.
17. Once any necessary retitering of primary antibodies is accomplished, immunostaining results obtained with the automated stainer must be comparable to or better than those obtained with that laboratory's currently used manual methods.

1.4. Reagent Requirements

Standardized staining protocols require standardized reagents to ensure reproducible results. Manufacturers of automated immunostaining instruments may supply proprietary buffers and wash solutions for their instrument, but the properties of these should be within the range of parameters that are used in manual staining. (The primary additives to manufacturer-supplied reagents are surfactants.)

The titer of the antisera is critical and must be determined experimentally for each antigen–antibody pair on each automated stainer. The ideal titer may be affected by a number of factors, such as adsorption to surfaces (effectively reducing the titer), denaturation caused by excess temperature, excessive bubbling/foaming during application to the slide, and evaporative loss. Addition of chemicals to increase shelf life of reagents or to affect their wetting properties may also influence the avidity of an antibody for its antigen. The influence of each of these factors for each particular automated stainer depends on the specific design of the instrument and on the supplied reagents for that instrument.

1.5. Evaluation of Results

The results of immunohistochemical staining are assessed microscopically. All sites of chromogen deposition, however, may not result from specific antigen–antibody interaction. There may be nonspecific adsorption of any of the various immune reagents used in the procedure to tissue sites owing to electrostatic, hydrophilic, or hydrophobic interactions. In addition, normal tissue processes may result in unexpected distribution of reactants, such as the localization of antigens within the cytoplasm of tissue macrophages. Although this is a true localization of antigen, it is the result of passive acquisition of the antigen by the labeled cells, rather than by active manufacturing of the antigen. For these reasons, evaluation of immunohistochemical stains requires the experience of a trained observer, and cannot be interpreted simply on the presence or absence of chromogen in the completed preparation.

The studies described here were undertaken to assess the functionality of current automated immunohistology stainers in a working histology laboratory. Each instrument was made available, and training in its use provided, by the manufacturer. The specific tests for electrical or mechanical function that were used for each instrument were designed to explore the unique design features of each instrument, and to test for any idiosyncrasies that might cause test failures or erratic results in routine use. This chapter will not discuss staining results as they pertain to specificity or sensitivity on tissue specimens because, although such tests were performed, they are valid only in terms of the chemistries used, and because reagent chemistries are continually evolving, any comments on these might be rapidly dated.

2. Automated Immunohistology Stainers

2.1. The Instrumentation Laboratory Code-On Immuno/DNA Slide Stainer

2.1.1. Description

The Code-On immunochemistry stainer was developed from the Histomatic XYZ slide stainer by David Brigatti. The Code-On has 11 solution vessels with a capacity of 200 mL each, six stations that hold reagent isolator pads, a high-temperature oven, an incubator, and a water bath. The reagent isolator pads are used for holding manually aliquoted antisera, detection reagents, and chromogens. The slides are held in pairs in a single carrier that has a total capacity of 60 slides. The slide carrier is moved from position to position by a robotic arm and can be programmed to visit any position on the instrument for any length of time. Staining protocols can be written by the user, and the total number of different protocols maintained by the instrument is limited only by the size of the hard drive used in the IBM PC computer that controls the staining instrument.

The Code-On is a capillary gap stainer. The capillary gap is formed between two slides, each of which may bear specimens. The slides of a pair are kept approx 150 μm apart by the clamping action of the slide holder, and by raised areas that exist on each slide at the label area and at each of the opposite corners at the end of the slide. The capillary gap, enclosed by the two slides, holds approx 150 μL of reagent, assuming that the gap fills to a height of approx 1.5 in. from the bottom of the slide pair.

This stainer dips the slides into solutions to remove paraffin and hydrate the slides to water. During each of these steps, the capillary gap fills in response to surface tension and is emptied between reagent steps by bringing the edge of each slide pair into contact with an absorbent pad. During immunostaining steps, the slide pairs are brought into contact with a drop of reagent. The reagent is pulled into the gap and is retained there until the incubation interval is complete. This blotting and refilling are done for each reagent change.

The efficiency with which a stainer, using the capillary gap technology, functions is directly related to the ability of the gap to fill or empty. This depends, in part, on the surface tension of the fluid being used. The surface tension of aqueous fluids is much greater than that of alcoholic solutions, making it much more difficult to fill or empty the gap with an aqueous solution unless the appropriate surfactants are added. The capillary gap system is intolerant of thick, loose, or folded sections, since these will either eliminate the gap or will interfere with staining of the facing section. The Code-On requires the use of special slides to create the precise spacing required by the technique. If the dimensions of the gap are not precisely maintained, the gap is prone to trap air bubbles that prevent it from properly filling or emptying.

Reagents are contained in any or all of the 30 wells of the "reagent isolator pads," which are manufactured from silicon polymer. Individual aliquots of antisera or other reagents are manually pipeted into the wells prior to starting the staining procedure or during the staining run if reagent stability is an issue. Because of the precise mechanical alignment necessary to bring specific slide pairs into contact with the appropriate reagents on the isolator pads, the Code-

On requires calibration of the robotic arm prior to the beginning of each staining run.

The computer program that runs the Code-On provides great flexibility for modifying any one or more steps in an existing staining protocol, and permits previous programs to be copied and subsequently modified. The computer comes with a dot-matrix printer to provide printed program definitions, solution station locations, and listings of reagents to be placed in specific individual isolator wells. Once the operator has selected or defined a program, loaded solutions, placed slides into the carrier that is clipped to the robotic arm, and the instrument calibration is satisfactory, then the run can be initiated.

The Code-On stainer has facilities for washing slides in running water, for incubating in a closed chamber, and for heating in an oven. The incubation chamber serves to slow evaporation of reagent from the capillary gaps during staining. The oven is a necessity for the high-temperature steps required during *in situ* staining protocols.

2.1.2. Performance Evaluation

A series of experiments were done to evaluate the efficiency with which the capillary gaps fill and empty. Aqueous solutions of normal swine serum proteins were prepared at various protein concentrations. Various concentrations of a surfactant, Brij-35, were added to each solution (0.0625, 0.125, 0.25, or 0.50 vol %). Test solutions were placed in reagent isolators and picked up between blank Probe-On Plus slides. To avoid contamination of the slides, they were handled with powderless rubber gloves. The Code-On was programmed to dip the slide pairs into the test solutions for 1 min. After the arm lifted, the ascent of the fluid into the gaps was measured in millimeters. The rise of fluid into the capillary gap was unaffected by either protein or Brij-35 concentrations, and bubbles were found randomly distributed. This test was repeated using identical protein and Brij-35 concentrations to assess draining of solutions from capillary gaps. After picking up solutions, the robotic arm moved the slide pairs to a fresh blotter pad with which it stayed in contact for 1 min. At the conclusion of this interval, the lower level of the fluid was measured in millimeters. The lower the concentration of protein, the more efficiently the gap emptied. The Brij-35 did not appear to affect the draining efficiency consistently. The pattern of draining differed among the slide pairs with fluid being trapped in some gaps, whereas others drained almost fully.

The efficiency of reagent removal from the capillary gap was assessed by filling capillary gaps between blank slides with peroxidase–streptavidin complex. After picking up the solution, the slide pairs were blotted for 1 min. The slide pairs were then subjected to three changes in buffer, followed by three, 1-min drains on clean pads. At the conclusion of the wash steps, the individual pads were exposed to 3-amino-9-ethylcarbazole (AEC) to assess peroxidase activity. The majority of reagent was drained from the capillary gap during the initial pad contact. A small residual amount of reagent was removed in the first wash step, but after the second and third washes, no appreciable levels of reagent could be detected.

To assess evaporative loss of reagent from the surface of reagent isolator pads, distilled water was added to a series of reagent isolator pads. The temperature of the laboratory was 25.7°C, and the relative humidity was 29% during these tests. (This humidity is typical during a Michigan winter because of prolonged periods of forced air heating.) The instrument cover (an accessory) was not available for this test. A weighed aliquot of 200 µL of water was pipeted into each isolator well. Over a 2-h period (the time for a typical immunostaining run), fluid was removed from the individual isolator wells and weighed on a Mettler balance. There was a 20% loss of volume at 40 min, 28% at 80 min, and 47.5% at 120 min. Other than the low relative humidity of the laboratory, there was no obvious air flow from ducts, doors, or windows to contribute to this rapid evaporation.

To assess evaporative loss of fluid from filled capillary gaps, clean slide pairs were filled with colored, distilled water. After a 1-min fill, the maximum height of fluid in the gap was marked on each slide pair. Slides were then segregated into two groups: one remained exposed to room air, and the other was placed into the closed incubation chamber. At timed intervals, fluid loss in millimeters was recorded. Laboratory temperature was a constant 25°C with a relative humidity of 28%. After 40 min, air-exposed slides had an average loss of 33% of their volume, whereas slides in the incubation chamber lost only 9% in the same time interval.

To assess reagent binding to reagent isolator pads, a series of previously used Isolons® were cleaned according to company-supplied protocols. For each isolon, either primary antibody, secondary antibody, or streptavidin–peroxidase complex was added to the wells. Saline-filled wells served as controls. The filled Isolon wells were incubated at room conditions for 20-, 40-, or 60-min intervals, after which they were sequentially exposed to the remainder of the reagents necessary to develop color in a routine immunostaining protocol. At the conclusion of the experiment, the amount of AEC color developed was visually assessed. In general, the experiments indicated that there was some adsorption of primary antibody to isolons. There was minimal adsorption of secondary antibody or of streptavidin–peroxidase complex. In a repeat experiment, Isolons were coated with Sigma-Cote silanation solution prior to beginning the experiment. There was no significant difference between silanated Isolons and those not silanated.

2.1.3. Performance Discussion

The advantages of the Code-On are the presence of an oven capable of heating slides to temperatures necessary to perform *in situ* hybridization and the ability to deparaffinize sections on the stainer. The capillary gaps performed variably and somewhat unpredictably, bubbles appearing after filling in some slide pairs and lakes of fluid being left after draining in others. This occurred with both blank slides filled with colored solutions and during actual immunostaining runs. Aqueous solutions tended to fill only the lower half of the gap reliably, requiring tissue placement to be in this region. It is advisable therefore, during clinical staining runs, to use double-mounted slides, with the control tissue toward the top, in order to verify the upper limit of reagent ascent. The filling of the gaps was exquisitely sensitive to and disrupted by the presence of powder from rubber gloves or finger oil on the slides. To guarantee removal of reagents from the gap, a minimum of three buffer washes should be used.

Because the 150-μL aliquots of antisera that are placed in the isolons are exposed to the room environment and suffer a potential 50% evaporative loss over 2 h, the optional hood should be required as standard equipment, a beaker of water should be placed in the oven, heated, and allowed to act as a humidity source, slide staining runs should be kept as short as possible, and the three-well reagent isolator should be used whenever possible because of its larger fluid capacity. To prevent excessive evaporative loss of fluids from the filled capillary gaps, it is imperative that the slides be placed in the incubation chamber having a proper seal. If left exposed to room environmental conditions, fluid evaporates from the capillary gaps with resultant reagent concentration.

2.2. Shandon's Cadenza

2.2.1. Description

Shandon's Cadenza is also a capillary gap device. Unlike the Code-On, the capillary gap is formed between a glass slide and a disposable, plastic "coverplate." The slide and coverplate are clamped together, and reagents are automatically pipeted into the top of the coverplate-slide assembly. The top of the coverplate is molded to form a funnel to receive reagents. When approx 100 μL of reagent are pipeted into the assembly, the reagent fills the capillary gap where it is retained by surface tension. Washing is accomplished by the automatic addition of 2 mL of wash buffer/programmed buffer rinse. This effectively provides 20 fluid changes within the gap. Excess fluid runs out the bottom of the assembly into a waste container. When a staining run ends, the machine continually reapplies buffer in order to prevent slide drying until the operator halts the process and removes the slides.

The Cadenza incorporates a robotic pipet system that picks up reagents from a reagent carousel and delivers them to the appropriate slide(s). The instrument is controlled by a built-in computer, and has an LCD display and a keyboard for programming a staining run. The machine can store up to 10 programs of the user's design with a capacity of 20 slides/staining run. Each run is preceded by a slide setup step in which individual slides are programmed for specific primary antisera. Antisera or other reagents are set into numbered positions in the reagent carousel, and the appropriately numbered reagent is programmed into each slide's staining sequence.

The robotic pipet sequentially dispenses each reagent. Between reagents, the pipet is placed into a wash position and thoroughly rinsed with buffer. Excess buffer in this wash stage flushes the exterior surface of the pipet. The pipet is paired with a sensor that detects the presence of fluid in the reagent carousel. This sensor activates an alarm and halts the machine if insufficient reagent is present to complete a step. This might occur if a reagent vial was empty, deficient in fluid, or absent because of a technician's error.

Like the Leica/Jung and the Ventana, only aqueous solutions can be used on the Cadenza, so slides must be deparaffinized and hydrated prior to loading onto the machine. Slides are already wet when loaded, and although care must be taken to avoid introducing bubbles into the gap during this process, this was rarely a problem. Additionally, it is inadvisable to use H_2O_2 on the Cadenza, since small bubbles, generated during this step, may adhere to the plastic coverplate and interfere with subsequent fluid flow through the capillary gap.

Fluid that is flushed from the capillary gap is collected in a waste tray beneath the slide-coverplate assemblies. This tray has a limited capacity and must be drained prior to a staining run. The entire slide-coverplate and reagent carousel area are enclosed within a Plexiglas™ cover during operation, which protects reagents from laboratory conditions (temperature, humidity) and prevents inadvertent contact with the robotic pipet arm during movement.

2.2.2. Performance Evaluation

Because the Cadenza has a cover that encloses the slides and reagents, a test was performed to evaluate conditions within the closed cover during operation. A prewarmed Cadenza, in standby mode with 50 mL of water in the waste tray, was observed over a 4-h period. With the lid closed, recordings of temperature and humidity were taken from a sensor placed above the slide coverplate assemblies. After 2.5 h, the temperature within the instrument rose +4°C, and the humidity dropped –3%. After this interval, the interior temperature and humidity remained stable.

To evaluate evaporative loss of fluid from the capillary gap, the instrument was programmed to dispense 100 μL of colored tap water into the gaps created between coverplates and clean, blank slides. The waste tray was filled with 50 mL of water, and the instrument lid was closed. Slides were observed for a 2-h period, and fluid loss was assessed. For the first 60 min, there was no appreciable fluid loss. At 90 min, an average of 7 mm (fluid height in the gap) was lost. At 120 min, an average of 12 mm of fluid was lost. These measurements correspond to a total loss of fluid from the gap of 24% over the 2-h period. Whether the fluid loss was the result of evaporation or slow, gravity-induced leakage from the capillary gap was not assessed.

In a similar evaporative loss study, a defined volume of distilled water was placed in reagent vials that were weighed on a Mettler balance and placed in the reagent carousel. The instrument's lid was kept closed, and at intervals, the measurements were repeated. During the test, the temperature within the machine averaged 25°C, and the humidity was 28%. After 2 h, the reagent vials had lost 6% of their original volume.

To assess antibody adsorption to vials, a series of vials were filled with primary antisera. The vials were capped and stored for up to 1 mo at 4°C. At the end of the storage period, the primary antibody was removed, and the vials were exposed to a standard immunostaining reagent sequence with 5-min time intervals for each reagent. There was no evidence of any antibody adsorption to the vials, suggesting that reagent can be stored in the Cadenza vials between staining runs. Similar studies showed there was no detectable binding of antisera to coverplates in tests that ranged up to overnight in length.

To assess washout from the capillary gap, streptavidin–peroxidase complex was automatically dispensed into the gap between clean slides and coverplates. After filling the capillary gap, a buffer rinse was dispensed. As the 2 mL of buffer flowed across the slide surface, effluent was collected in 100-μL aliquots. The series of tubes containing the effluent were reacted with AEC chromogen for 5 min. At the end of the reaction, the tubes were photographed. The first four tubes collected generally had a strong chromogenic reaction, the next four had a reduced intensity, and the remaining tubes had no detectable reaction. The majority of the complex was apparently washed out in the first 400 μL of the 2000 μL rinse.

To test the precision with which 100 μL were pipeted, individually dispensed aliquots were collected into preweighed weigh boats, which were then reweighed on a Mettler balance. In a series of tests, the largest aliquot recorded was 74.6 μL and the smallest recorded was 4.0 μL. The average aliquot dispensed was 35.9 μL Since this was obviously outside the instrument's specifications, the company performed a calibration service, and the experiment was repeated. After service, the largest aliquot dispensed was 90.6 μL, and the lowest was 86.1 μL with the average being 87.7 (12.3% error).

To assess the efficiency with which the pipet is washed between reagent steps, two vials were placed into the reagent carousel. One contained distilled water containing dark blue food color, and the other contained 1 mL of tap water. Two slides with coverplates were placed onto the machine. The machine was programmed to pick up the blue water and dispense it onto slide 1. After dispensing, the droplets emerging from the bottom of the slide were collected into a small vial. The machine then automatically washed the pipet, per the computer program, prior to entering the vial containing tap water. The machine then picked up water from the tap water vial and dispensed it onto slide 2. The effluent from slide 2 assesses carryover of reagents from the internal surface of the probe. The solution in the tap water vial, after probe entry, assesses reagent carryover on the external surface of the probe. Using a spectrophotometer, the blue dye concentration was determined for both cases. The effluent from slide 2 had a visible light blue color indicating some internal carryover of the initial reagent, even after the pipet had been purged with buffer. This amounted to 0.004% of the original solution concentration. The tap water vial also had a light blue color, indicating some carryover of dye on the external surface of the pipet. This amounted to 0.012% of the original dye concentration.

A test was conducted to determine if the coverplate plastic had any deleterious effect on the AEC chromogen. Coverplates were finely ground, and the resulting plastic powder was added to AEC. Various concentrations of plastic were used. After 10 min of incubation with the plastic, streptavidin–peroxidase complex was added. After another 5 min, the tubes were photographed. Over a range of added plastic from 0.01 to 0.04 g, there was no detectable decrease in the chromogenic reaction as compared to a tube with no added plastic.

2.2.3. Performance Discussion

The advantages of the Cadenza are the efficient filling of the capillary gap because of the assistance of gravitational forces, the extremely efficient wash steps because of the 20 times capillary gap fluid exchange, and the lack of technician's attention needed once a run is in progress. There was detected a minimal amount of reagent carryover that could theoretically induce a false-positive result in a slide treated with ultrasensitive detection reagents and preceded on the immunostainer by one receiving a concentrated antibody. Although we have not experienced such a problem, the following steps would help minimize the chance of it occurring at all:

1. Place a negative control sera slide immediately following one receiving a concentrated antibody;
2. Carefully follow the manufacturer's pipet cleaning protocols; and
3. Use maximal dilutions of primary antibodies.

As part of a periodic maintenance schedule, the volume of fluid dispensed by the pipeter should be quantitatively assessed to detect any restricted flow that might occur owing to blockage of the pipeter. If double-mounted slides are used, with the control tissue placed at the bottom of the slide, the technologist will be made aware of any insufficient reagent delivery.

2.3. The Leica/Jung Histostainer Ig

2.3.1. Description

The Leica/Jung stainer closely mimics manual staining procedures. Slides are maintained in a flat position, with tissue facing up, and reagents are dispensed onto the slide surface. The unit has a capacity of up to 20 slides/run and can permanently store 10 separate staining protocols. Each protocol draws from a list of reagents that cannot be altered or added to by the user. In addition, the program automatically inserts a wash step between every reagent application. The slides are loaded onto a flat platen that rotates over a water bath whose temperature is adjustable from ambient to 45°C. In addition to providing heat, the water bath also provides humidity within the staining environment.

Reagents are placed in the center of the slide platen and are picked up by a special pipet. The reagent is dispensed onto the slide through the pipet tip's spray nozzle, which atomizes the reagent as it is placed on the slide. The operator can select the area of the slide (in squares of 1 in.) to receive reagent, but this must be programmed separately for each slide. It is possible to apply reagent to one, two, or all three squares, the latter totally covering a standard microscope slide. For each third of the slide selected, the stainer dispenses 100 µL, 300 µL being dispensed when coverage of all three squares has been indicated. After completion of a reagent step, reagent is removed from the slide by an air knife that is carried just in front of the spray nozzle. As one reagent is blown from the slide surface, the next reagent is applied so there is no opportunity for slide drying to occur. The spray nozzle is replaceable, and its replacement is recommended after each slide run.

Only aqueous solutions can be used on the Leica/Jung stainer since it is inadvisable to add organic solvents to the waste-containing area of the instrument. All reagents that are removed from the slide surface go into this area, which also is the water bath and contains the heating coils. In addition, the structure of the water bath is plastic, and the manufacturer does not recommend exposure to organic solvents. Deparaffinization and hydration of slides, therefore, must be done manually before placing slides on the stainer. Counterstaining on the instrument with standard dyes after the immunostaining steps is also not recommended because of blockage of the nozzle by the dyes and because the counterstain is washed into the water bath where it stains the instrument.

The computer, which controls the stainer, is an integral part of the unit. Programming is performed using a restricted set of buttons (not a full keyboard), and menu selections are displayed on an LCD screen. The slides are individually programmed to receive the appropriate reagents. After programming the instrument to recognize reagent vial contents and locations, the computer calculates the required volume for each reagent for the entire run. An external printer can be attached to print out a number of different reports, including one that lists all actions performed on each individual slide in real time. A total of 46 error codes are recognized and reported by the instrument. When a staining run ends, the machine continually reapplies buffer in order to prevent slide drying until the operator halts the process.

The pipet draws up buffer prior to aspirating reagent. Buffer and reagent are separated by an air bubble in the pipet. This additional buffer within the pipet is used to assist in washing the pipet between the dispensing of aliquots of different reagents. There are also two dipping washes of the external surface of the nozzle between aliquot dispensing events. Buffer rinses of slides occur in sprays of 450 µL.

2.3.2. Performance Evaluation

To assess the reproducibility of spray volume delivery, a series of preweighed weigh boats were placed on the instrument's carousel, and the instrument was programmed to spray a single time into each. After spray delivery, the boats were reweighed, and the volume of liquid calculated. The maximum amount of fluid dispensed was 105.1 µL, and the minimum was 65.7 µL The average of 20 separate determinations was 88.8 µL, yielding an error of 11.2%.

Carryover of reagents was investigated by placing a vial of distilled water, colored deep blue with food dye, in reagent position 1. A second vial of clear distilled water was placed in position 2 of the carousel. The stainer was programmed to pick up solution from vial 1 and spray it into a weigh boat (equivalent to one square). The stainer then automatically rinsed its nozzle with two changes of buffer prior to loading solution from the second reagent vial. After the rinses, the instrument entered the second vial, loaded 100 µL of distilled water, and sprayed this into a second weigh boat. Vial 2, which the probe had entered, was saved for analysis. The second weigh boat effluent contained any carryover of reagent present on the internal surfaces of the spray mechanism. The second vial contained only carryover of reagent present on the external surfaces of the nozzle and its connections. There was a light blue color present in the second weigh boat indicating some internal reagent carryover. This was determined spectrophotometrically as being 0.038% of the original solution concentration. Vial 2, which was used to assess the external nozzle carryover, was colorless, indicating that the exterior of the nozzle was thoroughly washed.

To assess the reproducibility of the spray pattern, a series of 1 in. square absorbent pads were placed on glass slides, which were then loaded onto the platen of the instrument. The instrument was carefully leveled and was programmed to spray reagent (distilled water with red dye) into designated positions: either upper, center, or lower thirds of the slide. The spray pattern was circular, and was generally applied evenly and centered on the slide. The edges of the paper, measuring approx 4 mm in each corner, did not receive reagent.

For a second test of spray pattern, special slides were obtained from Erie Scientific, Inc. (Portsmouth, NH) that were evenly coated over their entire surface with a paint-like coating that does not make fluids bead up, nor does it allow bleeding of applied fluids. These slides were loaded onto the platen, and the instrument was programmed to spray dye onto all three positions of each slide. The slides were then kept on the machine until all water had evaporated, leaving the dye residue on the coating in the original area sprayed. This test resulted in circular spray patterns over the slide surface that were superimposed. In general, dye residue was absent from the outer 5 mm of the slide. In addition, there was less dye residue on the "inner" third of the slide, involving approx the distal 1–1.5 cm of glass surface. It should be noted that this test was conducted with dry glass slides, and in normal use, slides containing tissues would be wet prior to applying reagent.

The efficiency of the buffer wash steps was determined by programming the instrument to spray streptavidin–peroxidase complex onto slides previously treated with aminosilane. The instrument was then programmed so that one series of slides received a single buffer wash, one series received two cycles of buffer wash, and one series received three cycles of buffer wash. After the final buffer wash, the slides were sprayed with AEC chromogen. The presence of any chromogen color on the slides would indicate insufficient washing. All slides were visually colorless, indicating that the washing steps function efficiently.

To assess any antibody adsorption to Leica manufactured vials, a primary antibody was stored in the reagent vials for up to 1 mo at 4°C. Subsequently, the antibody was removed, and the remaining reagents of an immunostaining run were sequentially added, as in the Cadenza vial and Code-On isolon experiments. Slight vial staining was seen after the 1-mo incubation period, indicating minimal antibody adsorption.

An experiment was run to assess the control of temperature and humidity during a staining run. After filling the water bath, the stainer was prewarmed for one-half hour. The instrument was programmed to run at ambient temperature, and for the next 4 h, temperature and humidity readings were taken. The average temperature over the time interval was 25.4°C with a 1.2°C temperature rise from the initial measurement. The average humidity during the course of the experiment was 69%. There was a 6% rise in humidity during the 4-h incubation period.

2.3.3. Performance Discussion

The advantages of the Leica/Jung instrument are that incubation temperatures can be elevated for shorter runs, humidity is controlled, and the amount of antibody applied to a slide can be customized for the amount of tissue present. There was, however, a low level of reagent carryover similar to that observed with the Cadenza. *See* **Subheading 2.2.3.** for recommendations for detection and prevention of this potential problem.

In order to avoid placing tissues in the area of the slide that the spray pipeter did not predictably reach, they should be placed in the center of the slide. The nozzles of the spray pipeter occasionally had improper spray patterns and had to be rejected. This was detected prior to run initiation using a quality-control test recommended by the manufacturer. In addition, the amount of fluid dispensed by the pipeter varied as much as 11% less than the 100 µL that should be delivered to one square on a slide. This should be regularly monitored using techniques described in **Subheading 2.3.2.**

2.4. The Ventana 320

2.4.1. Description

The Ventana 320 holds the slides horizontally, much as in manual staining. The self-contained unit has a capacity of 40 slides/run, but because all staining runs are conducted at elevated temperatures, a single immunostaining run is completed in approx 1 h. The Ventana is unique among the immunostainers because it utilizes a set of reagents in bar-coded vials. The user places bar-coded labels on slides, and the instrument then dispenses an appropriate primary reagent to each slide, based on the bar-coded label. The unit also uses bar codes on the individual reagent vials to distinguish one from the other.

The use of heat during the staining process raises the possibility of drying of reagent onto the slide surface. In the Ventana 320, this is avoided by the use of a proprietary Liquid Coverslip. This material is a type of aliphatic oil, which floats on the surface of the aqueous reagents and prevents evaporation. This oil is inert to the various reagents used in immunostaining. Reagents are removed from the slide surface by jets of buffer wash.

In the Ventana, all reagents are carried in unique dispensers. These are similar to repetitive dispenser pipets. They consist of a reagent reservoir and a dispensing plunger. To dispense reagent, an air-driven plunger simply presses on the top of the dispenser plunger. The accuracy of each dispenser is solely controlled by the accuracy of the repetitive dispensers. All secondary reagents, from secondary antibody through hematoxylin counterstain, are supplied by Ventana in bar-coded dispensers. These bar codes, as are those supplied for labeling slides and primary antibodies, are proprietary to ensure the use of Ventana-supplied secondary reagents. The instrument reads the bar code the first time it sees a dispenser and assumes that the dispenser is full. The bar code identifies the reagent and also contains a serial number (not accessible to the user). The computer records the serial number and tracks the dispenser so that it may be used a fixed number of times. After the number of fixed aliquots has been dispensed, that particular serial number is permanently locked out of the machine. Should the user attempt to refill and reuse a dispenser, the computer will see it as empty and reject it.

The Ventana is controlled by a built-in computer system. Since it only uses company-supplied secondary reagents, the computer programs are supplied as "recipe" files. Recipe files are computer programs designed for processing either paraffin-embedded tissue or frozen section/cytospin samples, or for titering a primary antibody. Each recipe file consists of a specific sequence of rinses, enzyme inhibitors, antibodies, detection complexes, chromogens, counterstains, and if necessary, pretreatment enzymes. Although the Ventana can utilize either 3,3'-diaminobenzidine (DAB) or AEC as the chromogen, only one can be used per slide run.

All waste solutions are flushed into the heating water bath so this instrument can be used for only the aqueous steps in immunostaining. Deparaffinization and hydration of slides must be done manually, prior to loading slides onto the stainer.

The computer system of the Ventana provides a number of convenience and quality-assurance checks. Prior to beginning a staining run, the labeled slides are checked, and the reagent carousel is compared to ensure that there is a reagent available that corresponds to each slide present. This ensures that the bar codes on each slide are readable and verifies the number of slides loaded. At the conclusion of a staining run, a variety of reports can be printed using the built-in printer. These reports detail the reagents used, the number of slides run, the date, the elapsed time, the temperature set points, and other information about conditions of the staining run. The instrument also reports any error conditions that may occur during a staining run. There are a total of 45 different error codes listed in the user manual. The instrument also provides the capability of performing overnight staining runs in the absence of technical supervision. The stained slides are then allowed to dry out overnight. The next morning, these dried slides are rehydrated in aqueous buffer and coverslipped.

2.4.2. Performance Evaluation

To evaluate the bar-coded slide labels supplied by Ventana, a number of tests were done by placing bar-code-labeled slides in vials containing eosin, hematoxylin, 100% ethyl alcohol, or xylene for 5 min. All labels subjected to these tests remained firmly affixed to the slides. In a second test of the bar codes, a number 2 pencil was used to place marks onto the bar codes in an attempt to cause the machine to misread the labels. The instrument was confused only by pencil marks made parallel to the bars of the code. To simulate sloppy application of bar codes that might occur in a routine laboratory, bar codes were applied to the slides at considerable angles. Slides were then read by the Ventana. The machine correctly read all such angled labels unless the labels were so badly positioned that portions of the bar code were not scanned by the reading beam. When the machine could not read a badly positioned label, an error code was generated. Bar codes were also placed on slides with small bubbles or minor folds beneath the label. Again, the machine correctly read these labels, as long as they were not distorted completely out of the reading light beam. To assess the effect of vibration on the bar code reading mechanism, a shaking water bath running at maximum speed was situated adjacent to the stainer. In spite of obvious mechanical shaking of the laboratory bench and the Ventana, no misreading of the labels occurred.

To assess the effect of sloppy loading of slides into the platen of the Ventana, a series of bar-coded slides were placed into the slide holders on the platen at a variety of angles or placed on the instrument unclipped into the slide holders. In all cases, when the instrument scanned the slides, it reported error conditions and would not allow the staining run to proceed.

To assess the reproducibility of reagent dispensing, a disposable reagent dispenser (intended for primary antibody) was loaded with distilled water. Aliquots were manually dispensed into preweighed weighing boats by depressing the dispenser plunger. After dispensing, the weigh boats were reweighed on a Mettler balance. The largest volume dispensed was 103.26 µL, and the lowest was 93.55 µL. The average volume dispensed for 20 trials was 98.2 µL, yielding an error rate of 1.8%.

The area of the Ventana 320 in which the slides are held during staining is totally enclosed and inaccessible during a staining run. Therefore, it was not possible to measure the temperature directly within the staining chamber in our laboratory. The manufacturer's data collected from "on slide thermistors" that measure glass-surface temperatures show a variation of only 0.5°C.

The Ventana uses wash jets to remove reagents between staining steps. To assess the efficiency of the washes, one of the two double wash ports was totally blocked. Blank slides were then programmed for standard immunochemistry runs and stained. Even total blockage of one of the double wash ports caused only a low level of glass slide staining by DAB. Error codes were not displayed. Under normal circumstances, dispensing functions are manually verified during machine quality-control procedures.

The manufacturer supplied a statement that for over 80 antisera tested, the Liquid Coverslip (oil) did not diminish antisera titer. The only exception to this might occur if antisera and Liquid Coverslip were emulsified, a process not expected to occur under normal conditions. We tested four antisera for possible deleterious effects of Liquid Coverslip. Aliquots of pretitered, primary antisera were incubated at 25°C for 32 min with equal amounts of Liquid Coverslip. The aqueous phase was then removed. These antisera were then placed into Shandon's Cadenza immunostainer, as were the same antisera not incubated with oil. Positive control tissues were stained with each set of reagents, and the resulting slides compared. This qualitative comparison detected no difference between antisera exposed to Liquid Coverslip and the same antisera that had not been so exposed.

To assess possible adsorption of antisera to the Ventana dispensers, a Ventana pipeter filled with L26 antisera was stored at 4°C for up to 1 mo. At the end of the storage period, the antisera was decanted, and the pipeter thoroughly rinsed with PBS. The pipeter was then stained to assess antibody adsorption using sequential application of standard immunostaining reagents. After 4 wk of storage, there was no detectable immunoreactivity in the Ventana pipeter.

2.4.3. Performance Discussion

The advantages of the Ventana 320 are several. Because it uses elevated incubation temperatures and its reagents are designed for maximal performance at that temperature, the Ventana 320 has a total run time of 121 min allowing for 4 runs/8-h d, resulting in 160 slides/d. It requires very little setup time because all reagents are supplied in prepackaged dispensers that are simply "snapped" into place on the carousel. It requires little or no technician's attention once a run has begun and can be left to do an overnight run. Additionally, slides left to dry after an overnight run show no artifact attributable to this procedure.

The Ventana 320 is designed to be used only with detection reagents, enzymes, chromogens, and counterstains supplied by Ventana, but it allows the use of primary antisera not marketed by the company. To use such antisera, however, the company requires purchase of Ventana pipeters and bar codes for each individual antisera. Although this limits a laboratory's ability to be flexible, it assists in assuring quality control of reagents.

This immunostainer does not allow for user-designed protocols, and can use only a single detection system and chromogen per run. Multiple runs per day can be used, however, because of the shortened run time. The elevated temperature that makes this rapid process time available may have deleterious effects on some of the more heat-labile antibodies. We did not have the opportunity to test this in our laboratory.

3. Conclusions

Each instrument evaluated in this study was capable of producing excellent immunostains, and all are superior to manual methods when properly maintained and operated. The choice of an immunostainer for a given laboratory should involve a number of factors that must be taken into consideration in addition to those performance issues addressed here. An individual laboratory must assess its work load, its expected increment in immunostaining requests over the life of the staining instrument, and the expertise required to successfully operate the chosen instrument. It is highly recommended that any laboratory contemplating the acquisition of an automated immunohistology stainer discuss its particular needs with all suppliers prior to making a final decision.

Acknowledgments

We thank the following technicians for their invaluable assistance: Ann Marie Boback, BS, MT (ASCP), Christina Genaw, HTL, (ASCP), Michelle Carney, HTL, (ASCP), and Dayna M. Elfont (BA).

Index